The Probability Tutoring Book

AN INTUITIVE COURSE
FOR ENGINEERS AND SCIENTISTS
(AND EVERYONE ELSE!)

Revised Printing

Carol Ash

*University of Illinois
at Urbana–Champaign*

IEEE
PRESS

The Institute of Electrical and Electronics Engineers, Inc., New York

This book may be purchased at a discount from the publisher when ordered in bulk quantities. For more information contact:

IEEE PRESS Marketing
Attn: Special Sales
PO Box 1331
445 Hoes Lane
Piscataway, NJ 08855-1331
Fax: (908) 981-8062

Printed in the United States of America

10 9 8 7 6 5 4 3 2

ISBN 0-7803-1051-9 [pbk]
IEEE Order Number: PP0288-1 [pbk]

ISBN 0-87942-293-0
IEEE Order Number: PC0288-1

Library of Congress Cataloging-in-Publication Data

Ash, Carol (date)
    The probability tutoring book : an intuitive course for engineers and scientists (and everyone else!) / Carol Ash.
        p.   cm.
    "IEEE order number: PC0288-1"–T.p. verso.
    Includes index.
    ISBN 0-87942-293-9
    1.   Engineering mathematics.   2.   Probabilities.   I.   Title.
TA340.A75   1993
620'.001'51—dc20   92-53183
                                                                CIP

Contents

CONTINUOUS PROBABILITY

Preface

This is the second in a series of tutoring books,[1] a text in probability, written for students in mathematics and applied areas such as engineering, physics, chemistry, economics, computer science, and statistics. The style is unlike that of the usual mathematics text, and I'd like to describe the approach and explain the rationale behind it.

Mathematicians and consumers of mathematics (such as engineers) seem to disagree as to what mathematics actually is. To a mathematician, it's important to distinguish between rigor and informal thinking. To an engineer, intuitive thinking, geometric reasoning, and physical argument are all valid if they illuminate a problem, and a formal proof is often unnecessary or counterproductive.

The typical mathematics text includes applications and examples, but its dominant feature is *formalism*. Theorems and definitions are stated precisely, and many results are proved at a level of rigor that is acceptable to a working mathematician. This is bad. After teaching many undergraduates, most quite competent and some, in fact, blindingly bright, it seems entirely clear to me that most are not ready for an abstract presentation. At best, they will have a classroom teacher who can translate the formalism into ordinary English ("what this really means is ... "). At worst, they will give up. Most will simply learn to read around the abstractions so that the textbook at least becomes useful as a source of examples.

This text uses informal language and thinking whenever possible. This is the appropriate approach even for mathematics majors: Rigorous probability

---

[1]The first is *The Calculus Tutoring Book* by Carol Ash and Robert Ash (New York: IEEE Press, 1986).

isn't even *possible* until you've had a graduate-level course in measure theory, and it isn't *meaningful* until you've had this informal version first.

In any textbook, problems are as important to the learning process as the text material itself. I chose the problems in this book carefully, and much consideration was given to the *number* of problems, so that if you do most of them you will get a good workout. To be of maximum benefit to students, the text includes detailed solutions (prepared by the author) to all problems.

I'd like to thank the staff at the IEEE PRESS, Dudley Kay, Executive Editor; Denise Gannon, Production Supervisor; and Anne Reifsnyder, Associate Editor. I appreciate the time and energy spent by the reviewers, Dr. Robert E. Lover and members of the IEEE Press Board.

Most of all I owe my husband, Robert B. Ash, for patiently and critically reading every word and for being better than anyone else at teaching mathematics in general, and probability in particular, to students and wives.

Carol Ash

Introduction

Probability is a hard subject. Students find it hard, teachers find it hard, textbook writers find it hard. And you can't put the blame on the "theory" because this text will not be theoretical. It emphasizes informal ideas and problem solving, and the solutions will seem simple in retrospect, but you may find them hard to figure out on your own. This book will help you all it can, but in the end the only way to learn is to do *many* problems.

On the plus side, probability is useful and intensely interesting. Many courses and some entire disciplines have a probability prerequisite, for example, stochastic processes, information theory, signal detection, signal processing, control systems, quantum mechanics, system theory, and of course statistics.

There are two types of probability problems, discrete and continuous. Experiments with a "limited" number of outcomes are called discrete. There are only finitely many poker hands, so finding the probability of a royal flush in poker is a discrete problem. The discrete category includes experiments with an infinite number of outcomes as long as it is a "countably" infinite number. If the experiment is to toss a coin until a head turns up, the possible outcomes are

outcome 1	never get a head
outcome 2	H on 1st toss
outcome 3	TH
outcome 4	TTH
outcome 5	TTTH
⋮	

The list of outcomes is countably infinite, so finding the probability that it takes at most 10 tosses to get the first head is a discrete problem.

Here's an example of a continuous problem:

> John will be passing the corner of Main and First at some time between 1:00 and 2:00.
>
> Mary will be passing the same corner between 1:30 and 2:00.
>
> Each agrees to wait 5 minutes for the other.
>
> Find the probability that they meet.

There is an infinite number of possible arrival times for John (all the numbers between 1 and 2), and it is an "uncountably" infinite number since there is no way to list all the times and label then 1st, 2nd, 3rd, ... Similarly for Mary. So the problem is continuous.

Discrete probability has more charm, but if you are in engineering, continuous probability will most likely be more useful for you. Discrete probability has no special prerequisite—high school algebra is enough. For continuous probability, you'll need integral calculus. Techniques of integration are not important—a computer can always do the hard integrals—but you will have to remember how to set up a double integral; a review of double integrals is included before they're used in Chapter 5.

The text begins with discrete probability in Chapters 1–3. The rest of the book, Chapters 4–9, covers continuous probability with occasional flashbacks to the discrete case. Discrete and continuous probability have certain basic ideas in common but in practice they will seem quite different. I hope you enjoy them both.

CHAPTER 1

Basic Probability

SECTION 1-1 PROBABILITY SPACES

We want to answer the questions "What are probabilities?" and "How does an event get a probability?"

Sample Space of an Experiment

A sample space corresponding to an experiment is a set of outcomes such that exactly one of the outcomes occurs when the experiment is performed. The sample space is often called the *universe*, and the outcomes are called *points* in the sample space.

There is more than one way to view an experiment, so an experiment can have more than one associated sample space. For example, suppose you draw one card from a deck. Here are some sample spaces.

> *sample space 1* (the most popular)   The space consists of 52 outcomes, 1 for each card in the deck.
> *sample space 2*   This space consists of just 2 outcomes, black and red.
> *sample space 3*   This space consists of 13 outcomes, namely, 2, 3, 4, . . . , 10, J, Q, K, A.
> *sample space 4*   This space consists of 2 outcomes, picture and non-picture.

Any outcome or collection of outcomes in a sample space is called an *event*, including the *null* (empty) set of outcomes and the set of *all* outcomes.

In the first sample space, "black" is an event (consisting of 26 points). It is also an event in sample space 2 (consisting of 1 point). It is *not* an event in sample spaces 3 and 4, so these spaces are not useful if you are interested in the outcome black.

Similarly, "king" is an event in sample spaces 1 and 3 but not in 2 and 4.

Probability Spaces

Consider a sample space with n points.

Probabilities are numbers assigned to events satisfying the following rules.

(1) Each outcome is assigned a non-negative probability such that the sum of the n probabilities is 1.

This axiom corresponds to our intuitive understanding of probabilities in real life. The weather reporter never predicts a negative chance of snow, and the chance of snow plus the chance of rain plus the chance of dry should be 100%, that is, 1.

(2) If A is an event and $P(A)$ denotes the probability of A, then

$P(A) = $ sum of the probabilities of the outcomes in the event A

A sample space together with an assignment of probabilities to events is called a *probability space*. Note that probabilities are always between 0 and 1.

Figure 1 shows a probability space with six outcomes a, b, c, d, e, f and their respective probabilities. The indicated event B contains the three outcomes d, e, f and

$$P(B) = .1 + .2 + .3 = .6$$

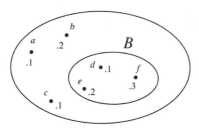

Figure 1

Probabilities may be initially assigned to outcomes any way you like, as long as (1) is satisfied. Then probabilities of events are determined by (2).

To make our probabilities useful, we try to assign initial probabilities to make a "good" model for the experiment.

Suppose you put slips of paper labeled a, b, c, d, e, f in a bag, shake it thoroughly, put on a blindfold, and draw one out; that is, you pick one of a, b, c, d, e, f at random. The appropriate model should have the six outcomes equally likely, so instead of the probabilities in Fig. 1, you should assign each outcome the probability $1/6$. Then

$$P(B) = \frac{1}{6} + \frac{1}{6} + \frac{1}{6} = \frac{3}{6} = \frac{\text{number of outcomes in the event}}{\text{total number of outcomes}}$$

This suggests an important special case. Suppose a sample space has n points and we make the special assignment

(1′) $$P(\text{each outcome}) = \frac{1}{n}$$

(Note that axiom (1) is satisfied as required.) Then the outcomes are equally likely. In this case it follows that

(2′)
$$\boxed{\begin{aligned} P(\text{event}) &= \frac{\text{number of outcomes in the event}}{\text{total number of outcomes}} \\ &= \frac{\text{favorable outcomes}}{\text{total outcomes}} \end{aligned}}$$

Use (1′) and (2′) if an experiment is "fair," in particular if an outcome is picked at random.

For the problems in this book, you may assume that *dice and coins and decks of cards are fair* unless specifically stated otherwise. If an experiment is not fair (e.g., toss a biased coin), you will be given the initial probabilities. In real life you might select the initial probabilities by playing with the coin: If you toss it many times and 63% of the tosses are heads, then it is reasonable to make the initial assignment $P(\text{heads}) = .63$.

How do you decide if your mathematical model (the probability space) is a good one? Suppose you assign initial probabilities so that $P(\text{event } B)$ turns out to be .37. If many people each perform the experiment many times and "most" of the people find that B happens "close to" 37% of the time, then you have a good model.

Example 1   (a fair deck)

Draw one card from a deck. Consider sample space 1 (containing 52 points). In the absence of any special information to the contrary we always choose to assign probabilities using (1′) and (2′). Then

$$P(\text{ace of spades}) = \frac{1}{52}$$

$$P(\text{ace}) = \frac{\text{favorable}}{\text{total}} = \frac{4}{52} = \frac{1}{13}$$

$$P(\text{card}) \geq 10) = \frac{20}{52}$$

> In this text, the ace is considered to be a picture, along with jack, queen, king.
>   Unless otherwise stated, the ace is high, so that
>
> $$\text{ace} > \text{king} > \text{queen} > \text{jack} > 10$$

Example 2   (a biased deck)

Suppose that the 16 pictures in sample space 1 are assigned probability $1/32$ and the 36 non-pictures are assigned prob $1/72$. (Note that the sum of the probs is 1, as required.) This corresponds to a deck in which the pictures are more likely to be drawn (maybe the pictures are thicker than the non-pictures). Then

$$P(\text{ace of spades}) = \frac{1}{32}$$

$P(\text{ace}) = \text{sum of probs of the aces}$

$$= \frac{1}{32} + \frac{1}{32} + \frac{1}{32} + \frac{1}{32} = \frac{4}{32}$$

$P(\text{card} \geq 10) = \text{probs of the 10's} + \text{probs of the pictures}$

$$= 4 \cdot \frac{1}{72} + 16 \cdot \frac{1}{32}$$

$P(\text{card} \geq 2) = \text{sum of all the probs} = 1$

$P(\text{red spade}) = 0$ since the event red spade contains no outcomes

Impossible Events and Sure Events

Probabilities are always between 0 and 1.
   A *sure* event is one that contains *all* points in the sample space (e.g., card ≥ 2). The probability of a sure event is 1.

An *impossible* event is an event that contains *no* points in the sample space (e.g., red spade). The probability of an impossible event is 0.

The converses are not necessarily true: There are possible events that have probability 0 and non-sure events with probability 1. (See the footnote in Section 2.1 and see (8) in Section 4.1.)

Complementary (Opposite) Events

If A is an event, then its complement $\bar{A}$ is the set of outcomes *not* in A.

For example, if A is red card, then $\bar{A}$ is black card; if $\bar{A}$ is king, then A is non-king.

It follows from (1) and (2) that in any probability space

(3)
$$\boxed{P(A) = 1 - P(\bar{A})}$$

Example 3   (tossing two fair dice)

Let's find the probability of getting an 8 (as a sum) when you toss a pair of fair dice.

The most useful sample space consists of the following 36 points; think of one die as red and the other as blue—each point indicates the face value of the red die and the face value of the blue die.

(4)
$$
\begin{array}{cccccc}
1,1 & 2,1 & 3,1 & 4,1 & 5,1 & 6,1 \\
1,2 & 2,2 & 3,2 & 4,2 & 5,2 & 6,2 \\
1,3 & 2,3 & 3,3 & 4,3 & 5,3 & 6,3 \\
1,4 & 2,4 & 3,4 & 4,4 & 5,4 & 6,4 \\
1,5 & 2,5 & 3,5 & 4,5 & 5,5 & 6,5 \\
1,6 & 2,6 & 3,6 & 4,6 & 5,6 & 6,6 \\
\end{array}
$$

There are five outcomes favorable to 8, namely, $(2,6)$, $(6,2)$, $(5,3)$, $(3,5)$, $(4,4)$. So

$$P(8) = \frac{5}{36}$$

Example 4

Toss two dice. Find the probability that they show *different* values, for example, $(4,6)$ and $(2,3)$ but *not* $(2,2)$.

You can count the favorable outcomes directly, or better still, by (3),

$$P(\text{non-matching dice}) = 1 - P(\text{matching dice}) = 1 - \frac{6}{36} = \frac{5}{6}$$

Problems for Section 1-1

Toss two dice. Find the probability of each of the following events.

1. sum is 7

2. 7 or 11

3. second die $>$ first die

4. at least one of the dice is a 6

5. both dice are ≥ 5

6. at least one die is ≥ 5

7. neither die is over 4

8. both dice are even

9. at least one die is odd

SECTION 1-2   COUNTING TECHNIQUES

In a probability space where the outcomes are equally likely,

$$P(\text{event}) = \frac{\text{number of favorable outcomes}}{\text{total number of outcomes}}$$

In order to take advantage of this rule you must be able to find the total number of outcomes in an experiment and the number that are favorable to your event. For the dice problems in the last section, they were easy to find by inspection. But it usually isn't that simple, so we'll first derive some counting procedures before continuing with probability.

The Multiplication Principle

Suppose you have 3 shirts (blue, red, green) and 2 pairs of pants (checked, striped). The problem is to count the total number of outfits.

The tree diagram (Fig. 1) shows all possibilities: there are $2 \times 3 = 6$ outfits.

Instead of drawing the tree, which takes a lot of space, think of filling a pants slot and a shirt slot (Fig. 2). The pants slot can be filled in 2 ways and the shirt slot in 3 ways, and the total number of outfits is the *product* 2×3.

You'll get the same answer if the tree is drawn with 3 shirt branches first, each followed by 2 pants branches. Equivalently, it doesn't matter if you name the first slot pants and the second shirts as in Fig. 2, or vice versa.

(1)
> If an event takes place in successive stages (slots), decide in how many ways each slot can be filled, and then multiply to get the total number of outcomes

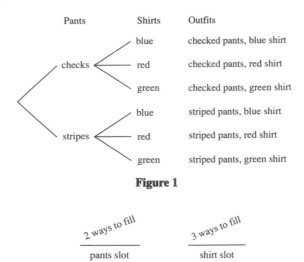

Figure 1

Figure 2   Number of outfits $= 3 \times 2$

Example 1

The total number of 4-letter words is

$$26 \cdot 26 \cdot 26 \cdot 26$$

(Each spot in the word can be filled in 26 ways.)

Example 2

The total number of 4-letter words that can be formed from 26 different scrabble chips is

$$26 \cdot 25 \cdot 24 \cdot 23$$

(The first spot can be filled in 26 ways, the second in only 25 ways since you have only 25 chips left, etc.)

$7 \times 7 \times 7$ versus $7 \times 6 \times 5$

The answer $7 \times 7 \times 7$ is the number of ways of filling 3 slots from a pool of 7 objects where each object can be used over and over again; this is called sampling *with replacement*. The answer $7 \times 6 \times 5$ is the number of ways of filling 3 slots from a pool of 7 objects where an object chosen for one slot cannot be used again for another slot; this is called sampling *without replacement*.

Review

Let's review the product $n!$ (n factorial), which turns up frequently in applications of the multiplication principle. By definition,

$$n! = n(n-1)(n-2)\ldots 1$$

Another definition sets

$$0! = 1$$

(This turns out to be a convenient choice for 0!, as you'll soon see.)
For example,

$$6! = 6 \cdot 5 \cdot 4 \cdot 3 \cdot 2 \cdot 1 = 720$$

$$\frac{10!}{7!} = \frac{10 \cdot 9 \cdot 8 \cdot 7 \cdot 6 \cdot 5 \cdot 4 \cdot 3 \cdot 2 \cdot 1}{7 \cdot 6 \cdot 5 \cdot 4 \cdot 3 \cdot 2 \cdot 1} = 10 \cdot 9 \cdot 8 = 720$$

$$5! \times 6 = 6!$$

Permutations (Lineups)

Consider 5 objects $A_1, \ldots, A_5$. To count all possible lineups (permutations) such as $A_1 A_5 A_4 A_3 A_2$, $A_5 A_3 A_1 A_2 A_4$, and so on, think of filling 5 slots, one for each position in the line, and note that once an object has been picked, it can't be picked again. The total number of lineups is $5 \cdot 4 \cdot 3 \cdot 2 \cdot 1$, or 5!.

In general,

$$\boxed{n \text{ objects can be permuted in } n! \text{ ways.}}$$

Suppose you want to find the number of permutations of size 5 chosen from the 7 items $A_1, \ldots, A_7$, for example, $A_1 A_4 A_2 A_6 A_3$, $A_1 A_6 A_7 A_2 A_3$. There are 5 places to fill, so the answer is

$$7 \cdot 6 \cdot 5 \cdot 4 \cdot 3$$

(Some authors will write the answer in the more compact notation 7!/2!.)

Combinations (Committees)

Now that we've counted permutations, let's try counting committees. I'll illustrate the idea by finding the number of committees of size 4 that can be chosen from the 17 objects $A_1, \ldots, A_{17}$.

Note how committees differ from permutations: A_1, A_{17}, A_2, A_{12} is the *same* committee as A_{12}, A_1, A_{17}, A_2, but $A_1 A_{17} A_2 A_{12}$ and $A_{12} A_1 \; A_{17} A_2$ are *different* permutations. Order doesn't count in a committee; it does count in a permutation.

It isn't correct to fill four committee slots and get "answer" $17 \cdot 16 \cdot 15 \cdot 14$ because a committee doesn't have a first member, a second member, and so on—the order of the members doesn't matter. The number $17 \cdot 16 \cdot 15 \cdot 14$ counts permutations, not committees. But you can get the committee answer by comparing the list of permutations with the list of committees.

List of Committees	**List of Lineups**	
(a) P_1, P_7, P_8, P_9	$(a_1) \quad P_1 P_7 P_8 P_9$	
	$(a_2) \quad P_7 P_1 P_9 P_8$	There are 4!
	$\vdots$	of these.
	$(a_{24}) \quad P_9 P_7 P_1 P_8$	
(b) P_3, P_4, P_{12}, P_6	$(b_1) \quad P_3 P_4 P_{12} P_6$	
	$(b_2) \quad P_4 P_{12} P_3 P_6$	There are 4!
	$\vdots$	of these.
	$(b_{24}) \quad P_{12} P_3 P_4 P_6$	

etc.

Each committee gives rise to 4! lineups, so

$$\text{number of committees} \times 4! = \text{number of lineups}$$

$$\text{number of committees} = \frac{\text{number of lineups}}{4!} = \frac{17 \cdot 16 \cdot 15 \cdot 14}{4!}$$

$$= \frac{17!}{4! \, 13!}$$

The symbol $\binom{17}{4}$ stands for the number of committees of size 4 from a population of size 17 (it's pronounced 17 on 4 or 17 choose 4). It is also written as $C(17, 4)$. We've just shown that

$$\binom{17}{4} = \frac{17!}{4! \, 13!}$$

Here's the general result:

The symbol $\binom{n}{r}$, called a *binomial coefficient*, stands for the number of committees of size r that can be chosen from a population of size n or, equivalently, the number of combinations of n things taken r at a time.

Its value is given by

(2)
$$\binom{n}{r} = \frac{n!}{r! \, (n-r)!}$$

For example, the number of 4-person committees that can be formed from a group of 10 is

$$\binom{10}{4} = \frac{10!}{4!\,6!} = \frac{10 \cdot 9 \cdot 8 \cdot 7 \cdot 6 \cdot 5 \cdot 4 \cdot 3 \cdot 2 \cdot 1}{4 \cdot 3 \cdot 2 \cdot 1 \cdot 6 \cdot 5 \cdot 4 \cdot 3 \cdot 2 \cdot 1}$$

$$= \frac{10 \cdot 9 \cdot 8 \cdot 7}{4 \cdot 3 \cdot 2 \cdot 1} = 10 \cdot 3 \cdot 7 = 210$$

(Cancel as much as you can before doing any arithmetic.)

Some Properties of $\binom{n}{r}$

(3)
$$\boxed{\binom{n}{r} = \binom{n}{n-r}}$$

For example,

$$\binom{17}{4} = \binom{17}{13}$$

This holds because picking a committee of size 4 automatically leaves a committee of 13 leftovers and vice versa so the list of committees of size 4 and the list of committees of size 13 are the same length. Alternatively, $\binom{17}{4}$ and $\binom{17}{13}$ are equal because by the formula in (2) each is 17!/4! 13!.

(4)
$$\boxed{\binom{n}{1} = \binom{n}{n-1} = n}$$

This follows because there are clearly n committees of size 1 that can be chosen from a population of size n. It also follows from the formula in (2) since

$$\binom{n}{1} = \binom{n}{n-1} = \frac{n!}{1!\,(n-1)!}$$

which cancels down to n.

(5)
$$\boxed{\binom{n}{n} = \binom{n}{0} = 1}$$

We have $\binom{n}{n} = 1$ because there is just one way to form a committee of size n from a population of size n. If you try to use the formula in (2), you get

$$\binom{n}{n} = \frac{n!}{n!\, 0!}$$

and it will be 1, provided you define $0! = 1$. In other words, you can define $0!$ any way you like, but it is convenient to call it 1 because then the formula in (2) continues to hold even when r is n or 0.

Once you have $\binom{n}{n} = 1$, it follows from (3) that $\binom{n}{0} = 1$ also. If you want to interpret $\binom{n}{0}$ as the number of committees with no members, look at it like this: There is 1 committee that can be formed with no members, namely, the null (empty) committee.

Example 4

I'll find the probability of getting the queen of spades (denoted Q_S) in a poker hand.

A poker hand is a committee of 5 cards drawn from 52.
The *total* number of poker hands is $\binom{52}{5}$.

Finding *favorable* hands amounts to selecting a committee of size 4 (the rest of the hand) from 51 (the rest of the deck). So there are $\binom{51}{4}$ favorable hands and

$$P(Q_S) = \frac{\binom{51}{4}}{\binom{52}{5}} = \frac{51!}{4!\, 47!} \cdot \frac{5!\, 47!}{52!} = \frac{5}{52}$$

Example 5

I'll find the probability of *not* getting the queen of spades in poker.

Method 1 (directly)   Again, the total number of poker hands is $\binom{52}{5}$. Each favorable hand contains 5 cards chosen from the 51 non-Q_S's. So there are $\binom{51}{5}$ favorable hands, and

$$P(\text{not getting the queen of spades}) = \frac{\binom{51}{5}}{\binom{52}{5}}$$

Method 2 (indirectly)

$$P(\overline{Q_S}) = 1 - P(Q_S) = 1 - \text{ answer to example 4} = 1 - \frac{5}{52}$$

The two answers agree:

$$\text{method 1 answer} = \frac{51!}{5!\, 46!} \frac{5!\, 47!}{52!} = \frac{47}{52} = \text{ method 2 answer}$$

$9 \times 8 \times 7$ versus $\binom{9}{3}$

Both count the number of ways in which 3 things can be chosen from a pool of 9. But $9 \times 8 \times 7$ corresponds to choosing the 3 things to fill labeled slots (such as president, vice president, secretary), while $\binom{9}{3}$ corresponds to the case where the 3 things are not given different labels or distinguished from one another in any way (such as 3 co-chairs).

From a committee/lineup point of view, $9 \times 8 \times 7$ counts lineups of 3 from a pool of 9, while $\binom{9}{3}$ counts committees.

Poker Hands versus Poker Lineups

Consider the probability of getting a poker hand with all hearts.

A poker hand is a committee of cards, and

$$(6) \qquad P(\text{all hearts}) = \frac{\binom{13}{5}}{\binom{52}{5}}$$

(For the numerator, pick 5 cards from the 13 hearts.)

If we consider a poker *lineup*, that is, a *lineup* of 5 cards drawn from 52, then

$$(7) \qquad P(\text{all hearts}) = \frac{13 \cdot 12 \cdot 11 \cdot 10 \cdot 9}{52 \cdot 51 \cdot 50 \cdot 49 \cdot 48}$$

The answers in (6) and (7) agree since

$$\frac{\binom{13}{5}}{\binom{52}{5}} = \frac{13!}{5! \, 8!} \cdot \frac{5! \, 47!}{52!} = \frac{13 \cdot 12 \cdot 11 \cdot 10 \cdot 9}{52 \cdot 51 \cdot 50 \cdot 49 \cdot 48}$$

In (6) the underlying sample space is the set of *unordered* samples of size 5 chosen without replacement from a population of 52. In (7), the sample space is the set of *ordered* samples so that $A_S K_H J_D 2_S 3_C$ and $K_H A_S J_D 2_S 3_C$ are different outcomes. The probability of all hearts is the *same* in both spaces so it's OK to use poker lineups instead of poker hands. (But (6) seems more natural to most people.)

Warning

In some instances, such as the probability of all hearts in a poker hand, you can use ordered samples or unordered samples as long as you are *consistent* in the numerator and denominator. *If you use order in one place, you must use it in the other place as well.*

Problems for Section 1-2

1. A committee of size 5 is chosen from $A_1, \ldots, A_9$. Find the probability that
   (a) the committee contains A_6
   (b) the committee contains neither A_7 nor A_8

2. Find the probability of a bridge hand (13 cards) with the AKQJ of spades and no other spades.

3. If 2 people are picked from 5 men, 6 women, 7 children, find the probability that they are not both children.

4. Find the probability of a poker hand with
   (a) no hearts
   (b) the ace of spades and the ace of diamonds (other aces allowed also)
   (c) the ace of spades and the ace of diamonds and no other aces

5. Four women check their coats and the coats are later returned at random. Find the probability that
   (a) each woman gets her own coat
   (b) Mary gets her own coat

6. Compute
   (a) $7!/5!$ (c) $\binom{8}{5}/\binom{4}{3}$ (e) $\binom{12345}{0}$ (g) $\binom{12345}{12344}$
   (b) $\binom{7}{5}$ (d) $\binom{12345}{1}$ (f) $\binom{12345}{12345}$

7. In a certain computer system you must identify yourself with a password consisting of a single letter or a letter followed by as many as 6 symbols which may be letters or digits, for example, Z, ZZZZZZ6, RUNNER, JIMBO, R2D2. Assuming that any password is as likely to be chosen as any other, what is the probability that John and Mary choose the same password?

8. Consider samples of size 3 chosen from $A_1, \ldots, A_7$.
   (a) Suppose the samples are drawn *with replacement* so that after an item is selected, it is replaced before the next draw (e.g., A_1 can be drawn more than once) and *order counts* (e.g., $A_2 A_1 A_1$ is different from $A_1 A_2 A_1$). How many samples are there?
   (b) Suppose the samples are drawn *without replacement* and *order counts*. How many are there?
   (c) How many samples are there if the sampling is *without replacement* and *order doesn't count*?
   (d) What is the fourth type of sampling? (Counting in this case is tricky and won't be needed in this course.)

9. Find the probability of a royal flush in poker (a hand with AKQJ 10 all in the same suit).

10. There are 7 churches in a town. Three visitors pick churches at random to attend. Find the probability that

    (a) they all choose the same church
    (b) they do not all choose the same church
    (c) they choose 3 different churches
    (d) at least 2 of them choose the same church

11. In the state lottery the winning ticket is decided by drawing 6 different numbers from 1 to 54. The order in which the numbers are drawn is irrelevant so that the draws 2, 54, 46, 37, 1, 6 and 54, 2, 37, 46, 1, 6 are the same.

    For $1 you get two tickets, each with your choice of 6 numbers on it. What is the probability that you win the lottery when you spend $1?

SECTION 1-3   COUNTING TECHNIQUES CONTINUED

Here are some examples that are more intricate but can still be done using the multiplication principle and permutation and combination rules from the last section.

Example 1

A committee of size 7 is chosen from 6 men, 7 women, 8 children. Let's find the probability that the committee contains

(a) 2 men, 4 women, 1 child

(b) 2 men

   (a) The total number of committees is $\binom{21}{7}$.

   For the favorable outcomes, the 2 men can be picked in $\binom{6}{2}$ ways, the 4 women in $\binom{7}{4}$ ways, and the child in 8 ways. (Think of filling three slots: a 2-man subcommittee, a 4-woman subcommittee, and a 1-child subcommittee.) So

   $$P(2\text{M}, 4\text{W}, 1\text{C}) = \frac{\binom{6}{2}\binom{7}{4} \cdot 8}{\binom{21}{7}}$$

   (b) For the favorable outcomes, the 2 men can be picked in $\binom{6}{2}$ ways, the 5 others in $\binom{15}{5}$ ways. So

   $$P(2 \text{ men}) = \frac{\binom{6}{2}\binom{15}{5}}{\binom{21}{7}}$$

Example 2

Find the probability that a poker hand contains only one suit.

To count the favorable hands, pick the suit in 4 ways. Then pick the 5 cards from that suit in $\binom{13}{5}$ ways. So

$$P(\text{only one suit}) = \frac{4 \cdot \binom{13}{5}}{\binom{52}{5}}$$

Example 3

A box contains 40 white, 50 red, 60 black balls. Pick 20 without replacement. Find the prob of getting

(a) 10 white, 4 red, 6 black

(b) 10 white

(a) The total number of outcomes is $\binom{150}{20}$. For the favorable, pick 10 white out of 40, pick 4 red out of 50, pick 6 black out of 60:

$$P(10\text{W}, 4\text{R}, 6\text{B}) = \frac{\binom{40}{10}\binom{50}{4}\binom{60}{6}}{\binom{150}{20}}$$

(b) For the favorable, pick 10 white out of 40, pick 10 others from the 110 non-white:

$$P(10\text{W}) = \frac{\binom{40}{10}\binom{110}{10}}{\binom{150}{20}}$$

Indistinguishable versus Distinguishable Balls

You may not have realized it, but as we picked committees and used fav/total in example 3, we assumed that balls of the same color could be distinguished from one another; for example, we assumed that balls were named $W_1, \ldots, W_{40}$; $R_1, \ldots, R_{50}$; and $B_1, \ldots, B_{60}$. The probability of getting 10W, 4R, 6B is the same whether or not the balls have names painted on them, so our assumption is not only convenient but legal.

> If an experiment involves n white balls, to find any probabilities we assume the balls are named $W_1, \ldots, W_n$.

Example 4

Consider strings of digits and letters of length 7 without repetition. Find the probability that a string contains 2 digits, 4 consonants, and 1 vowel.

Method 1   For the favorable outcomes, pick 2 digits, 4 consonants, and 1 vowel and then line them up.

$$\text{prob} = \frac{\binom{10}{2}\binom{21}{4} \cdot 5 \cdot 7!}{36 \cdot 35 \cdot 34 \cdot 33 \cdot 32 \cdot 31 \cdot 30}$$

Method 2   For the favorable outcomes, pick 2 positions in the string for the digits, 4 places for the consonants, leaving 1 for the vowel. Then fill the spots.

$$\text{prob} = \frac{\binom{7}{2}\binom{5}{4} \cdot 10 \cdot 9 \cdot 21 \cdot 20 \cdot 19 \cdot 18 \cdot 5}{36 \cdot 35 \cdot 34 \cdot 33 \cdot 32 \cdot 31 \cdot 30}$$

Double Counting

I'll find the probability of a poker hand with two pairs.

If I regard a poker hand as a committee, then the total number of hands is $\binom{52}{5}$.

Now I need the number of favorable hands.

To get started, it often helps to write out some favorable outcomes—the steps involved in constructing an outcome may suggest the slots to be filled in the counting process. Here are some poker hands with two pairs:

Hand 1: Q_H, Q_D, J_S, J_H, A_H
Hand 2: 4_C, 4_H, K_D, K_C, 2_H

First, let's analyze an *incorrect* method.

Each outcome involves a first face (e.g., queen), two suits in that face (e.g., heart and diamond), a second face (e.g., jack), two suits in that face (e.g., spade and heart), and a fifth card not of either face (to avoid a full house). So use these slots:

Step 1. Pick a face value.
Step 2. Pick 2 of the 4 cards in that face.
Step 3. Pick another face value.   *WRONG*
Step 4. Pick 2 out of the 4 cards in that face.
Step 5. Pick a fifth card from the 44 of neither face.

Filling these 5 slots we get "answer"

$(*)$ $13 \cdot \dbinom{4}{2} \cdot 12 \cdot \dbinom{4}{2} \cdot 44$

This is wrong because it counts the following as different outcomes when they are really the same hand:

Outcome 1	Outcome 2
Pick queen, hearts and spades.	Pick jack, hearts and clubs.
Pick jack, hearts and clubs.	Pick queen, hearts and spades.
Pick ace of clubs.	Pick ace of clubs.

Step 1 implicitly fills a slot named first face value and step 3 fills a slot named second face value. But in a poker hand, the two faces for the two pairs can't be distinguished as first and second so the slots are illegal.

The "answer" in (∗) counts every outcome twice. (Once you notice this, you can divide by 2 to get the right answer.)

Here's a correct version (from scratch):

Pick a committee of 2 face values for the pairs.

Pick 2 cards from each face value.

Pick a fifth card from the 44 not of either face.

The number of favorable hands is $\binom{13}{2}\binom{4}{2}\binom{4}{2} \cdot 44$.

And the probability of two pairs is

$$\frac{\binom{13}{2}\binom{4}{2}\binom{4}{2} \cdot 44}{\binom{52}{5}}$$

In general, an "answer" that *counts some outcomes more than once* is referred to as a *double count* (the preceding double count happens to count *every* outcome exactly *twice*). Double counts can be hard to resist.

Warning

> A common mistake is to use, say, $7 \cdot 6$ instead of $\binom{7}{2}$, that is, to fill slots implicitly named first thingo and second thingo when you should pick a committee of two things.

Symmetries

Here are some typical symmetries.

Draw cards either with or without replacement.

(1)
> $P(\text{ace on 1st draw}) = P(\text{ace 2nd}) = P(\text{ace 3rd})$, etc.
>
> $P(\text{ace 3rd, king 10th}) = P(\text{A 1st, K 2nd}) = P(\text{A 2nd, K 1st,})$, etc.

Intuitively, each position in the deck of cards has the same chance of harboring an ace; each *pair* of positions has the same chance of containing an ace and king, and so on.

Similarly, draw with or without replacement from a box containing red, white, and black balls.

(2)
$$\boxed{\begin{aligned} P(\text{RWBW drawn in that order}) &= P(\text{WWRB}) \\ &= P(\text{RBWW}), \text{ etc.} \end{aligned}}$$

Whatever the distribution of colors you draw, you're just as likely to get them in one order as another.

It isn't safe to rely on intuition, so I'll do one proof as a justification. I'll show that

$$P(\text{A on 3rd draw, K on 5th draw}) = P(\text{K on 1st, A on 2nd})$$

This is immediate if the drawings are with replacement: Each probability is

$$\frac{4 \cdot 4}{52 \cdot 52}$$

Suppose the drawing is with*out* replacement. Then

$$P(\text{K on 1st, A on 2nd}) = \frac{\text{fav}}{\text{total}} = \frac{4 \cdot 4}{52 \cdot 51}$$

$$P(\text{A on 3rd, K on 5th}) = \frac{\text{fav}}{\text{total}} = \frac{\text{fav}}{52 \cdot 51 \cdot 50 \cdot 49 \cdot 48}$$

For the fav, there are 5 slots to fill. The 3rd slot can be filled in 4 ways, the 5th slot in 4 ways, and then the other 3 slots in $50 \cdot 49 \cdot 48$ ways. So

$$P(\text{A on 3rd, K on 5th}) = \frac{4 \cdot 4 \cdot 50 \cdot 49 \cdot 48}{52 \cdot 51 \cdot 50 \cdot 49 \cdot 48} = \frac{4 \cdot 4}{52 \cdot 51}$$

The "other 3 slots" canceled out, leaving the same answer as $P(\text{K 1st, A 2nd})$, QED.

Example 5

Draw without replacement from a box with 10 white and 5 black balls.

To find the prob of W on the 1st and 4th draws (no information about 2nd and 3rd), take advantage of symmetry and switch to an easier problem:

$$P(\text{W on 1st and 4th}) = P(\text{W on 1st and 2nd}) = \frac{10 \cdot 9}{15 \cdot 14}$$

Problems for Section 1-3

1. Find the prob that a poker hand contains
   (a) 3 diamonds and 2 hearts
   (b) 2 spades, one of which is the ace
   (c) 4 black and 1 red
   (d) 2 aces
   (e) the ace of spades but not the king of spades

2. Three Americans A_1, A_2, A_3, 7 Russians $R_1, \ldots, R_7$ and 8 Germans $G_1, \ldots, G_8$ try to buy concert tickets. Only 5 tickets are left. If the tickets are given out at random, find the prob that
   (a) R_3 gets a ticket and so do 2 of the 3 Americans
   (b) only 1 of the Germans gets a ticket

3. If 3 people are picked from a group of 4 married couples, what is the prob of not including a pair of spouses?

4. If a 12-symbol string is formed from the 10 digits and 26 letters, repetition not allowed, what is the prob that it contains 3 even digits?

5. Find the prob of getting 3 whites and 2 reds if you draw 11 balls from a box containing 25 white, 30 red, 40 blue, and 50 black.

6. If four people are assigned seats at random in a 7-seat row, what is the prob that they are seated together?

7. Find the prob that a 3-card hand contains 3 of a kind (i.e., 3 of the same value).

8. (a) Find the prob that a 4-card hand contains 2 pairs.
   (b) Find the prob that a 5-card hand contains a full house (3 of a kind and a pair).

9. Find the prob that a poker hand contains
   (a) a flush (5 cards of the same suit)
   (b) 4 aces
   (c) 4 of a kind (e.g., 4 jacks)
   (d) a pair (and nothing better than one pair)

10. Suppose b boys and g girls are lined up at random. Find the prob that there is a girl in the ith spot.

11. Here are some counting problems with proposed answers that double count. Explain *how* they double count (produce specific outcomes which are counted as if they are distinct but are really the same) and then get correct answers.
    (a) To count the number of poker hands with 3 of a kind:

    Pick a face value and 3 cards from that value.
    Pick one of the remaining 48 cards not of that value (to avoid 4 of a kind).

Pick one of the remaining 44 not of the first or second value (to avoid 4 of a kind and a full house).

Answer is $13 \cdot \binom{4}{3} \cdot 48 \cdot 44$. WRONG

(b) To count 7-letter words with 3 A's:

Pick a spot for the first A.
Pick a spot for the second A.
Pick a spot for the third A.
Pick each of the remaining 4 places with any of the non-A's.

Answer is $7 \cdot 6 \cdot 5 \cdot 25^4$. WRONG

(c) To count 2-card hands not containing a pair:

Pick any first card.
Pick a second card from the 48 not of the first face value.

Answer is $52 \cdot 48$. WRONG

SECTION 1-4   ORS AND AT LEASTS

OR versus XOR

The word "or" has two different meanings in English. If you've seen Boolean algebra you know that engineers have two different words, OR and XOR, to distinguish the two meanings:

A OR B means A or B or both, called an *inclusive* or.

Similarly, A OR B OR C means one or more of A, B, C (exactly one of A, B, C or any two of A, B, C or all three of A, B, C). On the other hand,

A XOR B means A or B but not both, an *exclusive* or.

In this book, *"or" will always mean the inclusive OR* unless specified otherwise.

In the real world you'll have to decide for yourself which kind is intended. If a lottery announces that any number containing a 6 or a 7 wins, then you win with a 6 or 7 or both; that is, 6 OR 7 wins (inclusive or). On the other hand, if you order a Coke or a 7-Up, you really mean a Coke or a 7-Up but not both; that is, Coke XOR 7-Up.

OR Rule (Principle of Inclusion and Exclusion)

For 2 events,

(1)
$$P(A \text{ or } B) = P(A) + P(B) - P(A \text{ and } B)$$

For 3 events,

(2)

$$P(A \text{ or } B \text{ or } C) = P(A) + P(B) + P(C)$$
$$- [P(A \& B) + P(A \& C) + P(B \& C)]$$
$$+ P(A \& B \& C)$$

Here's the general pattern for n events:

(3)

$$P(A_1 \text{ or } A_2 \text{ or } \ldots \text{ or } A_n)$$
$$= \text{1-at-a-time terms}$$
$$- \text{2-at-a-time terms}$$
$$+ \text{3-at-a-time terms}$$
$$- \text{4-at-a-time terms}$$
$$\text{etc.}$$

For example,

$$P(A \text{ or } B \text{ or } C \text{ or } D)$$
$$= P(A) + P(B) + P(C) + P(D)$$
$$- [P(A \& B) + P(A \& C) + P(A \& D) + P(B \& C) + P(B \& D) + P(C \& D)]$$
$$+ [P(A \& B \& C) + P(A \& B \& D) + P(B \& C \& D) + P(A \& C \& D)]$$
$$- P(A \& B \& C \& D)$$

Proof of (1)

Suppose event A contains the 4 outcomes indicated in Fig. 1 with respective probs $p_1, \ldots, p_4$. And suppose B contains the indicated 3 outcomes.

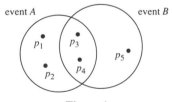

Figure 1

The event "A or B" contains all 5 outcomes in Fig. 1, so

$$P(A \text{ or } B) = p_1 + p_2 + p_3 + p_4 + p_5$$

On the other hand,

$$P(A) + P(B) = p_1 + p_2 + p_3 + p_4 \quad + \quad p_3 + p_4 + p_5$$

This is *not* the same as $P(A \text{ or } B)$ because it counts the probs p_3 and p_4 *twice*. We *do* want to count them since this is an inclusive or, but we don't want to count them twice. So to get $P(A \text{ or } B)$, start with $P(A) + P(B)$ and then subtract the probs in the intersection of A and B, that is, subtract $P(A \text{ and } B)$ as in (1).

Warning

The "or" in rule (1) is *inclusive*; it means A or B *or both*. We subtract away $P(A\&B)$ not because we want to throw away the both's but because we don't want to count them twice.

In other words,

$$P(A \text{ or } B) = P(A \text{ or } B \text{ or both}) = P(A) + P(B) - P(A\&B)$$

Proof of (2)

Suppose A contains the 5 outcomes in Fig. 2 with indicated probs, B contains the indicated 6 outcomes, and C contains the indicated 4 outcomes. Then

$$P(A \text{ or } B \text{ or } C) = p_1 + \cdots + p_9$$

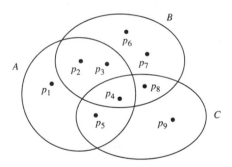

Figure 2

Look at $P(A) + P(B) + P(C)$. It does include $p_1, \ldots, p_9$, but it counts p_2, p_3, p_5, p_8 twice each and counts p_4 three times.

If we subtract away

$$\underbrace{P(A\&B),}_{p_2+p_3+p_4} \quad \underbrace{P(A\&C),}_{p_4+p_5} \quad \underbrace{P(B\&C)}_{p_4+p_8}$$

then p_2, p_3, p_5, p_8 will be counted once each, not twice, but now p_4 isn't counted at all. So formula (2) adds $P(A\&B\&C)$ back in to include p_4 again.

Example 1

I'll find the probability that a bridge hand (13 cards) contains 4 aces or 4 kings.

By the OR rule,

$$P(4 \text{ aces or } 4 \text{ kings}) = P(4 \text{ aces}) + P(4 \text{ kings}) - P(4 \text{ aces and } 4 \text{ kings})$$

The total number of hands is $\binom{52}{13}$.

When you count the number of ways of getting 4 aces, don't think about kings at all (the hand may or may not include 4 kings—you don't care). The other 9 cards can be picked from the 48 non-aces so there are $\binom{48}{9}$ hands with 4 aces.

Similarly, for outcomes favorable to 4 kings, pick the other 9 cards from the 48 non-kings. And for the outcomes favorable to the event 4 kings and 4 aces, pick the other 5 cards from the 44 remaining cards. So

$$P(4 \text{ aces or } 4 \text{ kings}) = \frac{\binom{48}{9}}{\binom{52}{13}} + \frac{\binom{48}{9}}{\binom{52}{13}} - \frac{\binom{44}{5}}{\binom{52}{13}}$$

Unions and Intersections

Many books use the union symbol instead of "or" and the intersection symbol instead of "and" so that the OR rule for two events looks like this:

$$P(A \cup B) = P(A) + P(B) - P(A \cap B)$$

I use "or" and "and" because they seem more natural. Most people would refer to 3 aces *or* 3 kings rather than to the *union* of 3-ace hands and 3-king hands.

Mutually Exclusive (Disjoint) Events

Suppose events A, B, C, D are mutually exclusive, meaning that no two can happen simultaneously (A, B, C, D have no outcomes in common). Then all the "and" terms in (1), (2), and (3) drop out, and we have

(1′) $$P(A \text{ or } B) = P(A) + P(B)$$

(2′) $$P(A \text{ or } B \text{ or } C) = P(A) + P(B) + P(C)$$

(3′) $$P(A_1 \text{ or } \ldots \text{ or } A_n) = P(A_1) + \cdots + P(A_n)$$

Example 2

Consider poker hands containing all spades or all hearts. The events "all spades" and "all hearts" are mutually exclusive since they can't happen simultaneously. So by (1′),

$$P(\text{all spades or all hearts}) = P(\text{all spades}) + P(\text{all hearts})$$

$$= \frac{\binom{13}{5}}{\binom{52}{5}} + \frac{\binom{13}{5}}{\binom{52}{5}}$$

Warning

$P(A \text{ or } B)$ is not $P(A) + P(B)$ *unless* A and B are mutually exclusive. If they are not, *don't forget to subtract $P(A\&B)$*.

At Least One

To illustrate the general idea I'll find the probability that a poker hand contains at least one ace.

Method 1 (the best for this particular example)

$$P(\text{at least one ace}) = 1 - P(\text{no aces}) = 1 - \frac{\binom{48}{5}}{\binom{52}{5}}$$

Method 2

$$P(\text{at least one ace}) = P(1A \text{ or } 2A \text{ or } 3A \text{ or } 4A)$$

The events one ace (meaning *exactly* one ace), two aces, three aces, four aces are mutually exclusive, so we can use the abbreviated OR rule.

$$P(\text{at least one ace}) = P(1A) + P(2A) + P(3A) + P(4A)$$

$$= \frac{4\binom{48}{4} + \binom{4}{2}\binom{48}{3} + \binom{4}{3}\binom{48}{2} + 48}{\binom{52}{5}}$$

Method 3

$$P(\text{at least one ace}) = P(A_S) + P(A_H) + P(A_C) + P(A_D)$$
$$- [P(A_S \& A_H) + \text{ other 2-at-a-time terms}]$$
$$+ [P(A_S \& A_H \& A_C) + \text{ other 3-at-a-time terms}]$$
$$- P(A_S \& A_H \& A_C \& A_D)$$

This long expansion isn't as bad as it looks.

The first bracket contains 4 terms all having the same value, namely, $\binom{51}{4}/\binom{52}{5}$.

The second bracket contains $\binom{4}{2}$ terms; all have the value $\binom{50}{3}/\binom{52}{5}$.

The third bracket contains $\binom{4}{3}$ terms, each with the value $\binom{49}{2}/\binom{52}{5}$.

So

$$P(\text{at least one ace}) = \frac{4\binom{51}{4} - \binom{4}{2}\binom{50}{3} + \binom{4}{3}\binom{49}{2} - 48}{\binom{52}{5}}$$

In this example, method 1 was best, but you'll see examples favoring each of the other methods.

Warning

Here is a wrong way to find the probability of at least one ace in a poker hand. The denominator is $\binom{52}{5}$ (right so far). For the numerator:

WRONG

Pick 1 ace to be sure of getting at least one.
Pick 4 more cards from the remaining 51.
So the numerator is $4\binom{51}{4}$.

The numerator is wrong because it counts the following as different outcomes when they are the same:

Outcome 1	Outcome 2
Pick the A_S as the sure ace.	Pick the A_H as the sure ace.
Then pick Q_H, A_H, 2_D, 6_C.	Then pick Q_H, A_S, 2_D, 6_C.

So the numerator *double counts*.

> Don't try to count at least n thingos by presetting n thingos to be sure and then going on from there. It just won't work.

Example 3

Here's how to find the prob that a bridge hand contains at most 2 spades.

Method 1

$$P(\text{at most 2 spades}) = 1 - P(3S \text{ or } 4S \text{ or } \ldots \text{ or } 13S)$$
$$= 1 - [P(3S) + P(4S) + \cdots + P(13S)]$$

OK, but too slow!

Method 2

$$P(\text{at most 2 spades}) = P(\text{no S}) + P(1S) + P(2S)$$

$$= \frac{\binom{39}{13} + 13\binom{39}{12} + \binom{13}{2}\binom{39}{11}}{\binom{52}{13}}$$

Exactlies Combined with At Leasts

I'll find the probability of a poker hand with 2 spades and at least 1 heart.

Method 1

Use a variation of the rule

$$P(\text{at least 1 heart}) = 1 - P(\text{no hearts})$$

to get

$$P(\text{2 spades and at least 1 heart}) = P(2S) - P(\text{2S and no H})$$

$$= \frac{\binom{13}{2}\binom{39}{3} - \binom{13}{2}\binom{26}{3}}{\binom{52}{5}}$$

Method 2

$$P(\text{2 spades and at least 1 heart})$$
$$= P(\text{2S and H}) + P(\text{2S and 2H}) + P(\text{2S and 3H})$$
$$= \frac{\binom{13}{2} \cdot 13 \cdot \binom{26}{2} + \binom{13}{2}\binom{13}{2} \cdot 26 + \binom{13}{2}\binom{13}{3}}{\binom{52}{5}}$$

At Least One of Each

Form committees of 6 from a population of 10 Americans, 7 Russians, and 5 Germans.

$$P(\text{at least 1 of each European nationality})$$
$$= P(\text{at least one R and at least one G})$$
$$= 1 - P(\text{no R or no G})$$
$$= 1 - [P(\text{no R}) + P(\text{no G}) - P(\text{no R and no G})]$$
$$= 1 - \frac{\binom{15}{6} + \binom{17}{6} - \binom{10}{6}}{\binom{22}{6}}$$

Warning

The complement of

at least 1 of each European nationality

is *not* no Europeans; that is, the complement is *not*

no R   *and*   no G

Rather, the complement is

no R   *or*   no G

Some Basic Pairs of Complementary Events

Here is a brief list of some complementary events. (You should understand the logic of the list rather than force yourself to memorize it.)

Event	**Complement**
A or B	not A and not B
A and B	not A or not B
at least 1 king	no kings
at least 1 king and at least 1 queen	no kings or no queens
at least 1 king or at least 1 queen	no kings and no queens
at least 1 of each suit	no S or no H or no C or no D

Problems for Section 1-4

1. Find the prob that a poker hands contains
   (a) 2 aces or 2 kings
   (b) 3 aces or 3 kings
   (c) the ace or king of spades

2. A box contains 10 white balls, 20 reds, and 30 greens. Draw 5 without replacement. Find the prob that
   (a) the sample contains 3 white or 2 red or 5 green
   (b) all 5 are the same color

3. Consider computing $P(A_1$ or A_2 or ... or $A_8)$.
   (a) Eventually, you have to subtract away the 2-at-a-time terms such as $P(A_5 \& A_3)$. How many of these terms are there?
   (b) Eventually, you will have to add in the 3-at-a-time terms such as $P(A_1 \& A_4 \& A_5)$. How many are there?

4. A jury pool consists of 25 women and 17 men. Among the men, 2 are Hispanic, and among the women, 3 are Hispanic. If a jury of 12 people is picked at random what is the prob that it

   (a) contains no women or no Hispanics
   (b) contains no women and no Hispanics

5. Find the prob that a poker hand contains the jack of spades XOR the queen of spades (the jack or queen but not both).

6. Find the prob that a poker hand contains

   (a) at least 1 spade
   (b) at least 3 spades
   (c) at most 2 aces
   (d) 4 pictures including at least 2 aces

7. Find the prob that a bridge hand contains

   (a) at least one royal flush (AKQJ 10 in the same suit)
   (b) at least one 4-of-a-kind (e.g., 4 kings)

8. A hundred people including the Smith family (John, Mary, Bill, Henry) buy a lottery ticket apiece. Three winning tickets will be drawn without replacement. Find the prob that the Smith family ends up happy. (Find three methods if you can, for practice.)

9. Find the prob that a 4-person committee chosen from 6 men, 7 women, and 5 children contains

   (a) 1 woman
   (b) at least 1 woman
   (c) at most 1 woman
   (d) at least 1 of each category
   (e) no women and at least 1 man

10. There are 50 states and 2 senators from each state. Find the prob that a committee of 15 senators contains

   (a) at least 1 from each of Hawaii, Massachusetts, and Pennsylvania
   (b) at least 1 from the three-state region composed of Hawaii, Massachusetts, and Pennsylvania
   (c) 1 from Hawaii and at least 1 from Massachusetts

11. (the game of rencontre—the matching game)   Seven husbands $H_1, \ldots,$ H_7 and their wives $W_1, \ldots, W_7$ are matched up at random to form 7 new coed couples.

   (a) Find the prob that H_3 is matched with his own wife.
   (b) Find the prob that H_2, H_5, and H_7 are all matched with their own wives.

(c) Find the prob that at least one husband is matched with his wife and simplify to get a pretty answer.

Suggestion: The only feasible method is to use

$$P(\text{at least one match}) = P(H_1 \text{ is matched or } H_2 \text{ or } \ldots \text{ or } H_7)$$

(d) Find the prob that no husband is matched with his wife.

Review

Here's one way to draw the graph of the inequality $x - y < 2$. First, draw the graph of the equation $x - y = 2$, a line. The line divides the plane into two regions (Fig. 1). One of the regions corresponds to $x - y < 2$ and the other to $x - y > 2$. To decide which region goes with which inequality, test a point. Point $(100, 0)$ for instance is in region 2, and it satisfies the inequality $x - y > 2$. So it is region 1 that must be the graph of $x - y < 2$.

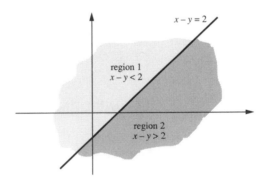

Figure 1

Another way to graph inequalities is to solve for y. The graph of

$$y < x - 2$$

is the region *below* line $y = x - 2$ and the graph of

$$y > x - 2$$

is the region *above* line $y = x - 2$.

Here are some more graphs of inequalities.

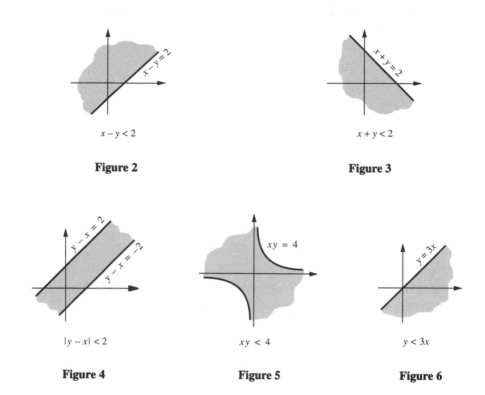

Figure 2

Figure 3

Figure 4

Figure 5

Figure 6

SECTION 1-5   CONTINUOUS UNIFORM DISTRIBUTIONS

This is supposed to be a chapter on *discrete* probability. But there is a type of continuous problem that is so like the discrete case that I'm going to tell you about it now.

One-Dimensional Continuous Uniform Distributions

Trains stop at your station at 8:00 A.M., 8:15 A.M., and 8:30 A.M. You arrive at the station at random between 8:00 A.M. and 8:30 A.M. We'll find the probability that you have to wait only 5 minutes or less before a train arrives.

We choose as our mathematical model a sample space consisting of all points in the interval $[0, 30]$ (or we could think in hours and use the interval $[0, 1/2]$). Figure 1 shows the total set of outcomes and the favorable arrival times (the times that lead to a wait of 5 minutes or less). All outcomes are equally likely (because you arrive at random) but we can't use

$$\frac{\text{favorable } number \text{ of outcomes}}{\text{total } number \text{ of outcomes}}$$

because there are an uncountably infinite number of each (that's why the problem is continuous rather than discrete). But in the same spirit we'll use

$$P(\text{wait 5 minutes or less}) = \frac{\text{favorable } \textit{length}}{\text{total } \textit{length}} = \frac{5+5}{30} = \frac{1}{3}$$

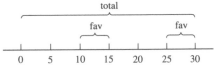

Figure 1

> Suppose a point is chosen at random in an interval—the outcome of the experiment is said to be *uniformly distributed* on (or in) the interval. We choose as the mathematical model a sample space (universe) consisting of all points in the interval, where
>
> $$P(\text{event}) = \frac{\text{favorable length}}{\text{total length}}$$

Two-Dimensional Continuous Uniform Distributions

John arrives at random between time 0 and time t_1, and Mary arrives at random between time 0 and t_2, where $t_1 < t_2$. Find the probability that John arrives before Mary.

Let x be John's arrival time and y be Mary's arrival time. Each outcome of the experiment is a *pair* of arrival times. The universe is the set of points in the rectangle in Fig. 2. The experiment amounts to picking a point at random from the rectangle.

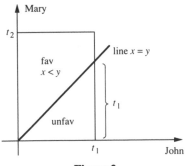

Figure 2

The favorable points are those in the rectangle where $x < y$ (see the review section if you don't remember how to graph inequalities).

For the 2-dimensional continuous analog of fav/total, we'll use

$$P(\text{John before Mary}) = \frac{\text{favorable area}}{\text{total area}}$$

Perhaps the easiest way to compute the fav area in Fig. 2 is indirectly, by first finding the unfavorable area:

$$P(\text{John before Mary}) = \frac{\text{entire rectangle} - \text{unfavorable triangle}}{\text{entire rectangle}}$$

$$= \frac{t_1 t_2 - \frac{1}{2} t_1^2}{t_1 t_2}$$

Suppose a point is chosen at random in a region in the plane—the outcome of the experiment is said to be *uniformly distributed* on (or in) the region.

As a special case, if a number x is chosen at random from the interval $[a, b]$ and a second number y is chosen (independently from the x choice) at random from the interval $[c, d]$, then the point (x, y) is uniformly distributed in a 2-dimensional rectangle.

We choose as the mathematical model a sample space (universe) consisting of all points in the region, where

$$P(\text{event}) = \frac{\text{favorable area}}{\text{total area}}$$

Warning

If *one* number is picked at random in an interval $[a, b]$, then probabilities involving that number are found using fav length/total length in the 1-dimensional universe $[a, b]$.

Suppose *two* numbers (e.g., his arrival time and her arrival time) are picked at random in $[a, b]$ and $[c, d]$, respectively. The probability of an event involving *both* numbers is found using fav area/total area in the 2-dimensional rectangle $a \leq x \leq b, c \leq y \leq d$.

Make sure you use the right dimension.

Events of Probability 0

Suppose an arrival time is uniformly distributed on $[0, 30]$. The probability of arriving *between* times 6 and 10 is 4/30. But the probability of arriving *at* the one particular time 7 (or 3/4 or 8 or π) is 0, even though the event is *not* impossible. (This is different from most discrete situations.) Only events that have "length" have positive probabilities. Furthermore, when you ask for the probability of arriving between 6 and 10, it doesn't matter whether you include the 6 and 10 or not (because the individual endpoints carry no probability). The following probabilities are all the same, namely, 4/30:

$$P(6 \leq \text{ arrival } \leq 10)$$
$$P(6 < \text{ arrival } < 10)$$
$$P(6 < \text{ arrival } \leq 10)$$
$$P(6 \leq \text{ arrival } < 10)$$

Similarly, suppose a point is uniformly distributed in a region in 2-space. Then the probability that it lies in a given subregion is positive, namely, subregion area/total area. But the probability that the point specifically is $(3, 7)$ or (π, e) or $(\frac{2}{3}, 0)$ is 0, even though these events are not impossible. Only events that have "area" have positive probabilities.

Example 1   (Buffon's needle problem—a marvelous problem)

A needle is tossed onto a planked floor. The needle has length L, and the width of each plank is D where $L < D$. I'll find the probability that the needle hits a crack.

The needle can land angled in any one of four ways (Fig. 3). If I find that 40% of the needles in one particular orientation hit the crack, then by symmetry, 40% of the needles in any other orientation also hit the crack, and the probability in general that a needle hits the crack would be .4. So all I have to do is work with one particular orientation.

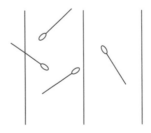

Figure 3

Let's use the orientation in Fig. 4 where the needle makes an acute angle θ with the horizontal. Let y be the distance from the eye of the needle to the right-hand crack. The position of the needle is determined by θ and y. Tossing the needle onto the floor at random is equivalent to picking θ at random between 0 and $\pi/2$ and picking y at random between 0 and D. The universe is a rectangle in θ, y space.

Figure 5 shows that a needle hits a crack if

$$L \cos \theta \geq y$$

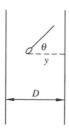

Figure 4

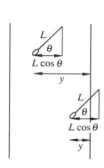

Figure 5

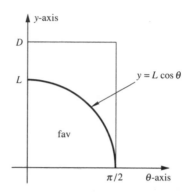

Figure 6

So the favorable region is the graph of $y \leq L \cos \theta$, the region under the graph of $y = L \cos \theta$. Figure 6 shows the favorable region inside the universe. Then

$$P(\text{needle hits a crack}) = \frac{\text{favorable area}}{\text{total area}}$$

$$= \frac{\int_0^{\pi/2} L \cos \theta \, d\theta}{(\pi/2)D}$$

$$= \frac{L}{(\pi/2)D} = \frac{2L}{\pi D}$$

(As expected, the probability goes down if L gets smaller or if D gets larger.)

Problems for Section 1-5

1. A number x is chosen at random between -1 and 1 so that x is uniformly distributed in $[-1, 1]$. Find the prob that

(a) $-\frac{1}{2} < x < 0$ (c) $|x - .5| \le .1$

(b) $-\frac{1}{2} \le x \le 0$ (d) $3x^2 > x$

2. A point is chosen at random in a circle of radius 9. Find the prob that it's within distance 2 of the center.

3. Consider the quadratic equation $4x^2 + 4Qx + Q + 2 = 0$, where Q is uniformly distributed on $[0, 5]$. Find the prob that the roots are real.

4. Suppose θ is uniformly distributed on $[-\pi/2, \pi/2]$. Find the prob that $\sin \theta > \frac{1}{3}$.

5. Consider a circle with radius R. Here are two ways to choose a chord in the circle. In each case find the prob that the chord is longer than the side of an inscribed equilateral triangle

(a) One end of the chord is point Q. The other end is determined by rotating the needle in the diagram by θ degrees where θ is uniformly distributed on $[0, 180]$.

(b) Pick a point at random on the radius AB in the diagram (so that the indicated distance d is uniformly distributed on $[0, R]$). Draw a chord through the point perpendicular to AB.

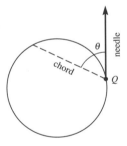

Figure P.5(a)

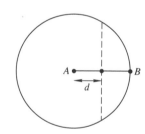

Figure P.5(b)

6. Choose a number at random between 0 and 1 and choose a second number at random between 1 and 3. Find the prob that

(a) their sum is ≤ 3 (b) their product is > 1

7. John and Mary agree to meet and each arrives at random between 10:00 and 11:00. Find the prob that

(a) the first to arrive has to wait more than 10 minutes for the other

(b) Mary arrives at least 20 minutes before John

(c) they arrive at the same time

REVIEW FOR THE NEXT CHAPTER

Series for e^x

$$1 + x + \frac{x^2}{2!} + \frac{x^3}{3!} + \frac{x^4}{4!} + \cdots = e^x \text{ for all } x$$

Series for e^{-x} (replace x by $-x$ in the e^x series)

$$1 - x + \frac{x^2}{2!} - \frac{x^3}{3!} + \frac{x^4}{4!} - \cdots = e^{-x} \text{ for all } x$$

Geometric Series

$$1 + x + x^2 + x^3 + x^4 + \cdots = \frac{1}{1-x} \quad \text{for } -1 < x < 1$$

$$a + ar + ar^2 + ar^3 + \cdots = \frac{a}{1-r} \quad \text{for } -1 < r < 1$$

Differentiated Geometric Series

$$1 + 2x + 3x^2 + 4x^3 + \cdots = D\left(\frac{1}{1-x}\right) = \frac{1}{(1-x)^2}$$

$$\text{for } -1 < x < 1$$

Finite Geometric Series

$$a + ar + ar^2 + \cdots + ar^n = \frac{a - ar^{n+1}}{1-r}$$

$$1 + x + x^2 + x^3 + \cdots + x^n = \frac{1 - x^{n+1}}{1-x}$$

Differentiated Finite Geometric Series

$$1 + 2x + 3x^2 + \cdots + nx^{n-1} = D\left(\frac{1 - x^{n+1}}{1-x}\right)$$

$$= \frac{nx^{n+1} - (n+1)x^n + 1}{(1-x)^2}$$

CHAPTER 2

Independent Trials and 2-Stage Experiments

**SECTION 2-1   CONDITIONAL PROBABILITY AND
                  INDEPENDENT EVENTS**

The aim of the chapter is to look at experiments with 2 (or more) stages. There
are two types:

1. The second stage depends on the first stage. For example, toss a coin,
    and if it comes up heads draw one card, but if it comes up tails draw
    two cards.

2. The second stage doesn't depend on the first stage. For example, toss
    a coin twice (or more).

  For the first type we need the idea of conditional probability, and for the
second type we need the idea of independent events.

Conditional Probability

Draw a card from a deck. The probability that the card is a king is 4/52.
But suppose you get a glimpse of the card, enough to tell that it's a picture.
The probability that the card is a king *given that it is a picture* is denoted by
$P(\text{king}|\text{picture})$ and is called a *conditional* probability. To find it, switch to a
new universe consisting of the 16 pictures:

$$P(\text{king}|\text{picture}) = P(\text{king in the new universe}) = \frac{4 \text{ kings}}{16 \text{ pictures}} = \frac{1}{4}$$

Here's the informal definition of conditional probability.

> (1) $P(B|A)$ is the probability that B occurs given that A has occurred. To find $P(B|A)$, cut the universe down to A, and then find the probability of B within the new universe.

For example, if 2 cards are drawn from a deck (without replacement) then

$$P(\text{spade on 2nd draw}|\text{spade on 1st draw}) = \frac{12 \text{ spades left in deck}}{51 \text{ cards left in deck}}$$

and

$$P(\text{spade on 2nd draw}|\text{heart on 1st draw}) = \frac{13 \text{ spades left}}{51 \text{ cards left}}$$

Formal Definition of Conditional Probability

If the condition A is more complicated than heart on 1st draw, it might not be so easy to work directly within the new universe. So we'll get a formula for conditional probabilities.

Here's an example to show the rationale behind the formula. Consider the universe in Fig. 1 where

$$P(A) = .6 \quad \text{and} \quad P(B) = .3$$

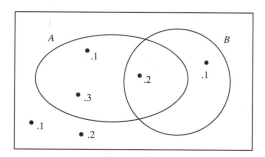

Figure 1

To assign a value to $P(B|A)$, consider A as the new universe. Within that new universe we see just one B-outcome, with probability .2. But a probability of .2 (w.r.t. the old universe where the total probability is 1) counts for more in the new universe, which only has a total probability of .6. It seems natural to choose

$$P(B|A) = \frac{.2}{.6}$$

Here's the idea in general:

$$(2) \qquad \boxed{P(B|A) = \frac{P(A \text{ and } B)}{P(A)} = \frac{\text{favorable prob within the } A \text{ world}}{\text{total prob of the } A \text{ world}}}$$

Most books call (2) the formal definition of conditional probability.

If you can easily picture the new universe determined by the condition A, then use (1) to find $P(B|A)$ directly. Otherwise, use (2), a safe, mechanical method.

The new A universe together with conditional probabilities assigned via (1) and/or (2) is a legitimate new probability space with all the usual properties; for example,

$P(B|A) = 1 - P(\bar{B}|A)$

$P(\text{at least one of them}|A) = 1 - P(\text{none of them}|A)$

$P(B \text{ or } C|A) = P(B|A) + P(C|A) - P(B \text{ and } C|A)$

etc.

Example 1

Find the prob that a poker hand contains 2 jacks if you already know it contains 1 ace (exactly 1 ace).

By (2),

$$P(2J|1A) = \frac{P(2J \text{ and } 1A)}{P(1A)} = \frac{\binom{4}{2} 4 \binom{44}{2}/\binom{52}{5}}{4\binom{48}{4}/\binom{52}{5}} = \frac{\binom{4}{2}\binom{44}{2}}{\binom{48}{4}}$$

Example 2

Draw 2 cards without replacement. Find the prob of 2 jacks given that at least one of the cards is a picture.

$$P(2J|\text{at least one picture}) = \frac{P(2J \text{ and at least one picture})}{P(\text{at least one picture})}$$

The event 2J and at least one picture is the same as the event 2J since a hand with 2J automatically has at least one picture. So

$$P(2J|\text{at least one picture}) = \frac{P(2J)}{1 - P(\text{no pictures})} = \frac{\binom{4}{2}/\binom{52}{2}}{1 - \binom{36}{2}/\binom{52}{2}}$$

$$= \frac{\binom{4}{2}}{\binom{52}{2} - \binom{36}{2}}$$

Chain Rules for AND

The definition in (2) can be rewritten as

(3)
$$P(A \text{ and } B) = P(A)\, P(B|A)$$

and in this form it can be thought of as a rule for AND.

The AND rule says that if A occurs, say, 20% of the time and B occurs 30% *of those times*, then A and B occur simultaneously 6% of the time.

By symmetry, we also have

(3')
$$P(A \text{ and } B) = P(B)\, P(A|B)$$

Similarly,

$$P(A \,\&\, B \,\&\, C) = P(A)\, P(B|A)\, P(C|A \,\&\, B)$$

and so on for as many events as you like.

(The event "A and B" is often denoted by AB.)

Example 3

Inside each box of Whizzo cereal is a card with 8 paint-covered circles. Underneath the paint, 3 of the circles contain the words "you," "win," and "prize," respectively. Figure 2 shows a typical card. If you can return the card to the Whizzo company with the paint scraped off precisely those three circles, you'll get a refund. (If you scrape the paint off a blank circle and try to paint over it again to cover your mistake, they'll be able to tell you cheated.) To what percentage of customers should the company be prepared to send refunds?

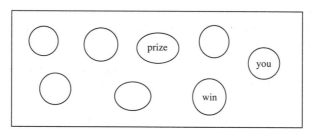

Figure 2

The experiment amounts to drawing 3 balls without replacement from a box containing 3 good balls and 5 bad balls. We want the probability of getting 3 goods.

Method 1

$$P(3 \text{ good}) = \frac{\text{favorable committees}}{\text{total committees}} = \frac{1}{\binom{8}{3}} = \frac{3!\,5!}{8!} = \frac{3!}{8 \cdot 7 \cdot 6} = \frac{1}{56}$$

Be prepared for about 2% of the customers to win refunds.

Method 2   Use the chain rule for AND.

$$P(3 \text{ good}) = P(\text{good on 1st draw \& good on 2nd draw \& good on 3rd draw})$$
$$= P(\text{G on 1st}) \, P(\text{G on 2nd}|\text{G on 1st}) \, P(\text{G on 3rd}|\text{G on 1st \& 2nd})$$
$$= \frac{3}{8} \cdot \frac{2}{7} \cdot \frac{1}{6}$$

Independent Events

If two events are unrelated so that the occurrence (or non-occurrence) of one of the events doesn't affect the likelihood of the other event, the events are called independent. We want to express this idea mathematically.

Here's the definition that's commonly given for independent events:

(4) $\quad\boxed{A \text{ and } B \text{ are independent iff } P(A \text{ and } B) = P(A)P(B).}$

To see why (4) captures the intuitive idea of independence, combine it with (3) to get

(5) $\quad\boxed{\begin{array}{l} A \text{ and } B \text{ are independent} \quad \text{iff } P(B|A) = P(B) \\ \qquad\qquad\qquad\qquad\qquad\qquad\quad \text{iff } A \text{ has no effect on } B; \end{array}}$

and similarly combine (4) with (3') to get

(6) $\quad\boxed{\begin{array}{l} A \text{ and } B \text{ are independent} \quad \text{iff } P(A|B) = P(A) \\ \qquad\qquad\qquad\qquad\qquad\qquad\quad \text{iff } B \text{ has no effect on } A. \end{array}}$

For example, suppose you toss a coin and draw a card. The sample space consists of $2 \cdot 52 = 104$ points such as heads and king of hearts, heads and queen of spades, and so on. Since the coin toss and card draw are intended to be independent, use (4) to assign probabilities to events:

$$P(\text{head and king of hearts}) = P(\text{head})P(\text{king of hearts}) = \frac{1}{2} \cdot \frac{1}{52} = \frac{1}{104}$$

$$P(\text{tail and heart}) = P(\text{tail})P(\text{heart}) = \frac{1}{2} \cdot \frac{13}{52} = \frac{1}{8}$$

The *three* events A, B, C are called independent *if all the following hold*:

(7)
$$P(A \text{ and } B \text{ and } C) = P(A)P(B)P(C)$$
$$P(A \text{ and } B) = P(A)P(B)$$
$$P(A \text{ and } C) = P(A)P(C)$$
$$P(B \text{ and } C) = P(B)P(C)$$

Similarly the *four* events A, B, C, D are independent if *all* the following hold:

(8)
$$P(A \text{ and } B \text{ and } C \text{ and } D) = P(A)P(B)P(C)P(D)$$
$$P(\text{any 3 at a time}) = \text{product of separate probs}$$
$$P(\text{any 2 at a time}) = \text{product of separate probs}$$

> If balls (or cards) are drawn *with* replacement, then any event associated with one drawing and any event associated with another drawing are physically independent; we refer to the drawings themselves as independent.
>
> Similarly, successive coin (or die) tosses are independent.
>
> In such cases, use (4) and its generalizations in (7) and (8), and so on, to assign probabilities.

Example 4

Toss a biased coin 3 times where $P(H) = 1/3$. Find the prob of HHT in that order.

The tosses are independent, so

$$P(\text{HHT}) = P(H)P(H)P(T) = \frac{1}{3} \cdot \frac{1}{3} \cdot \frac{2}{3}$$

Example 5

John and Mary take turns tossing one die; John goes first. The winner is the first player to throw a 4. Find the prob that John wins and the prob that Mary wins.

Method 1   Use the notation $\overline{4}\overline{4}4$ to represent the event

J throws a non-4, then M throws a non-4, and then J throws a 4.

Then

$$P(\text{J wins}) = P(4 \text{ or } \bar{4}4 \text{ or } \bar{4}\bar{4}\bar{4}\bar{4}4 \text{ or } \dots)$$
$$= P(4) + P(\bar{4}\bar{4}4) + P(\bar{4}\bar{4}\bar{4}\bar{4}4) + \cdots$$

(events are mutually exclusive)

$$= P(4) + [P(\bar{4})]^2 P(4) + [P(\bar{4})]^4 P(4) + \cdots$$

(tosses are independent)

$$= \frac{1}{6} + \left(\frac{5}{6}\right)^2 \frac{1}{6} + \left(\frac{5}{6}\right)^4 \frac{1}{6} + \left(\frac{5}{6}\right)^6 \frac{1}{6} + \cdots$$

This is a geometric series (see the review before Chapter 2) with $a = 1/6$, $r = (5/6)^2$, so

$$P(\text{J wins}) = \frac{1/6}{1 - (5/6)^2} = \frac{6}{11}$$

And

$$P(\text{no one wins}) = P(\bar{4} \text{ forever}) = \left(\frac{5}{6}\right)^\infty = 0$$
$$P(\text{Mary wins}) = 1 - P(\text{John wins}) = \frac{5}{11}$$

Method 2 (slick)   Let p be the prob that John wins. Then

(9)
$$p = P(4) + P(\bar{4}\bar{4} \text{ and then J throws the first 4 in the rest of the game})$$
$$= P(4) + P(\bar{4}\bar{4})P(\text{J throws the first 4 in the rest of the game}|\bar{4}\bar{4})$$

But the situation as the rest of the game begins (after $\bar{4}\bar{4}$) is the same as the situation at the beginning of the game where the prob that John wins is p. So

$$P(\text{J throws the first 4 in the rest of the game } |\bar{4}\bar{4}) = p$$

and (9) becomes

$$p = P(4) + P(\bar{4}\bar{4})p$$
$$p = \frac{1}{6} + \left(\frac{5}{6}\right)^2 p$$

Solve the equation to get the answer $p = 6/11$.

Probability of *A* Before *B*

Draw from a deck with or without replacement. Let's find the probability that a deuce is drawn before a picture.

Here's an intuitive argument. (Intuition can't always be trusted but in this case a more formal proof could be given to back it up.)

For all practical purposes, you must eventually draw a deuce or picture, that is, there *will* be a winning round: If the drawing is without replacement, this round will surely occur; if the drawing is with replacement, this round is not sure but occurs with probability 1 since the opposite event, 3–10 forever, has probability $\left(\frac{32}{52}\right)^{\infty} = 0$.[1] The most the other cards can do is delay the inevitable winning round; they have no effect on *who* wins, deuce or picture. So you might as well assume that the deck contains only the 4 deuces and 16 pictures to begin with, in which case the probability that you draw a deuce first is 4/20. In other words,

P(deuce before picture as you draw repeatedly from the original deck)
   $= P$(D before P on the round a winner is decided)
   $= P$(D before P|no D's or P's have been drawn yet but one is drawn *now*)
   $= P$(D on one draw from a new 20 card universe of D's and P's)
   $= \dfrac{4}{20}$

Here's a general rule for $P(A$ before $B)$ where A and B are two of the possible (disjoint) outcomes of an experiment and the experiment is performed over and over independently.

Drawings *with* replacement are one instance of repeated independent experiments so they are covered by the rule. Drawings with*out* replacement are not in this category, but it just so happens that the rule holds in this case too (as the deuce/picture example showed).

$P(A$ occurs before $B)$
   $=$ prob of A in *one* trial in a new universe where *only* A and B can occur
Equivalently,

$$P(A \text{ occurs before } B) = \frac{P(A)}{P(A) + P(B)}$$

$$= \frac{\text{fav prob in an } A, B \text{ only world}}{\text{total prob in the } A, B \text{ world}}$$

Problems for Section 2-1

1. Draw 2 cards from a deck without replacement. Find
   (a) P (second is a queen|first is a queen)
   (b) P(second is a queen|first is an ace)
   (c) P(first is higher or the 2 cards tie|first is a king)
       (Remember that aces are high.)
   (d) P(1 ace|first is an ace)

---

[1]The event 3–10 forever is *possible* because it *does* contain points, for example, 535353..., 34444..., 10^{∞} and so on. But it has probability 0.

2. Given $P(\text{rain}|\text{Jan. 7}) = .3$, find whichever of the following are possible.

  (a) $P(\text{not rain}|\text{Jan. 7})$ **(b)** $P(\text{rain}|\text{not Jan. 7})$

3. Toss 2 dice. Find the prob that the first is 6 given that the sum is 8.

4. Find the prob that a poker hand contains at least 1 king given that it contains at least 1 ace.

5. The North and South partners in bridge have 9 spades between them. Find the prob that the 4 spades held by the East-West pair split 3–1 (East has 3, West has 1, or vice versa).

6. A point is chosen at random from a unit square $ABCD$. Find the prob that it's in triangle ABD given that it's in triangle ABC.

7. A box contains 10 white, 9 black, and 5 red balls. Draw 4 without replacement.
  (a) Find the prob of BRBW (in that order).
  (b) Find $P(\text{BBRW})$.
  (c) Repeat part (a) if every time a red is drawn not only is it returned to the box but another red is added as well.
  (d) Find the prob of 2B, 1W, 1R.
  (e) Find the prob of W on the 4th draw.
  (f) Find the prob of W on the last two draws.

8. You send out three messages for help:

> a smoke signal that has prob .1 of being seen
> a message in a bottle that has prob .2 of being found
> a carrier pigeon that has prob .3 of arriving

Find the prob you are saved, assuming that smoke, bottles, and pigeons are independent.

9. Switches I and II in the diagram work independently. The prob is .7 that switch I is closed (allowing a signal to get through) and the prob is .2 that switch II is closed.

If a signal from A to B doesn't arrive, find the prob that switch II was open.

Figure P.9

10. (slippery)   Two balls are painted independently, white with prob 1/2 and black with prob 1/2, and then placed in a box. Find the prob that both balls are black if
  (a) you see a wet black paintbrush lying around
  (b) you draw a ball from the box and it's black

11. The prob that a missile hits its target is .8. Missiles are fired (independently) at a target until it is hit. Find the prob that it takes more than 3 missiles to get the hit.

12. A box contains 10 white balls and 5 black balls. Draw 4. Find the prob of getting W on the 1st and 4th draws if the drawing is

    (a) without replacement   **(b)** with replacement

13. Players A, B, C toss coins simultaneously. The prob of heads is p_a for A, p_b for B, and p_c for C.

    If the result is 2H and 1T or the result is 2T and 1H, then the player that is different from the other two is called the odd man out and the game is over. If the result is 3H or 3T, then the players toss again until they get an odd man out.

    Find the prob that A will be the odd man out.

14. Draw cards. Find the prob of getting a heart before a black card.

15. Keep tossing a pair of dice. Find the prob of getting 5 before 7.

16. There are 25 cars in the parking lot, with license numbers $C_1, C_2, \ldots,$ C_{25}. Assume the cars leave at random. Find the prob that
    (a) C_1 leaves before C_2
    (b) C_1 leaves before C_2 and C_3
    (c) C_1 leaves before C_2 which in turn leaves before C_3

17. A die is biased so that $P(1) = .2$ and $P(2) = .3$. Toss repeatedly. Find the prob of getting a 1 before a 2.

SECTION 2-2   THE BINOMIAL AND MULTINOMIAL DISTRIBUTIONS

Now we're ready to look at multi-stage experiments, beginning with the kind where the stages are independent. In particular we'll examine what happens when the same experiment is repeated over and over independently.

The Multinomial Distribution

Suppose an experiment has 4 mutually exclusive exhaustive outcomes $A, B, C,$ D (at each trial the result is exactly one of A, B, C, D).

Repeat the experiment, say, 9 times so that we have 9 independent trials. We'll show that

(1)

$$P(3A, 1B, 3C, 2D)$$
$$= \frac{9}{3!\,1!\,3!\,2!}\,[P(A)]^3\,[P(B)]^1\,[P(C)]^3\,[P(D)]^2$$

The general formula for n independent trials (instead of 9), where each trial has r possible outcomes (instead of 4), has the same pattern as (1). We say that the result of the n trials has a *multinomial distribution*.

Here's why the formula in (1) holds.

The sample space consists of 4^9 points, namely, all strings of length 9 using the letters A, B, C, D. The point $DDDDDDDDB$, for instance, corresponds to D on the first 8 trials followed by B on the 9th trial. Figure 1 shows some of the outcomes in the event "$3A, 1B, 3C, 2D$" (the favorable outcomes).

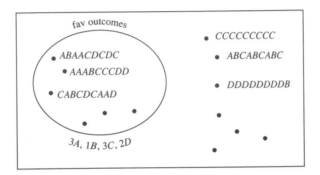

Figure 1   The universe

The 4^9 outcomes in the universe are not equally likely; for example, the outcome $CCCCCCCC$ has probability $[P(C)]^9$, while the outcome $AAAAAAAAA$ has probability $[P(A)]^9$. So we can't use fav/total. Instead, we'll find the probability of each favorable outcome and add them all up.

One of the *favorable* outcomes is $ABAACDCDC$. The trials are independent, so

(2)
$$P(ABAACDCDC) = P(A)P(B)P(A)P(A)P(C)P(D)$$
$$P(C)P(D)P(C)$$
$$= [P(A)]^3 \, [P(B)]^1 \, [P(C)]^3 \, [P(D)]^2$$

Similarly, *each* favorable outcome has the probability in (2).

Now we need to know how many favorable outcomes there are; that is, in how many ways can we arrange $3A, 1B, 3C, 2D$?

If the letters were $A_1, A_2, A_3, B, C_1, C_2, C_3, D_1,$ and D_2 we would have 9! permutations. Having 3 identical A's instead of A_1, A_2, A_3 means that the 9! is too large by a factor of 3!. Having 3 identical C's instead of C_1, C_2, C_3 makes the 9! too large by another factor of 3!. And having 2 identical D's instead of D_1, D_2 makes the 9! too large by a factor of 2!. So there are $9!/(3! \ 3! \ 2!)$ permutations.

(Here's another way to find the number of permutations. To line up the 9 letters, pick 3 places in the lineup for the 3 identical A's, pick one place for the B, pick 3 places for the 3 identical C's, then the last 2 places must go to the D's. This can be done in

$$\binom{9}{3} \cdot 6 \cdot \binom{5}{3} = \frac{9!}{3!\,6!} \cdot 6 \cdot \frac{5!}{3!\,2!} = \frac{9!}{3!\,3!\,2!}$$

ways.

So the event $3A, 1B, 2C, 2D$ consists of $9!/(3!\,3!\,2!)$ outcomes each with probability $[P(A)]^3 [P(B)]^1 [P(C)]^3 [P(D)]^2$. The probability of the event, the sum of these probabilities, is

$$\frac{9!}{3!\,3!\,2!} [P(A)]^3 [P(B)]^1 [P(C)]^3 [P(D)]^2$$

the formula in (1), QED.

(In order to show the pattern clearly, there's a 1! in the denominator in (1) to match the 1 B. But, of course, 1! is 1 so it's OK to leave it out.)

For example, if a die is tossed 100 times, following the pattern in (1),

$$P(25 \text{ 1's, } 30 \text{ 2's, and } 45 \text{ above } 2)$$

$$= \frac{100!}{25!\,30!\,45!} [P(1)]^{25} [P(2)]^{30} [P(\text{above } 2)]^{45}$$

$$= \frac{100!}{25!\,30!\,45!} \left(\frac{1}{6}\right)^{25} \left(\frac{1}{6}\right)^{30} \left(\frac{4}{6}\right)^{45}$$

$$= \frac{100!}{25!\,30!\,45!} \left(\frac{1}{6}\right)^{55} \left(\frac{4}{6}\right)^{45}$$

Some Typical Independent Trials

1. Coin tosses, die tosses
2. Tossing balls into boxes
3. Drawings *with* replacement
4. Drawings with*out* replacement from a *large* population (as in polling)

Actually, drawings with*out* replacement are *not* independent, but if the population is large, then one draw has such a slight effect on the next draw that for all practical purposes, they are independent and you can use the multinomial distribution to find probs.

The Classical Urn Problem

Draw 15 balls from a box containing 20 red, 10 white, 30 black, and 50 green. I'll find the probability of 7R, 2W, 4B, 2G.

If the drawing is *with replacement*, then there are 15 independent trials, and on any one trial, $P(R) = 20/110$, $P(W) = 10/110$, $P(B) = 30/110$, $P(G) = 50/110$. So

$$P(7R, 2W, 4B, 2G) = \frac{15!}{7!\ 2!\ 4!\ 2!} \left(\frac{20}{110}\right)^7 \left(\frac{10}{110}\right)^2 \left(\frac{30}{110}\right)^4 \left(\frac{50}{110}\right)^2$$

On the other hand, if the 15 balls are drawn *without replacement*, then they are a committee and

$$P(7R, 2W, 4B, 2G) = \frac{\binom{20}{7}\binom{10}{2}\binom{30}{4}\binom{50}{2}}{\binom{110}{15}}$$

Note that in the numerator, the "tops" must add up to 110 and the "bottoms" must add up to 15.

Example 1

Draw 15 with replacement from a box with 20R, 10W, 30B, 50G. The prob of 7R and 2W in the 15 draws is

$$P(7R, 2W) = P(7R, 2W, 6 \text{ others}) = \frac{15!}{7!\ 2!\ 6!} \left(\frac{20}{110}\right)^7 \left(\frac{10}{110}\right)^2 \left(\frac{80}{110}\right)^6$$

In a multinomial problem, make sure you remember the others if there are any.

On the other hand, the prob of 7R, 8W in the 15 draws is

$$P(7R, 8W) = P(7R, 8W, \text{no others}) = \frac{15!}{7!\ 8!\ 0!} \left(\frac{20}{110}\right)^7 \left(\frac{10}{110}\right)^8 \left(\frac{80}{110}\right)^0$$

$$= \frac{15!}{7!\ 8!} \left(\frac{20}{110}\right)^7 \left(\frac{10}{110}\right)^8$$

In this case, where there are no others, you might as well ignore them right from the beginning. There's no point in sticking in 0! and $(80/110)^0$ only to have them disappear again anyway.

Example 2

At the University of Illinois, 30% of the students are from Chicago, 60% are from downstate, and 10% are out of state. Find the probability that, if 10 students are picked at random by the Daily News to give their indignant reaction to the latest tuition increase, 6 are Chicagoans and 4 are out-of-staters.

The students are sampled with*out* replacement from a *large* population, so they can be treated as independent trials. Using the multinomial distribution,

$$P(6 \text{ Chicagos, 4 out-of-states}) = \frac{10!}{6!\ 4!}(.3)^6(.1)^4$$

Bernoulli Trials and the Binomial Distribution

Suppose an experiment has two outcomes, titled success and failure (e.g., tossing a coin results in heads or tails, testing a light bulb results in accept or reject, tossing a die results in 4 versus non-4). Independent repetitions of the experiment are called *Bernoulli trials*.

Suppose that in any one trial

$$P(\text{success}) = p, P(\text{failure}) = 1 - p = q.$$

Then, as a special case of the multinomial distribution,

$$P(k \text{ successes in } n \text{ trials}) = P(k \text{ successes}, n - k \text{ failures})$$

$$= \frac{n!}{k!\,(n - k)!}\, p^k q^{n-k}$$

The coefficient can be written as $\binom{n}{k}$ so that

(3)
$$P(k \text{ successes in } n \text{ trials}) = \binom{n}{k} p^k q^{n-k}$$

To repeat, here's *why* (3) holds. Each outcome consisting of k successes and $n - k$ failures has probability $p^k q^{n-k}$. The coefficient $\binom{n}{k}$ counts how *many* of these outcomes there are (as many as there are ways of picking k out of the n spots for the successes).

We say that the number of successes in n Bernoulli trials, where $P(\text{success}) = p$, has a *binomial distribution* with parameters n and p, meaning that (3) holds.

Example 3

If a fair coin is tossed 10 times, then, by (3),

$$P(2\text{H}) = \binom{10}{2} \left(\frac{1}{2}\right)^2 \left(\frac{1}{2}\right)^8 = \binom{10}{2} \left(\frac{1}{2}\right)^{10}$$

Example 4

A dealer sells 10 new cars. The prob that a new car breaks down is .3.

(a) Find the prob that 8 of the cars work and 2 break down.

(b) Find the prob that the first 2 cars sold break down and the others work.

The cars are Bernoulli trials. Each trial results in breaks or works.

(a) $P(8 \text{ work in 10 trials}) = \binom{10}{8}(.7)^8(.3)^2$

(b) $P(\text{BBWWWWWWWW}) = (.3)^2(.7)^8$

Example 5

Find the prob that a 6-letter word contains two Z's.

The positions in the words are independent trials since repetition is allowed. Each trial has two possible outcomes: Z with prob 1/26 and other with prob 25/26. The number of Z's has a binomial distribution, so

$$P(2Z) = \binom{6}{2} \left(\frac{1}{26}\right)^2 \left(\frac{25}{26}\right)^4$$

All or None Problems

An urn contains 20 white, 30 black, and 40 red balls. Draw 5 with replacement.
  If you want the probability that all 5 are white, you can use the binomial distribution to get

$$P(\text{all white}) = P(5\text{W}) = \binom{5}{5} \left(\frac{20}{90}\right)^5 \left(\frac{70}{90}\right)^0 = \left(\frac{20}{90}\right)^5$$

But it's faster to simply use

$$P(\text{all white}) = P(\text{WWWWW}) = \left(\frac{20}{90}\right)^5 \quad \text{(since the trials are ind)}$$

Similarly, the fastest way to get the probability of no whites is

$$P(\text{no whites}) = P(\bar{\text{W}}\bar{\text{W}}\bar{\text{W}}\bar{\text{W}}\bar{\text{W}}) = \left(\frac{70}{90}\right)^5$$

Example 6

A coin is biased so that $P(\text{H}) = .6$. If the coin is tossed 10 times,

$$P(\text{at least 2H}) = 1 - P(\text{no H}) - P(1\text{H}) = 1 - (.4)^{10} - \binom{10}{1}(.6)(.4)^9$$

The Geometric Distribution

Let $P(\text{heads}) = p$. Toss until a head turns up. Here's how to find the probability that it takes 10 tosses:

$$P(\text{it takes 10 tosses to get a head})$$
$$= P(\text{first 9 tosses are tails and 10th is head})$$
$$= (1 - p)^9 p$$

  The number of trials to get the first success in Bernoulli trials where $P(\text{success}) = p$ is said to have a *geometric* distribution with parameter p.

The Negative Binomial Distribution

Let $P(\text{heads}) = p$. Toss until there is a total of 17 heads. Here's how to find the probability that the game ends on the 30th toss:

$P(\text{it takes 30 tosses to get 17 heads})$
$= P(\text{16 heads in 29 tosses and heads on the 30th})$
$= P(\text{16 heads in 29 tosses})P(\text{H on 30th})$ (since tosses are ind)
$= \binom{29}{16} p^{16}(1 - p)^{13}p$

The number of trials needed to get r successes in Bernoulli trials where $P(\text{success}) = p$ is said to have a *negative binomial distribution* with parameters r and p.

Problems for Section 2-2

1. Of all the toasters produced by a company, 60% are good, 30% are fair, 7% burn the toast, and 3% electrocute their owners. If a store has 40 of these toasters in stock, find the prob that they have

   (a) 30 good, 5 fair, 3 burners, 2 killers   (c) no killers
   (b) 30 good, 4 fair

2. Toss 16 nickels. Find the prob of

   (a) no heads   (b) 7 heads   (c) at least 15 heads

3. Five boxes each contain 7 red and 3 green balls. Draw 1 ball from each box. Find the prob of getting more green than red.

4. Sixty percent of the country is Against, 30% is For, and 10% is Undecided. If 5 people are polled find the prob that
   (a) all are For
   (b) 1 is For, 2 are Against, and 2 are Undecided
   (c) a majority is Against

5. A couple has 6 children. Find the prob that they have

   (a) 3 girls and 3 boys   (c) GBGGBB in that order
   (b) 3 girls first and then 3 boys

6. A coin is biased so that $P(\text{H}) = .6$. Toss it 10 times.
   (a) Find the prob of 6 heads overall given that the second toss is tails.
   (b) Find the prob of at least 9 heads given that you got at least 8 heads.

7. Toss 10 balls at random into 5 boxes. Find the prob that

   (a) each box gets 2 balls   (c) box B_3 gets 6 balls
   (b) box B_2 is empty   (d) no box is empty

8. A machine has 7 identical independent components. The prob that a component fails is .2. In order for the machine to operate, 5 of its 7 components must work. Find the prob that the machine fails.

9. Find the prob that a 7-digit string contains

(a) two 4's   (b) exactly one digit > 5

10. (poker dice)   Toss 5 dice. Find the prob of getting a pair (and nothing better than a pair).

11. Ten pieces of candy are given out at random in a group of 5 boys, 7 girls, and 9 adults. Find the prob that 4 pieces of candy go to the girls if
(a) people are allowed to get more than one piece
(b) no one is allowed to get more than one piece

12. If 40% of marriages end in divorce and we assume that divorces are independent of one another, find the prob that of 8 couples
(a) only the Smiths and Joneses will stay married
(b) exactly 2 of the 8 couples will stay married

13. At a particular intersection with a stop sign you observe that 1 out of every 20 cars fails to stop. Find the prob that among the next 100 cars at least 3 don't stop.

14. Toss 6 balls at random into 10 boxes. Find the prob that
(a) they split 4–2 (4 go into one box and 2 into a second box)
(b) they split 3–3
(c) they all go into different boxes

15. A drawer contains 10 left gloves and 12 right gloves. If you pull out a handful of 4 gloves, what's the prob of getting 2 pairs (2L and 2R)?

16. Fifteen percent of the population is left-handed. If you stop people on the street what's the prob that
(a) it takes at least 20 tries to get a lefty
(b) it takes exactly 20 tries to get a lefty
(c) it takes exactly 20 tries to get 3 lefties
(d) it takes at least 20 tries to get 3 lefties (the more compact your answer, the better)
(e) the number of tries to get a lefty is a multiple of 5

17. A coin has $P(\text{head}) = p$. Find the prob that it takes 10 tosses to get a head and a tail (i.e., at least one of each).

SECTION 2-3   SIMULATING AN EXPERIMENT

A box contains 12 red balls and 8 black balls. Draw 10 times without replacement. Then

$$P(6 \text{ reds}) = \frac{\binom{12}{6}\binom{8}{4}}{\binom{20}{10}} \approx .35$$

The physical interpretation of the mathematical model is that if you do this 10-draw experiment many times, it is likely (but not guaranteed) that the percentage of times you'll get 6 reds will be close to (but not necessarily equal to) 35%.

(Notice how much hedging there has to be in the last paragraph. *Within* a mathematical model, theorems can be stated precisely and proved to hold. But when you try to apply your model to the real world, you are stuck with imprecise words like *many, very likely,* and *close to.*)

The random number generator in a computer can be used to simulate drawing balls from a box so that you can actually do the 10-draw experiment many times. The program that follows was done with Mathematica.

The subprogram enclosed in the box is a single 10-draw experiment: It draws 10 balls without replacement from an urn containing 12 reds and 8 blacks. When it's over, the counter named Red tells you the total number of reds in the sample. Here's how this part works.

- Start the Red counter at 0.
- Draw the first ball by picking an integer z at random between 1 and 20.
- The integers from 1 to 12 are the red balls; the integers from 13 to 20 are the blacks.
- Step up Red by 1 if z is a red ball, that is, if $z \le 12$ (in Mathematica this has to be written as $z <= 12$).
- On the second draw pick an integer z at random between 1 and 19.
- If the first draw was *red* then the integers from 1 to 11 are the red balls, the integers from 12 to 19 are the blacks.
- If the first draw was *black* then the integers from 1 to 12 are the red balls, the integers from 13 to 19 are the blacks.
- Again, the counter Red is stepped up by 1 if z is red.
- And so on through the 10 draws.

The program as a whole repeats the 10-draw experiment n times. The counter named SixReds keeps track of how often you get 6 reds in the 10 draws. The final output divides SixReds by n to get the fraction of the time that this happens, that is, the relative frequency of 6 red results.

Here's what happened when I entered the program and ran it twice.

```
In[1]

Percent6RedWO[n_] :=

(SixReds = 0;
```

```
Do[
```

```
  (Red = 0;
   total = 20;
   Do [z = Random[Integer,{1,total}];
       If [z <= 12 - Red, Red = Red + 1];
       total = total - 1,
       {10}
       ]
   );
```

```
    If[Red == 6, SixReds = SixReds + 1],

    {n}

    ];

SixReds/n//N    (*Here is the output;

                 The N makes it a decimal rather than a fraction*)
)

In[2]

Percent6RedWO[500] (*Repeat the ten-draw experiment 500 times*)

Out[2]

0.308
```

For that run, 31% of the time I got 6 reds in 10 draws.

```
In[3]

Percent6RedWO[1000]   (*Repeat the ten-draw experiment 1000 times*)

Out[3]

0.335
```

For the second run, 34% of the time I got 6 reds in 10 draws.

If the drawing is done *with* replacement, then

$$P(6 \text{ reds in } 10 \text{ draws}) = \binom{10}{3} \left(\frac{12}{20}\right)^6 \left(\frac{8}{20}\right)^4 \approx .25 \text{ (binomial distribution)}$$

Here's the program adjusted so that it simulates n 10-draw experiments where the drawing is *with* replacement. Only the boxed subprogram is changed: now it just picks 10 integers at random between 1 and 20 where on each draw the integers 1 to 12 are red.

```
In[4]

Percent6RedWITH[n_] :=

(SixReds = 0;

  Do[
```

```
    (Red = 0;
     total = 20;
     Do [z = Random[Integer,{1,total}];
         If [z <= 12, Red = Red + 1];
         {10}
         ]
    );
```

```
    If[Red == 6, SixReds = SixReds + 1],

    {n}

  ];

SixReds/n//N

)

In[5]

Percent6RedWITH[1000]

Out[5]

0.265
```

SECTION 2-4   THE THEOREM OF TOTAL PROBABILITY AND BAYES' THEOREM

This section is about 2-stage (or multi-stage) experiments where the second stage depends on the first.

The Theorem of Total Probability

Here's a typical 2-stage experiment.

(1)
A box contains 2 green and 3 white balls. Draw 1.
If the ball is green, draw a card from a fair deck.
If the ball is white, draw a card from a deck consisting of just the 16 pictures.

We'll find the probability of drawing a king.

The tree diagram in Fig. 1 describes the situation. The labels on the first set of branches are

$$P(\text{green}) = \frac{2}{5} \quad \text{and} \quad P(\text{white}) = \frac{3}{5}$$

The labels on the second set of branches are

$$P(\text{king}|\text{green}) = \frac{4}{52}, \qquad P(\text{non-king}|\text{green}) = \frac{48}{52}$$

$$P(\text{king}|\text{white}) = \frac{4}{16}, \qquad P(\text{non-king}|\text{white}) = \frac{12}{16}$$

Note that at each vertex of the tree, the sum of the probabilities is 1.

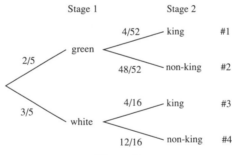

Figure 1

Since one of green and white has to occur at the first stage,

$$P(\text{king}) = P(\text{green and king or white and king})$$
$$= P(\text{green and king}) + P(\text{white and king})$$
$$\qquad \text{(by the OR rule for mutually exclusive events)}$$
$$= P(\text{green})P(\text{king}|\text{green}) + P(\text{white})P(\text{king}|\text{white})$$
$$\qquad \text{(by the AND rule)}$$
$$= \frac{2}{5} \cdot \frac{4}{52} + \frac{3}{5} \cdot \frac{4}{16}$$

If we use the notation

branch #1 = product of the probabilities along the branch $= \dfrac{2}{5} \cdot \dfrac{4}{52}$

then the answer can be written as

$$P(\text{king}) = \#1 + \#3 = \text{sum of favorable branches}$$

Here's the general rule, called the *theorem of total probability*.

If at the first stage the result is exactly one of A, B, C, then the probability of Z at the second stage is

(2) $\quad P(Z) = P(A)P(Z|A) + P(B)P(Z|B) + P(C)P(Z|C)$

And here's a restatement that makes it easy to use.

(3) $P(Z) = $ sum of favorable branches in Fig. 2 $= \#1 + \#3 + \#5$

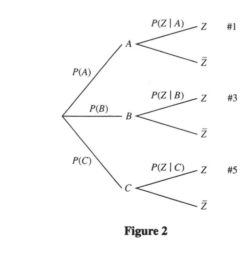

Figure 2

The theorem of total probability as stated in (2) can be thought of in a more general context, without reference to a 2-stage experiment: If a probability space can be divided into, say, 3 mutually exclusive exhaustive events A, B, C, then (2) holds for any event Z.

Bayes' Theorem

Let's use the same experiment again:

A box contains 2 green and 3 white balls. Draw 1 ball.

If the ball is green, draw a card from a fair deck.

If the ball is white, draw a card from a deck consisting of just the 16 pictures.

Suppose you draw a king on the second stage (Fig. 1 again). We'll go backward and find the probability that it was a green ball on the first stage. In other words, we'll find

$$P(\text{green on first stage}|\text{king on second})$$

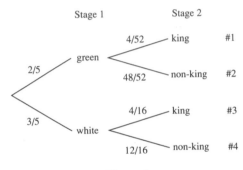

Figure 1

By the rule for conditional probability,

$$P(\text{green}|\text{king}) = \frac{P(\text{green and king})}{P(\text{king})}$$

The numerator is

$$P(\text{green})P(\text{king}|\text{green}) = \text{branch \#1}$$

By the theorem of total probability, the denominator is #1 + #3. So

$$P(\text{green}|\text{king}) = \frac{\#1}{\#1 + \#3} = \frac{\text{fav king branches}}{\text{total king branches}}$$

$$= \frac{\frac{2}{5} \cdot \frac{1}{13}}{\frac{2}{5} \cdot \frac{1}{13} + \frac{3}{5} \cdot \frac{1}{4}} = \frac{8}{47}$$

Here's the general rule, called *Bayes' Theorem*.

The a posteriori (backward conditional) probability of A at the *first* stage, given Z on the *second* stage (Fig. 3), is

$$P(A|Z) = \frac{P(A \text{ and } Z)}{P(Z)}$$

(4)
$$= \frac{Z\text{-branches that are favorable to } A}{\text{total } Z\text{-branches}}$$

$$= \frac{\#1}{\#1 + \#3 + \#5}$$

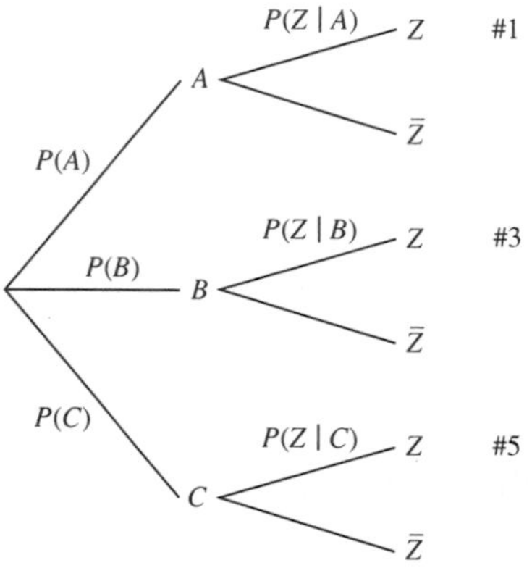

Figure 3

Example 1

Suppose $\frac{1}{2}\%$ of the population has a disease D. There is a test to detect the disease. A positive test result is supposed to mean that you have the disease, but the test is not perfect. For people *with* D, the test misses the diagnosis 2% of the time; that is, it reports a false negative. And for people with*out* D, the test incorrectly tells 3% of them that they have D; that is, it reports a false positive.

(a) Find the probability that a person picked at random will test positive.
(b) Suppose your test comes back positive. What is the (conditional) probability that you have D?

(a) Figure 4 shows the tree diagram. By the theorem of total probability,

$$P(\text{positive}) = \#1 + \#3 = (.005)(.98) + (.995)(.03)$$

(b) By Bayes' theorem,

$$P(D|pos) = \frac{\text{favorable pos branches}}{\text{total pos branches}} = \frac{\#1}{\#1 + \#3}$$

$$= \frac{(.005)(.98)}{(.005)(.98) + (.995)(.03)} \sim .14$$

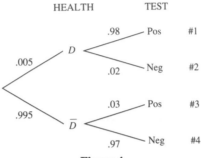

Figure 4

So even though the test seems fairly accurate (with success rates of 98% and 97%), if you test positive, the probability is only .14 that you actually have the disease. (The probability came out low because so few people have the disease to begin with.)

What we would really like to know in this situation is a *first stage* result: Do you have the disease? But we can't get this information without an autopsy. *The first stage is hidden.* But the second stage (the result of the test) is not hidden. The best we can do is make a prediction about the first stage by looking at the second stage. This illustrates why backward conditionals are so useful.

Problems for Section 2-4

1. The prob of color blindness is .02 for a man and .001 for a woman. Find the prob that a person picked at random is color blind if the population is 53% men.

2. Draw a card. If it's a spade, put it back in the deck and draw a second card. If the first card isn't a spade, draw a second card without replacing the first one. Find the prob that the second card is the ace of spades.

3. A multiple-choice exam gives 5 choices per question. On 75% of the questions, you think you know the answer; on the other 25% of the questions, you just guess at random. Unfortunately when you *think* you know the answer, you are right only 80% of the time (you dummy).

    (a) Find the prob of getting an arbitrary question right.

    (b) If you do get a question right, what's the prob that it was a lucky guess?

4. Box A has 10 whites and 20 reds, box B has 7 whites and 8 reds, and box C has 4 whites and 5 reds. Pick a box at random and draw one ball. If the ball is white, what's the prob that it was from box B?

5. Toss a biased coin where $P(H) = 2/3$. If it comes up heads, toss it again 5 times. If it comes up tails, toss it again 6 times. Find the prob of getting at least 4 heads overall.

6. Of 10 egg cartons, 9 contain 10 good eggs and 2 bad while a tenth carton contains 2 good and 10 bad. Pick a carton at random and pull out 2 eggs (without replacement). If both are bad, find the prob that you picked the tenth carton.

7. A box of balls contains 3 whites and 2 blacks.

    round 1   Draw a ball. Don't replace it.

    round 2   If the ball is white, toss a fair coin.

                      If the ball is black, toss a biased coin where $P(H) = .8$.

    round 3   If heads, draw 2 balls from the (depleted) box.

                      If tails, draw 1 ball.

Find the prob of getting at least 1 white at round 3.

8. An insurance company unofficially believes that 30% of drivers are careless and the prob that a driver will have an accident in any one year is .4 for a careless driver and .2 for a careful driver. Find the prob that a driver will have an accident next year given that she has had an accident this year.

9. Look at the accompanying tree diagram. What is each of the following the probability of?

    (a) .2   **(b)** .3   **(c)** (.2)(.3)

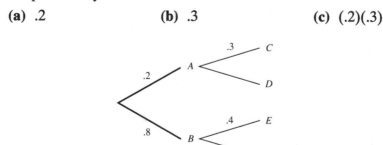

Figure P.9

10. Toss a die 3 times. Find the probability that the result of the third toss is larger than each of the first two.

Suggestion: Condition on the third toss and use the theorem of total probability.

SECTION 2-5   THE POISSON DISTRIBUTION

In Section 2.2 we found a formula, called the binomial distribution, for the probability of k successes in n Bernoulli trials. Now we'll look at a related formula called the Poisson distribution. Once you see how it's connected to the binomial, you can apply it in appropriate situations.

The Poisson Probability Function

Let λ be fixed. Consider an experiment whose outcome can be $0, 1, 2, 3, \ldots$. If

(1)
$$P(\text{outcome is } k) = \frac{e^{-\lambda}\lambda^k}{k!} \text{ for } k = 0, 1, 2, 3, \ldots$$

then we say that the outcome has a Poisson distribution with parameter λ. The formula in (1) is the *Poisson probability function*.

The Poisson prob function is legitimate because the sum of the probs is 1, as it should be:

$$\sum_{k=0}^{\infty} \frac{e^{-\lambda}\lambda^k}{k!} = e^{-\lambda} \underbrace{\sum_{k=0}^{\infty} \frac{\lambda^k}{k!}}_{e^{\lambda}} = 1$$

Here we used the standard series from calculus

$$e^x = \sum_{k=0}^{\infty} \frac{x^k}{k!} = 1 + x + \frac{x^2}{2!} + \frac{x^3}{3!} + \cdots$$

The Poisson Approximation to the Binomial

Consider n Bernoulli trials where $P(\text{success}) = p$.

Let $\lambda = np$, interpreted as the average number of successes to be expected in the n trials.

(To see the physical interpretation of λ, consider tossing a coin 200 times where $P(\text{heads}) = .03$, so that $n = 200$, $p = .03$. If many people toss 200 times each, one person might get no heads and another might get 199 heads, but it is very likely that the *average* number of heads per person is near $.03 \times 200 = 6$.)

Suppose n is large, p is small, their product λ is moderate, and k is much smaller than n. We'll show that

$$P(k \text{ successes in } n \text{ trials}) = \underbrace{\binom{n}{k} p^k (1-p)^{n-k}}_{\text{binomial dist}} \approx \underbrace{\frac{e^{-\lambda}\lambda^k}{k!}}_{\text{Poisson}}$$

So the Poisson can be used to approximate the binomial. The advantage of the Poisson, as you'll soon see, is that it has only the one parameter λ, while the binomial distribution has two parameters, n and p.

Here's why the approximation holds.

$$\binom{n}{k} p^k (1-p)^{n-k} = \frac{n(n-1)\cdots(n-k+1)}{k!} p^k (1-p)^{n-k}$$

Substitute

$$p = \frac{\lambda}{n}$$

and rearrange to get

(2) $$\binom{n}{k} p^k (1-p)^{n-k} = \frac{n(n-1)\cdots(n-k+1)}{n^k} \frac{\lambda^k}{k!} \frac{(1-\lambda/n)^n}{(1-\lambda/n)^k}$$

For large n, small p, moderate λ, and k much smaller than n,

$$\left(1-\frac{\lambda}{n}\right)^n \sim e^{-\lambda} \quad \text{(remember that } \lim_{x\to\infty}\left(1+\frac{a}{x}\right)^x = e^a\text{)}$$

$$\left(1-\frac{\lambda}{n}\right)^k = (1-\text{small})^k \sim 1$$

$$\frac{n(n-1)(n-1)\cdots(n-k+1)}{n^k} = \frac{n^k + \text{lower degree terms}}{n^k} \sim 1$$

So (2) becomes

$$\binom{n}{k} p^k (1-p)^{n-k} \sim 1 \cdot \frac{\lambda^k}{k!} \frac{e^{-\lambda}}{1} = \frac{e^{-\lambda}\lambda^k}{k!}$$

Application of the Poisson to the Number of Successes in Bernoulli Trials

Your record as a typist shows that you make an average of 3 mistakes per page. I'll find the probability that you make 10 mistakes on page 437.

Each symbol typed is an experiment where the outcome is either error or OK. The symbols on a page are typed independently, so they are Bernoulli trials. The number of mistakes on a page has a binomial distribution, where n is the number of symbols on a page and p is the probability of an error in a symbol. But we don't know n or p, so we can't use the binomial distribution. On the other hand, we do know that

$$\text{average number of mistakes per page } = 3$$

so the next best choice is the Poisson approximation to the binomial, with $\lambda = 3$:

$$P(10 \text{ mistakes on page 437}) = \frac{e^{-3} \, 3^{10}}{10!}$$

Suppose you want the probability of fewer than 4 mistakes in the 10-page introduction. If the typist averages 3 mistakes per page, then on the average there are 30 mistakes in the introduction, so use the Poisson with $\lambda = 30$:

$P(\text{fewer than 4 mistakes in Intro})$

$= P(0 \text{ mistakes in Intro}) + P(1) + P(2) + P(3)$

$= e^{-30} \left(1 + 30 + \dfrac{30^2}{2!} + \dfrac{30^3}{3!} \right)$ (Remember that $\lambda^0 = 1$ and $0! = 1$.)

Here's the general rule:

Suppose you have a bunch of Bernoulli trials.

You don't know n, the number of trials in the bunch, or the probability p of success on any one trial (if you did, you could use the binomial distribution).

But you do know that the average number of successes in a bunch is λ.

Then use the Poisson distribution to get

(3) $$P(k \text{ successes in a bunch}) = \frac{e^{-\lambda} \lambda^k}{k!}$$

(provided that it's a large bunch and successes are fairly rare).

Example 1

The police ticket 5% of parked cars. (Assume cars are ticketed indepen-dently.) Find the probability of 1 ticket on a block with 7 parked cars.

Each car is a Bernoulli trial with $P(\text{ticket}) = .05$, so

$$P(1 \text{ ticket on block}) = P(1 \text{ ticket in 7 trials}) = \binom{7}{1}(.95)^6(.05)$$

Example 2

On the average, the police give out 2 tickets per block. Find the probability that a block gets 1 ticket.

The cars are Bernoulli trials. We don't know the number of cars on a block or $P(\text{ticket})$, but we can use the Poisson with $\lambda = 2$:

$$P(1 \text{ ticket on block}) = 2e^{-2}$$

Example 2 continued

Find the probability that a 4-block strip gets a least 5 tickets.

On the average, a 4-block strip gets 8 tickets, so use the Poisson with $\lambda = 8$:

$$P(\text{at least 5 tickets on a 4-block strip})$$
$$= 1 - P(0) - P(1) - P(2) - P(3) - P(4)$$
$$= 1 - e^{-8} - 8e^{-8} - \frac{8^2 e^{-8}}{2!} - \frac{8^3 e^{-8}}{3!} - \frac{8^4 e^{-8}}{4!}$$

Warning

When you use (3) to find the probability of k successes in *a bunch*, you must use as λ the average number of successes in *the bunch*. In example 2, for parking tickets in *a block*, use $\lambda = 2$, but for parking tickets in *a 4-block strip*, use $\lambda = 8$.

Application of the Poisson to the Number of Arrivals in a Time Period

As a telephone call arrives at a switchboard, the arrival time is noted and the switchboard is immediately ready to receive another call.

Let λ be the average number of calls in an hour, the rate at which calls arrive. Assume that calls arrive independently. We'll show why it's a good idea to use the Poisson as the model, that is, to use

(4) $$P(k \text{ calls in an hour}) = \frac{e^{-\lambda}\lambda^k}{k!}$$

Divide the hour into a large number, n, of small time subintervals, so small that we can pretend that, at most, 1 call can arrive in a time subinterval. In other words, *during each subinterval, either no call arrives or 1 call arrives* (but it isn't possible for 2 calls to arrive). With this pretense, the n time subintervals are Bernoulli trials where success means that a call has arrived. We don't know n and p, but we do have λ, so it makes good sense to use the Poisson in (4).

Similarly, the Poisson distribution is the model for particles emitted, earthquakes occurring, and arrivals in general:

> If arrivals are independent, the number of arrivals in a time period has a Poisson distribution:
>
> (5) $$P(k \text{ arrivals in a time period}) = \frac{e^{-\lambda}\lambda^k}{k!}$$
>
> where the parameter λ is the arrival rate, the average number of arrivals in the time period.

Example 3

Suppose particles arrive on the average twice a second. Find the probability of at most 3 particles in the next 5 seconds.

The average number of particles in a 5-second period is 10, so use the Poisson distribution with $\lambda = 10$:

$$P(\text{at most 3 particles in 5 seconds}) = P(\text{none}) + P(1) + P(2) + P(3)$$

$$= e^{-10}\left(1 + 10 + \frac{10^2}{2!} + \frac{10^3}{3!}\right)$$

Warning

The λ in example 3 must be 10, the average number of particles arriving per *5 seconds*, not the original 2, which is the average number per second.

When you use (5) to find the probability of k arrivals in a time period, λ must be the average number of arrivals *in that time period*.

Problems for Section 2-5

1. On the average, a blood bank has 2 units of the rare type of blood, XYZ.
   (a) Find the prob that a bank can supply at least 3 units of XYZ.
   (b) If a community has two blood banks, find the prob that the community can supply at least 6 units of XYZ.

2. On the average there are 10 no-shows per airplane flight. If there are 5 flights scheduled, find the prob of
   (a) no no-shows   **(b)** 4 no-shows   **(c)** at most 4 no-shows

3. Assume drivers are independent.
   (a) If 5% of drivers fail to stop at the stop sign, find the prob that at least 2 of the next 100 drivers fail to stop.
   (b) If on the average 3 drivers fail to stop at the stop sign during each rush hour, find the prob that at least 2 fail to stop during tonight's rush hour.

4. On the average you get 2 speeding tickets a year.
   (a) Find the prob of getting 3 tickets this year.
   (b) Suppose you get 2 tickets in January. Find the prob of getting no tickets during the rest of the year (the other 11 months).

5. Phone messages come to your desk at the rate of 2 per hour. Find the prob that if you take a 15-minute break you will miss
   (a) no calls   **(b)** no more than 1 call

6. On the average, in a year your town suffers through λ_1 earthquakes, λ_2 lightning strikes, and λ_3 meteorites crashing to earth. Find the prob that there will be at least one of these natural disasters next year.

7. On the average, you get 3 telephone calls a day. Find the prob that in 5 years there will be at least one day without a call. (This takes two steps. First, find the prob of no calls in a day.)

8. If $P(H) = .01$, then the prob of 1 H in 1000 tosses is $\binom{1000}{1}(.99)^{999}(.01)$. What's the Poisson approximation to this answer?

9. The binomial distribution, and to a lesser extent the Poisson distribution, involves Bernoulli trials. Do you remember what a Bernoulli trial is?

Review Problems for Chapters 1 and 2

1. Draw 10 balls from a box containing 20 white, 30 black, 40 red, and 50 green. Find each prob twice, once if the drawing is with replacement and again if it is without replacement.
   (a) P(3W and 4R)
   (b) P(3W followed by 4R followed by 3 others)
   (c) P(4R followed by 3W followed by 3 others)

2. Find the prob that among the first 9 digits from a random digit generator, there are at least four 2's.

3. Draw from a deck without replacement. Find the prob that
   (a) the 10th draw is a king and the 11th is a non-king

(b) the first king occurs on the 10th draw
(c) it takes 10 draws to get 3 kings
(d) it takes at least 10 draws to get 3 kings

4. Find the prob that a bridge hand contains at least one card lower than 6 given that it contains at least one card over 9.

5. Form 12-symbol strings from the 26 letters and 10 digits. Find the prob that a string contains 3 vowels if repetition is

   (a) allowed   **(b)** not allowed

6. If the letters in ILLINOIS are rearranged at random, find the prob that the permutation begins or ends with L. (Just as you can assume in a probability problem that white balls are named $W_1, \ldots, W_n$, you can assume here that the word is $I_1L_1L_2I_2NOI_3S$, with all the letters distinguishable.)

7. Find the prob that John and Mary are next to one another if eight people are seated at random

   (a) on a bench   **(b)** around a circular table

8. Find the prob that a 3-letter word contains z (e.g., zzz, zab, bzc).

9. You notice that 1 out of every 10 cars parked in a tow-away zone is actually towed away. Suppose you park in a tow-away zone every day for a year. Find the prob that you are towed at most once.

10. At a banquet, m men and w women are introduced in random order to the audience. Find the prob that the last two introduced are men.

11. Given j married couples, k single men, and n single women, pick a man and a woman at random. Find the prob that
    (a) both are married
    (b) only one is married
    (c) they are married to each other

12. Find the prob of not getting any 3's when you toss a die

    (a) 10 times   **(b)** 100,000 times   **(c)** forever

13. Your drawer contains 5 black, 6 blue, and 7 white socks. Pull out 2 at random. Find the prob that they match.

14. If you get 6 heads and 4 tails in 10 tosses, find the prob that one of the heads was on the 8th toss.

15. Twenty-six ice cream flavors, A to Z, are available. Six orders are placed at random. Find the prob that the orders include
    (a) A and B once each
    (b) A and B at least once each
    (c) at least one of A and B (i.e., at least one A or at least one B)
    (d) two A's and at least two B's

    (e) all different flavors
    (f) all the same flavor

16. Consider the probability of getting a void in bridge, a hand with at least one suit missing.
    (a) What's wrong with the following answer?

        The total number of hands is $\binom{52}{13}$.

        For the favorable hands:

            Pick a suit to be missing. Can be done in 4 ways.

            Pick 13 cards from the other 3 suits. Can be done in $\binom{39}{13}$ ways.

$$\text{Answer is } \frac{4\binom{39}{13}}{\binom{52}{13}}$$

    (b) Find the right answer.

17. A box contains 6 black (named $B_1, \ldots, B_6$), 5 white, and 7 red balls. Draw balls. Find the prob that
    (a) B_3 is drawn before B_5
    (b) B_3 is drawn before any of the whites

18. One IRS office has three people to answer questions. Mr. X answers incorrectly 2% of the time, Ms. Y 3% of the time, and Ms. Z 4% of the time. Of all questions directed to this office, 60% are handled by X, 30% by Y, and 10% by Z.

What percentage of incorrect answers given by the office is due to Z?

19. Toss a coin 10 times. Find the prob of getting no more than 5 heads given that there are at least 3 heads.

20. Call the throw of a pair of dice lucky if the sum is 7 or 11.

Two players each toss a pair of dice (independently of one another) until each makes a lucky throw. Find the prob that they take the same number of throws.

21. A basketball player has made 85% of her foul shots so far in her career. Find the prob that she will make at least 85% of her next 10 foul shots.

22. Prizes are given out at random in a group of people. It's possible for a person to get more than one prize.
    (a) If there are 10 prizes and 5 people, find the prob that no one ends up empty-handed.
    (b) If there are 5 prizes and 10 people, find the prob that no one gets two (or more) prizes.
    (c) If there are 5 prizes and 6 men and 4 women, find the prob that all the prizes go to men.

23. Toss a penny and a nickel 20 times each. For each coin, $P(H) = .7$. If the overall result is 17H and 23T, find the prob that 11 of those 17 heads were from the penny.

24. Four hundred leaflets are dropped at random over 50 square blocks.
   (a) Find the prob that your block gets at least 3.
   (b) Find the Poisson approximation to the answer in (a).

25. Three players toss coins simultaneously. For each player, $P(H) = p$, $P(T) = q$. If the result is 2H and 1T or the result is 2T and 1H, then the player that is different from the other two is called the odd man out and the game is over. If the result is 3H or 3T, then the players toss again until they get an odd man out.

Find the prob that the game lasts at least 6 rounds.

26. Let $P(A) = .5$, $P(B) = .2$, $P(C) = 1$. Find $P(A \text{ or } B \text{ or } C)$ if
   (a) A, B, C are mutually exclusive
   (b) A, B, C are independent

27. On the average, there is a power failure once every four months.
   (a) Find the prob of a power failure during exam week.
   (b) Find the prob that it will be at least a month before the next failure.

28. A bus makes 12 stops and no one stop is more popular than another. If 5 passengers travel on the bus independently, what's the prob that 3 get off at one stop and 2 get off at another (so that you have a full house of stops).

29. Find   **(a)** $\binom{999}{1}$ **(b)** $\binom{999}{0}$ **(c)** $\binom{1000}{999}$ **(d)** $\dfrac{\binom{n+m-1}{n-1}}{\binom{n+m}{n}}$

30. Teams A and B meet in the world series (the first to win 4 games is the series winner). Assume the teams are evenly matched and the games are independent events. Find the prob that the series ends in 6 games.

31. Five shots are fired at random into a circle of radius R. The diagram shows an inscribed square and four other zones. Find the prob that the 5 shots end up
   (a) all in the same zone
   (b) in five different zones

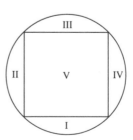

Figure P.31

32. Mary Smith has a 50-50 chance of carrying an XYZ gene. If she is a carrier, then any child has a 50-50 chance of inheriting the gene. Find the prob that her 4th child will not have the gene given that her first three children don't have it.

33. (the famous birthday problem) Find the prob that in a group of n people, at least two will have the same birthday.

34. You have 5 dice and 3 chances with each die to get a 6.

    For example, if you toss the third die and get a 6, then you move on to the fourth die. But if the third die is non-6, you get to try again and then again if necessary.

    Find the prob of getting two 6's overall with the 5 dice.

    Suggestion: First find the prob of getting a 6 from a single die in your 3 chances.

35. A message is sent across a channel to a receiver.

    > The probability is .6 that the message is $xxxxx$.
    > The probability is .4 that the message is $yyyyy$.

    For each letter transmitted, the probability of error (that an x will become y, or vice versa) is .1. Find the probability that the message was $xxxxx$ if 2 x's and 3 y's are received.

36. John's score is a number chosen at random between 0 and 3. Mary's score is chosen at random between 0 and 1. The two scores are chosen independently. Find the prob that
    (a) his score is at least twice hers
    (b) the max of the two scores is $\leq 1/2$
    (c) the min of the two scores is $\geq 1/2$

37. Five people are picked (without replacement) from a group of 20. Find the prob that John was chosen using these different methods.
    (a) fav committees/total committees
    (b) P(John was chosen 1st or 2nd or ... or chosen 5th)
    (c) $1 - P$(no John)

38. Find the probability that, in a group of 30 people, at least 3 were born on the fourth of July.

39. A class consists of 10 freshmen, 20 sophomores, 30 juniors, and 20 grads.
    (a) If grades A, B, C, D, E are assigned at random, find the prob that 4 freshmen get A's.
    (b) If 6 class offices (president, vice president, etc.) are assigned at random, find the prob that 4 freshmen get offices.

40. (*Computer Networks*, Tannenbaun, Prentice Hall, 1989)   A disadvantage of a broadcast subnet is the capacity wasted due to multiple hosts attempting to access the channel at the same time. Suppose a time period is divided into a certain number of discrete slots. During each time slot, the probability is p that a host will want to use the channel. If two or more hosts want to use a time slot, then a collision occurs, and the slot is wasted. If there are n hosts, what fraction of the slots is wasted due to collisions?

41. John will walk past a street corner some time between 10:00 and 11:00. Mary will pass the same street corner some time between 10:00 and 11:30. Find the prob that they meet at the corner if
(a) each agrees to wait 10 minutes for the other
(b) John will wait 10 (lovesick) minutes for Mary (but not vice versa)

42. Draw 20 times from the integers 1, 2, 3, ... , 100. Find the prob that your draws come out in increasing order (each draw is larger than the previous draw), if the drawing is

(a) with replacement   (b) without replacement

You can do it directly with fav/total (but most people get stuck on the fav).

CHAPTER 3

Expectation

SECTION 3-1   EXPECTED VALUE OF A RANDOM VARIABLE

You already know a lot about finding averages. If your grades for the year are

85% in a 3-point course
90% in a 4-point course
70% in a 5-point course

then their (weighted) average is

(1)
$$\frac{(85 \times 3) + (90 \times 4) + (70 \times 5)}{12}$$

In (1), each course grade in the numerator is weighted by its corresponding point value, and the denominator is the sum of the weights.

In a similar fashion, when the result of an experiment is a number, we'll find the average result.

Example 1

A couple decides to have 3 children. But if none of the 3 is a girl, they'll try again, and if they still don't get a girl, they'll try once more.

On the average, how many children do they end up having?

They can have 3 or 4 or 5 children. We'll average the three outcomes, a *weighted* average where the weights are the respective probabilities of the outcomes (an outcome with greater likelihood counts more in the average). The sum of the weights, that is, the sum of the three probabilities, is 1, so we can leave out the denominator:

average number of children (called the *expected number of children*)

$$= 3 \times P(3 \text{ children}) + 4 \times P(4 \text{ children}) + 5 \times P(5 \text{ children})$$

$$= 3P(\text{at least 1 G in 3 tries}) + 4P(\text{BBBG}) + 5P(\text{BBBB})$$

$$= 3[1 - P(\text{BBB})] + 4P(\text{BBBG}) + 5P(\text{BBBB})$$

$$= 3\left[1 - \left(\frac{1}{2}\right)^3\right] + 4\left(\frac{1}{2}\right)^4 + 5\left(\frac{1}{2}\right)^4$$

$$= \frac{51}{16} = 3\frac{3}{16}$$

We interpret this physically as meaning that if *many* couples have children according to this scheme, it is likely that the average family size would be near $3\frac{3}{16}$ children.

Random Variables

A numerical result of an experiment is called a *random variable* and is usually denoted with a capital letter like X, Y, Z.

In example 1 the number of children the couple has is a random variable. Other random variables associated with the experiment are the number of boys, number of girls, number of older brothers, and so on.

As another example, consider Bernoulli trials where $P(\text{success}) = p$.

Let X be the number of successes in n trials.

Let Y be the number of trials it takes to get the first success.

The random variable X has a binomial distribution, and Y has a geometric distribution. In particular,

(2)
$$P(X = k) = \binom{n}{k} p^k q^{n-k} \quad \text{for } k = 0, 1, 2, 3, \ldots, n$$

(3)
$$P(Y = k) = q^{k-1} p^k \quad \text{for } k = 1, 2, 3, \ldots$$

The formulas in (2) and (3) are called the *probability functions* for X and Y.

Expected Value of a Random Variable

If X is the number of children in example 1, then X takes on values 3, 4, 5 and expected value of $X = 3P(X = 3) + 4P(X = 4) + 5P(X = 5)$

In general,

> The *expectation* or *expected value* or *mean* of a random variable X is a weighted average of the values of X, where each value x is weighted by the probability of its occurrence. If the expectation is denoted by $E(X)$ or EX, then
>
> $$E(X) = \sum_x xP(X = x)$$

Note that if 0 is one of the values of X, the term $0 \times P(X = 0)$ can be left out of the sum since it's always 0 anyway.

Expectation of a Poisson Random Variable

Let X have a Poisson distribution with parameter λ, so that

$$P(X = k) = \frac{e^{-\lambda}\lambda^k}{k!} \quad \text{for } k = 0, 1, 2, 3, \ldots$$

We anticipate that

$$E(X) = \lambda$$

since we use the Poisson as the model for the number of arrivals in a time period where λ is physically the average number of arrivals per time period.

Here's the proof that the mathematical expectation agrees with the physical interpretation:

$$E(X) = \sum_{k=0}^{\infty} kP(X = k) = \sum_{k=0}^{\infty} k\frac{e^{-\lambda}\lambda^k}{k!}$$

$$= e^{-\lambda}\left(\frac{\lambda}{1!} + \frac{2\lambda^2}{2!} + \frac{3\lambda^3}{3!} + \cdots\right)$$

$$= e^{-\lambda}\lambda\left(\underbrace{1 + \lambda + \frac{\lambda^2}{2!} + \frac{\lambda^3}{3!} + \cdots}_{e^{\lambda}}\right)$$

$$= \lambda$$

Problems for Section 3-1

1. A box of 5 items is known to contain 3 good and 2 defective. If you test the items successively (meaning you draw without replacement), find the expected number of tests needed to identify the D's.

   Note that if you draw GGG, you are finished, since the remaining 2 items must be D's. If you draw GGD, then it will take one more draw to locate both D's. And it is never necessary to draw all 5 items.

2. An arrow is fired at random into a circle with radius 8.

   If it lands within 1 inch of the center, you win $10.
   If it lands between 1 and 3 inches from the center, you win $5.
   If it lands between 3 and 5 inches from the center, you win $2.
   Otherwise, you lose $4.

   Find your expected winnings.

3. A couple decides to have children until they get a girl, but they agree to stop with a maximum of 3 children even if they haven't gotten a girl yet. Find the expected number of

   (a) children   **(b)** girls   **(c)** boys

4. In roulette, a wheel stops with equal probability at any of the 38 numbers 0, 00, 1, 2, 3, ... , 36. If you bet $1 on a number, then you win $36 (net gain is $35) if the number comes up; otherwise, you lose your dollar. Find your expected winnings.

5. Let X be a random variable taking on values 0, 1, 2, 3, ... Show that

$$EX = \sum_{i=1}^{\infty} P(X \ge i)$$

   Suggestion: Write out the right-hand side in detail.

SECTION 3-2   THE METHOD OF INDICATORS

Some expectations are hard to find directly but are easy using the clever method of this section.

Expectation of a Sum

Before we get started we'll need this rule for $E(X + Y)$:

(1)
$$\boxed{E(X + Y) = E(X) + E(Y)}$$

   In other words, to find the expectation of a sum of random variables, just find all the separate expectations and add. (If the average number of girls in a family is 1.2 and the average number of boys is 1.3, then the average number of children is 2.5.)

Here's a justification.

If a value of the random variable X is denoted by x and a value of Y is denoted by y, then the value of $X + Y$ is $x + y$ and

$$E(X + Y) = \sum_{x,y} (x + y) \, P(X = x, Y = y)$$

$$= \sum_{x,y} x \, P(X = x, Y = y) + \sum_{x,y} y P(X = x, Y = y)$$

$$= \sum_{x} \left[x \underbrace{\sum_{y} P(X = x, Y = y)}_{P(X = x)} \right] + \sum_{y} \left[y \underbrace{\sum_{x} P(X = x, Y = y)}_{P(Y = y)} \right]$$

$$= \sum_{x} x \, P(X = x) + \sum_{y} y \, P(Y = y)$$

$$= E(X) + E(Y)$$

The Method of Indicators and the Mean of the Binomial Distribution

Let X have a binomial distribution with parameters n and p. You can think of X as the number of successes in n Bernoulli trials where $P(\text{success}) = p$.

We want to find EX, the expected number of successes in n trials.

If we try it directly, using the definition of expectation, we have

$$EX = \sum_{k=0}^{n} k \, P(X = k) = \sum_{k=0}^{n} k \binom{n}{k} p^k (1 - p)^{n-k}$$

$$= \binom{n}{1} pq^{n-1} + 2 \binom{n}{2} p^2 q^{n-2} + \cdots + n \binom{n}{n} p^n$$

This answer is correct, but it would take a lot of effort and combinatorial identities to simplify. So, instead, we'll try another method that works more smoothly in this and many other problems.

To find $E(X)$, define n new random variables X_i as follows. For $i = 1, 2, 3, \ldots, n$, let

$$X_i = \begin{cases} 1 & \text{if the } i\text{th trial is a success} \\ 0 & \text{otherwise} \end{cases}$$

The X_i's are called *indicator* random variables or *indicators*. There's an indicator for each trial, and each indicator signals (with a 1) if the trial is a success. For example, if $n = 5$ and the trials are

$$S \ S \ F \ F \ S$$

then

$$X = 3, \quad X_1 = 1, \quad X_2 = 1, \quad X_3 = 0, \quad X_4 = 0, \quad X_5 = 1$$

The total number of successes is the sum of all the signals, so

$$X = X_1 + \cdots + X_n$$

By (1),

$$EX = EX_1 + \cdots + EX_n$$

Now we need each of the EX_i's, which is simple because X_i takes on only two values. And, in this case, all the EX_i's are the same:

(2)
$$
\begin{aligned}
EX_i &= 0 \times P(X_i = 0) + 1 \times P(X_i = 1) \\
&= P(X_i = 1) \\
&= P(\text{success on } i\text{th trial}) \\
&= p
\end{aligned}
$$

So

$$
\begin{aligned}
EX &= EX_1 + \cdots + EX_n \\
&= \text{sum of } n \text{ terms each of which is } p \\
&= np
\end{aligned}
$$

(3)
> If X has a binomial distribution with parameters n and p, then
>
> $$E(X) = np$$

This is a very intuitive result (which we spotted back in Section 2-5). It says that if $P(\text{heads}) = 1/5$, then the expected number of heads in 20 throws is $\frac{1}{5} \times 20 = 4$.

Here's a summary of the *method of indicators*.

Suppose X can be written as

$$X = X_1 + X_2 + \cdots$$

where the X_i's are indicators (random variables that take on just the values 0 and 1). Then

$$EX = EX_1 + EX_2 + \cdots$$

Now all you have to do is find the expectation of each indicator and add. An indicator expectation is usually easy to find (which is why this is a useful method). As in (2), it boils down to

$$EX_i = P(X_i = 1)$$
$$= P(\text{the indicator found what it was assigned to look for})$$

Unfortunately, there is no rule for deciding whether or not to try indicators. And there is no rule for what to select as the indicators. Experience helps.

Expected Number of Reds in a Sample

Suppose 40% of the balls in a box are red. Draw n balls. Let X be the number of reds in the sample. Let's find EX.

If the drawings are *with* replacement, then they are Bernoulli trials where $P(\text{success}) = P(\text{red}) = .4$. By (3),

$$EX = np = n \times .4$$

Expect 40% of the sample to be red.

We'll show that the expectation is the *same* if the drawings are *without* replacement. Let

$$X_i = \begin{cases} 1 & \text{if the } i\text{th ball is red} \\ 0 & \text{otherwise} \end{cases}$$

Then

$$X = X_1 + \ldots + X_n$$

and

$$EX = EX_1 + \ldots + EX_n$$

Now we need the indicator expectations:

$$EX_i = P(X_i = 1)$$
$$= P(i\text{th ball is red})$$
$$= P(\text{1st ball is red}) \text{ (by symmetry)}$$
$$= .4$$

So

$$EX = \text{sum of } n \text{ terms each of which is } .4 = n \times .4 = 40\% \times n$$

> If, say, 40% of the balls in a box are red, then the mathematical expectation is that 40% of a sample will be red.
>   This holds for sampling with or without replacement.

Mean of a Geometric Random Variable

I'll continue to illustrate the method of indicators and at the same time find another famous expectation.

Consider Bernoulli trials where $P(\text{success}) = p$.

Let X be the number of tries to get a success. Let's find EX.

Let

$$X_i = \begin{cases} 1 & \text{if the first } i \text{ trials are F} \\ 0 & \text{otherwise} \end{cases}$$

Overall we're interested in the length of the F streak (if any) before the first S. For each trial there is an indicator assigned to see if the initial F streak is still alive. For example, if the outcome is

$$\text{F F F F S}$$

then

$$X = 5, \ X_1 = 1, \ X_2 = 1, \ X_3 = 1, \ X_4 = 1, \ X_5 = 0, \ X_6 = 0, \ X_7 = 0, \ldots$$

For this outcome, of the infinitely many indicators, only $X_1, \ldots, X_4$ kick in.

The sum of the indicators counts the initial streak of F's. Since X counts the number of wasted F's *plus* the first S, we have

$$X = 1 + X_1 + X_2 + X_3 + \cdots$$

and

$$E(X) = E(1) + E(X_1) + E(X_2) + E(X_3) + \cdots$$

By $E(1)$ we mean the expectation of a random variable that always takes the value 1, so

$$\boxed{E(1) = 1}$$

Now we need the expectations of the indicators:

$$EX_i = P(X_i = 1) = P(\text{first } i \text{ trials are F}) = q^i$$

So

$$EX = 1 + q + q^2 + q^3 + \cdots \qquad \text{(geometric series)}$$

$$= \frac{1}{1-q} = \frac{1}{p}$$

---

(4) | Let $P(\text{success}) = p$.

The expected number of Bernoulli trials to get the first success is $1/p$.

Again, this is intuitively reasonable: If $P(\text{head}) = 1/5$, then on the average it takes 5 tosses to get a head. If $P(\text{head}) = 2/5$, then on the average it takes $2\frac{1}{2}$ tosses to get a head.

Example 1   (mean of the negative binomial)

Suppose $P(\text{heads}) = p$. We'll find the expected number of tosses needed to get 3 heads.

Let X be the number of tosses to get 3 heads.

Let

X_1 = number of tosses needed to get the first H
X_2 = number of tosses needed after the first H to get the second H
X_3 = number of tosses needed after the second H to get the third H

If the outcome is

$$\text{T T H H T T T H}$$

then

$$X = 8, \quad X_1 = 3, \quad X_2 = 1, \quad X_3 = 4$$

In general,

$$X = X_1 + X_2 + X_3$$

The random variables X_1, X_2, X_3 are not indicators (they do not take on only the values 0,1) but we still have

$$EX = EX_1 + EX_2 + EX_3$$

The tosses are independent, so waiting for the second H after the first H and waiting for the third H after the second H are like waiting for the first H in the sense that X_2 and X_3 have the same distribution as X_1. All three have a geometric distribution with parameter p. So each has mean $1/p$ and

$$EX = \frac{1}{p} + \frac{1}{p} + \frac{1}{p} = \frac{3}{p}$$

For example, if $P(\text{heads}) = 1/5$, then on the average it takes 15 tosses to get 3 heads. More generally, if $P(\text{success}) = p$, then the expected number of Bernoulli trials to get k successes is k/p. (The result in (4) is the special case where $k = 1$.)

Example 2

A box contains 10 white balls and 14 black balls.

If balls are drawn *with* replacement, then the drawings are Bernoulli trials where $P(\text{W}) = 10/24$, so the expected number of trials to get a white is 24/10 (mean of the geometric distribution).

Find the expected number of trials needed if the drawings are with*out* replacement.

Let X be the number of draws needed to get a white. Name the black balls $\text{B}_1, \ldots, \text{B}_{14}$. For $i = 1, \ldots, 14$, let

$$x_1 = \begin{cases} 1 & \text{if } \text{B}_i \text{ is drawn before any white} \\ 0 & \text{otherwise} \end{cases}$$

This assigns an indicator to each black ball to see if it turns up before any white. If the outcome is

$$\text{B}_4 \, \text{B}_1 \, \text{B}_7 \, \text{B}_8 \, \text{W}$$

then

$$X = 5, \quad X_4 = 1, \quad X_1 = 1, \quad X_7 = 1, \quad X_8 = 1, \quad \text{other } X_i\text{'s are } 0$$

In general,

$$X = 1 + X_1 + X_2 + \cdots + X_{14}$$

and

$$\begin{aligned} EX &= 1 + EX_1 + EX_2 + \cdots + EX_{14} \\ &= 1 + P(\text{B}_1 \text{ before any W}) + P(\text{B}_2 \text{ before any W}) \\ &\quad + \cdots + P(\text{B}_{14} \text{ before any W}) \end{aligned}$$

The prob that B_i is drawn before any W is the same as the prob of getting B_i in one draw from a new box containing just the 10 W's and B_i, namely, 1/11. So

$$EX = 1 + 14 \cdot \frac{1}{11} = \frac{25}{11}$$

(less than the expected value when the drawing is *with* replacement).

Problems for Section 3-2

1. A box contains w whites and b blacks. Draw n without replacement. Let X be the number of whites in the sample.
   (a) Find EX immediately by quoting a known result.
   (b) For practice, find EX again using the following indicators: Let the white balls be named $W_1, W_2, \ldots, W_w$. Define indicators to watch each white ball: for $i = 1, 2, \ldots, w$, let

   $$X_i = 1 \text{ if } W_i \text{ is in the sample}$$

2. (the expected number of matches in the game of rencontre) A permutation of $a_1, \ldots, a_n$ has a fixed point if a_i appears in the ith place (its "natural" place) in the permutation. For example, the permutation $a_2 a_1 a_3 a_4 a_6 a_5$ has two fixed points, a_3 and a_4. If a permutation of $a_1, \ldots, a_n$ is chosen at random, what is the expected number of fixed points?

3. Ten married couples are seated (a) in a circle and (b) in a line. Find the expected number of wives sitting next to their husbands.

4. A consecutive string of heads is called a *run* of heads. For example, if the outcome of coin tossing is

   $$\boxed{\text{H H}} \quad \text{T T} \quad \boxed{\text{H H H H}} \quad \text{T} \quad \boxed{\text{H}} \quad \text{T} \quad \boxed{\text{H}}$$

   then there are 4 runs of heads (note that it's possible for a run to contain only one head). If $P(\text{heads}) = p$, find the expected number of runs of heads in n tosses.

5. Pick numbers at random between 0 and 1. The ith number chosen sets a record if it is larger than all the preceding choices.

   For example, if the numbers are .5, .2, .6, .34, .7, then three records are set, by the first number (which is always considered to set a record), and by .6 and .7. (The probability is 0 that a number is chosen twice, so ignore the possibility of tying a record.)

   Find the expected number of records in n trials. What happens to the expected number of records as $n \to \infty$?

6. Inside each box of Crunchies cereal the manufacturer has placed a picture of one of four Olympic athletes A, B, C, D. The pictures are distributed in equal numbers, so you are just as likely to get one as another.

   (Drawing cereal boxes without replacement from a *large* stock of boxes can be thought of as drawing *with* replacement. In other words, buying cereal is like drawing with replacement from an urn containing A's, B's, C's, D's in equal amounts.)

   (a) If you buy n boxes, find the expected number of different pictures you get.

   (b) Find the expected number of boxes you have to buy to get a complete set of pictures.

   Suggestion: Define random variables X_1, X_2, X_3, X_4 so that X_3, for instance, counts the number of draws to get a third letter after you already have two letters.

SECTION 3-3   CONDITIONAL EXPECTATION AND THE THEOREM OF TOTAL EXPECTATION

Definition of Conditional Expectation

Let X be a random variable. Then

$$E(X|\text{event } A) = \sum_x \times P(X = x|A)$$

The conditional expectation is really just an ordinary expectation, but within the new universe where event A has happened.

Example 1

Draw one card from a deck. Let X be its face value (an ace has value 11, and every other picture has value 10).

To find the (plain) expectation of X, average its 10 possible values:

$$EX = 2P(X = 2) + 3P(X = 3) + \cdots + 11P(X = 11)$$

$$= 2P(\text{deuce}) + \cdots + 10P(10, \text{ J, Q, or K}) + 11P(\text{ace})$$

$$= 2 \cdot \frac{1}{13} + \cdots + 9 \cdot \frac{1}{13} + 10 \cdot \frac{4}{13} + 11 \cdot \frac{1}{13} = \frac{95}{13}$$

To find the *conditional* expectation of X given that the card is a picture, average the *two* possible values:

$$E(X|\text{picture}) = 10P(X = 10|\text{picture}) + 11P(X = 11|\text{picture})$$

$$= 10 \cdot \frac{3}{4} + 11 \cdot \frac{1}{4} = \frac{41}{4}$$

The Theorem of Total Expectation

Box I has 20 reds, 10 whites, and box II has 10 reds, 10 whites.
Toss a coin where $P(\text{heads}) = .4$.
If H, draw 3 balls from box I.
If T, draw 5 balls from box II.
Let X be the number of reds in the sample. I'll find $E(X)$.

Remember that if $\frac{3}{4}$ of the balls in an urn are red, then, on the average, $\frac{3}{4}$ of any sample is red (whether the drawing is with or without replacement). So

$$E(\text{number of reds in a sample of 3 from urn I}) = \frac{20}{30} \cdot 3 = 2$$

and

$$E(\text{number of reds in a sample of 5 from urn II}) = \frac{10}{20} \cdot 5 = \frac{5}{2}$$

In other words, $E(X|\text{H}) = 2$ and $E(X|\text{T}) = 5/2$ (Fig. 1).

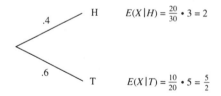

Figure 1

If 40% of the time you expect 2 reds and 60% of the time you expect 5/2 reds, then, overall,

$$EX = .4 \times 2 + .6 \times \frac{5}{2} = 2.3$$

Here's the general idea, called the *theorem of total expectation*.

Suppose a probability space can be divided into three mutually exclusive exhaustive events A, B, C (Fig. 2).
   Let X be a random variable defined on the space. Then

$$E(X) = P(A)\, E(X|A) + P(B)\, E(X|B) + P(C)\, E(X|C)$$

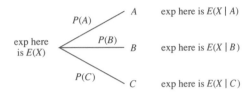

Figure 2

In other words, to get the overall expected value of X, find its expected value in the three new universes and then average *those* results using as weights the probabilities of the universes.
   Here's the proof.

$$EX = \sum_x xP(X = x) \qquad \text{(definition of } EX\text{)}$$

$$= \sum_x x[P(A)P(X = x|A) + P(B)P(X = x|B)$$

$$+P(C)P(X = x|C)] \qquad \text{(theorem of total prob)}$$

$$= P(A)\left[\sum_x xP(X = x|A)\right] + P(B)\left[\sum_x xP(X = x|B)\right]$$

$$+P(C)\left[\sum_x xP(X = x|C)\right] \qquad \text{(algebra)}$$

$$= P(A)\, E(X|A) + P(B)\, E(X|B) + P(C)\, E(X|C)$$

Warning

Here are some results from earlier sections:

Expected number of successes in n Bernoulli trials is np.
Expected number of Bernoulli trials to get the first success is $1/p$.
Expected number of reds in n draws from a box is $n \times$ percentage of reds in box.

You'll make extra work for yourself if you don't take advantage of them.

Example 1

In craps, a player makes a bet and then rolls a pair of dice.

If the resulting sum is 7 or 11, the player wins.

If the result is 2, 3, or 12, the player loses.

If another sum, r, comes up, it's called the player's point. The player keeps tossing, and if r comes up before 7, the player wins; if 7 turns up before r, the player loses.

Let X be the number of rolls in a game. I'll find $E(X)$ using the theorem of total expectation, conditioning on the first roll (Fig. 3).

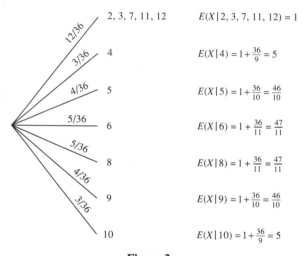

Figure 3

If the first roll is 2, 3, 7, 11, or 12, then the game is over with that 1 roll, so

$$E(X|2, 3, 7, 11, 12) = 1 \quad \text{(first branch in Fig. 3)}$$

Suppose the first roll is 4. Now that the point is 4, the game will continue until 4 or 7 comes up. The continuation consists of Bernoulli trials where

$$P(\text{success}) = P(\text{game-ending throw}) = P(4 \text{ or } 7) = \frac{9}{36}$$

Remember that if $P(S) = p$, then the number of trials to get the first S is a geometric random variable with expectation $1/p$.

So the expected number of *more* rolls, *after* the initial roll comes up 4, is 36/9 and

$E(X|4) =$ the one initial roll + expected number of rolls in the continuation

$$= 1 + \frac{36}{9} \quad \text{(second branch in Fig. 3)}$$

Similarly, if the first roll is 5, then the game continues until 5 or 7 comes up. The continuation consists of Bernoulli trials with

$$P(\text{success}) = P(\text{game-ending throw}) = P(5 \text{ or } 7) = \frac{10}{36}$$

The expected number of *more* rounds after the roll of 5 is 36/10, and

$E(X|5) =$ the one initial roll + expected number of rolls in the continuation

$$= 1 + \frac{36}{10} \quad \text{(third branch in Fig. 3)}$$

The other conditional expectations in Fig. 3 are found similarly. Finally,

$$E(X) = \frac{12}{36} \cdot 1 + \frac{3}{36} \cdot 5 + \frac{4}{36} \cdot \frac{46}{10} + \frac{5}{36} \cdot \frac{47}{11} + \frac{5}{36} \cdot \frac{47}{11} + \frac{4}{36} \cdot \frac{46}{10} + \frac{3}{36} \cdot 5 \sim 3.375$$

Example 2

A machine is erratic. When you push the button, one of three things happens.

> 60% of the time it goes into mode A and is done in 5 minutes.
> 30% of the time it goes into mode B, buzzes for 2 minutes, but produces nothing, so that you have to try again.
> 10% of the time it goes into mode C, buzzes uselessly for 3 minutes, after which time you must try again.

Find the expected amount of time it takes to get done.

Let X be the time it takes to get done. Let $E(X)$ be denoted by E. We'll find E using the theorem of total expectation, conditioning on the first try (Fig. 4).

> If the machine goes into mode A on the first try, then $X = 5$, so we have $E(X|A) = 5$.
> If the machine goes into mode B on the first try, then after 2 minutes we are ready to try again. The new situation is exactly the same as it was initially, so after the 2-minute waste of time, the new expected time to get done is E again. So $E(X|B) = 2 + E$.
> Similarly, $E(X|C) = 3 + E$.

So

$$E = (.6)(5) + (.3)(2 + E) + (.1)(3 + E)$$

Solve to get

$$E = \frac{39}{6} = 6\frac{1}{2}$$

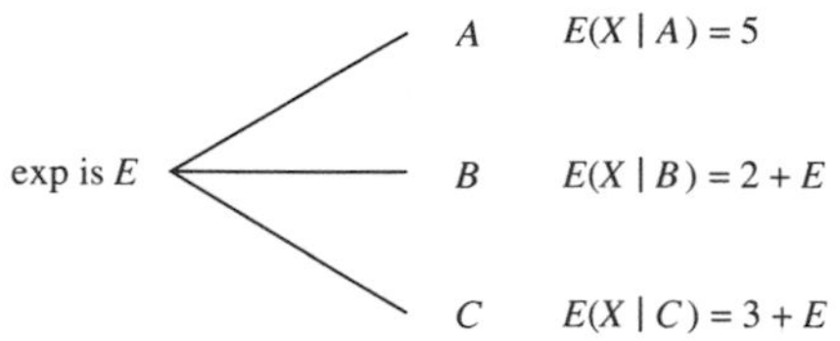

A $E(X \mid A) = 5$

exp is E B $E(X \mid B) = 2 + E$

C $E(X \mid C) = 3 + E$

Figure 4

Problems for Section 3-3

1. Roll a pair of dice. Find the expected value of the first die if the sum is 4.

2. A student working on a problem has three available methods.

   > Method A takes 5 hours but doesn't solve the problem.
   > Method B takes 2 hours and also leads nowhere.
   > Method C leads to a solution after 4 hours of work.

   If the student has no reason to favor one untried method over another (but naturally a method that fails will not be tried again), what is the expected length of time it takes to solve the problem?

3. Toss a die 10 times. If you get six 1's, find the expected number of 2's.

4. Spin the indicated arrow. If it lands in quadrant k, toss k dice. Find the expected number of 5's on the dice.

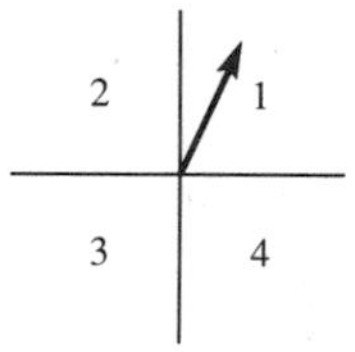

Figure P.4

5. A box contains 10 white and 20 black balls. Draw 8 balls without replacement. Find the expected number of whites in the first 3 draws given that you got 1 white in the last 5 draws.

6. Toss a coin n times where $P(\text{H}) = p$. Find the expected number of H given that you got at least 2H.

7. A coin has $P(\text{head}) = p$. Keep tossing until you finish a run of 5 heads (i.e., until you get 5 heads *in a row*). Find the expected number of tosses it takes.

Suggestion: Use the theorem of total expectation, conditioning on when the first tail occurs.

SECTION 3-4   VARIANCE

If John's grades are 51 and 49 and Mary's grades are 100 and 0, then each has a 50 average. But they seem like very different students (John is a dolt, while Mary is erratic). An average grade doesn't tell the whole story. We should also measure the spread around the average, called the variance. Here's an example to show how it's done.

Toss a 3-sided biased die where

$$P(1) = .5, \quad P(2) = .1, \quad P(3) = .4$$

Let X be the face value. Then

$$E(X) = 1(.5) + 2(.1) + 3(.4) = 1.9$$

To find the variance of X, denoted by Var X, look at the differences between the possible values of X and the mean of X:

$$1 - 1.9 = -.9 \quad \text{(happens with probability .5)}$$
$$2 - 1.9 = .1 \quad \text{(happens with probability .1)}$$
$$3 - 1.9 = 1.1 \quad \text{(happens with probability .4)}$$

Square the differences (so that a positive over-the-mean doesn't cancel a negative under-the-mean) and take a weighted average:

$$\text{Var } X = (-.9)^2 \, (.5) + (.1)^2 \, (.1) + (1.1)^2 \, (.4) = .89$$

Here's the general definition of the *variance* of a random variable:

Let μ_X denote the mean of X. Then

$$\text{Var } X = E(X - \mu_X)^2 = E(X - EX)^2$$

If Var X is large, then it is likely that X will be far from the mean. If Var X is small, it is likely that X will be close to its mean.

The *standard deviation* of X, denoted by σ, is defined as $\sqrt{\text{Var}X}$.

Variances are not usually computed directly as we just did for the die; there is a slicker method. All the details and theory and problems about variance will be postponed until the chapter on continuous expectation. Then we'll save time by doing discrete and continuous variance simultaneously.

Review Problems for Chapter 3

1. If 100 balls are tossed at random into 50 boxes, find the expected number of empty boxes.

2. Toss a coin 10 times. Find the expected number of heads in the first 5 tosses, given 6 heads in the 10 tosses.

3. In the game of chuck-a-luck, 3 dice are rolled. You can bet on any of the six faces 1, 2, ..., 6. Say you bet on 5.

   > If exactly one of the 3 dice is a 5, then you win \$1.
   > If two of the 3 dice are 5's, then you win \$2.
   > If all three of the dice are 5's, you win \$3.
   > If none of the dice are 5's, you lose \$1.

   Find your expected winnings.

4. If $x_1, \ldots, x_n$ are permuted, find the expected number of items *between* x_1 and x_2.

5. Toss a coin 10 times. Find the expected number of heads given that there are at least 9 heads. If you have time, try it twice, with and without indicators.

6. At the first stage of an experiment, a number X is selected so that X has a Poisson distribution with parameter λ. Then if $X = n$, toss a coin n times where $P(\text{heads}) = p$. (If $X = 0$, don't toss at all.) Find the expected number of heads.

7. You have n keys on your key ring. One of them unlocks the front door, but you forgot which one, so you keep trying keys until the door unlocks. Find the expected number of trials needed if

   (a) you sensibly try the keys without replacement (for practice, try it twice, with and without indicators)

   For reference:

   $$1 + 2 + 3 + \cdots + n = \frac{n(n+1)}{2}$$

   (b) you foolishly try the keys *with* replacement

8. Ten people each toss a coin 5 times. In each case, $P(\text{heads}) = p$. A person is shut out if she gets no heads. Find the expected number of shutouts.

9. (a) Toss 100 balls at random into 50 boxes. Find the expected number of balls in the third box.

   (b) Toss balls at random into 50 boxes. Find the expected number of tosses required to get a ball into the third box.

10. A hospital handles 20 births a day. Ten percent require a special fetal monitor. Find the expected number of days out of the year when the hospital will need at least two monitors.

11. The probability that a well is polluted is .1. Suppose 100 wells are tested as follows.

    Divide the wells into 5 groups of 20. For each group, water samples from each of the 20 wells are pooled and one test is performed. If a test is negative, then you know that all 20 wells are unpolluted. If a test is positive, then each of the 20 members is tested individually to see which ones are polluted.

    Find the expected number of tests needed with this method (as opposed to the 100 tests it would take to forget about pooling and just test each one).

12. Find the expected number of suits in a poker hand.

13. Toss a die.

    > If the face is even, you win $1.
    > If the face is 1 or 3, you lose $2.
    > If the face is 5, you go to prison. When in prison keep tossing.
    > If the next toss is even, you lose $1.
    > If the next toss is odd, you get out of prison and begin the game again.

    (a) Find the expected winnings.
    (b) Find the expected number of tosses in a game (until you either win or lose money).

14. Noah's ark contains a pair of A_1's, a pair of A_2's, ... , a pair of A_n's. Draw m times from the ark (without replacement). Find the expected number of pairs left.

15. An urn contains 3 red and 7 black balls. Draw one ball.

    Then toss a coin 4 times, but use a fair coin if the ball was red and use a biased coin with $P(\text{heads}) = 1/3$ if the ball was black.

    Find the expected number of heads.

16. For a couple planning to have children,

    $$P(\text{boy}) = p, P(\text{girl}) = 1 - p = q$$

    (a) If they keep trying until they have a girl, what is the expected number of children?
    (b) Suppose they plan to have a minimum of 6 children. If the first 6 are all boys, they will continue until they get a girl. What is

the expected number of children? (This can be done nicely with indicators or with the theorem of total expectation.)

17. A contest offers you (and 10 million others) the chance to win a million-dollar prize. No entry fee is required except for the price of a 25-cent stamp. Find your expected winnings to see if the contest is worth entering.

REVIEW FOR THE NEXT CHAPTER

Some Integrals for Reference

$$\int_0^1 x^n \, (1-x)^m \, dx = \frac{n!m!}{(n+m+1)!}$$

$$\int xe^{ax} \, dx = e^{ax} \left(\frac{x}{a} - \frac{1}{a^2} \right)$$

$$\int x^2 \, e^{ax} \, dx = e^{ax} \left(\frac{2}{a^3} - \frac{2x}{a^2} + \frac{x^2}{a} \right)$$

$$\int_0^\infty e^{-x^2} \, dx = \frac{1}{2} \sqrt{\pi}$$

$$\int_0^\infty e^{-ax^2} \, dx = \frac{1}{2} \sqrt{\frac{\pi}{a}} \qquad \text{for } a > 0$$

$$\int_{-\infty}^\infty e^{-x^2} \, dx = \sqrt{\pi}$$

$$\int_{-\infty}^\infty e^{-ax^2} \, dx = \sqrt{\frac{\pi}{a}} \qquad \text{for } a > 0$$

$$\int_0^\infty x^n \, e^{-x} \, dx = n! \qquad \text{for } n = 0, 1, 2, 3, \ldots$$

$$\int_0^\infty x^n \, e^{-ax} \, dx = \frac{n!}{a^{n+1}} \qquad \text{for } a > 0; \quad n = 0, 1, 2, 3, \ldots$$

Differentiating and Integrating Functions That Change Formula

Suppose

$$f(x) = \begin{cases} x & \text{for } 2 \leq x \leq 3 \\ x^2 & \text{for } 3 \leq x \leq 5 \end{cases}$$

Then

$$f'(x) = \begin{cases} 1 & \text{for } 2 \leq x \leq 3 \\ 2x & \text{for } 3 \leq x \leq 5 \end{cases}$$

and

$$\int_2^5 f(x) \, dx = \int_2^3 x \, dx + \int_3^5 x^2 \, dx = \left. \frac{x^2}{2} \right|_2^3 + \left. \frac{x^3}{3} \right|_3^5 = \frac{5}{2} + \frac{98}{3} = \frac{211}{6}$$

Similarly, if

$$g(x) = \begin{cases} x^2 & \text{if } 2 \le x \le 5 \\ 0 & \text{otherwise} \end{cases}$$

then

$$\int_{-\infty}^{\infty} g(x)\, dx = \int_{2}^{5} x^2\, dx = \left. \frac{x^3}{3} \right|_{2}^{3} = \frac{19}{3}$$

Integrating with Absolute Values

$$\int_{-2}^{3} |x|\, dx = \int_{-2}^{0} -x\, dx + \int_{0}^{3} x\, dx \quad \text{(Fig. 1)}$$

$$= \left. -\frac{x^2}{2} \right|_{-2}^{0} + \left. \frac{x^2}{2} \right|_{0}^{3} = \frac{13}{2}$$

$$\int_{-1}^{3} e^{|x|}\, dx = \int_{-1}^{0} e^{-x}\, dx + \int_{0}^{3} e^{x}\, dx \quad \text{(Fig. 2)}$$

$$= \left. -e^{-x} \right|_{-1}^{0} + \left. e^{x} \right|_{0}^{3} = e + e^3 - 2$$

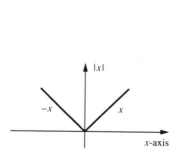

Figure 1

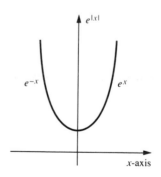

Figure 2

CHAPTER 4

Continuous Random Variables

Remember that a random variable is the numerical outcome of an experiment. So far (except for the uniform continuous case in Section 1.5), we've considered only random variables, which take on a *finite or countably infinite number of values* (e.g., the number of heads in 10 tosses). These random variables are called *discrete*. Now we're turning to random variables that can take an *uncountably infinite number of values*, such as all values in an interval. Typical examples are height, age, weight, distance, time, and so on. In this section, we'll show how *probability densities* are used to assign probabilities to events.

Densities

First, you have to know something about densities in general. Let's look at a familiar density, a mass density, to see the connection between the mass *density* and the *total* mass.

   If the mass density on a line is, say, 6 grams per centimeter, then an 8-cm interval contains 48 grams. In general, if the mass density is *constant*, then

(1)   total grams in an interval = length of interval × mass density

that is,

$$\text{total grams } = \text{centimeters } \times \frac{\text{grams}}{\text{centimeters}}$$

Suppose the mass density in an interval $[a, b]$ varies from point to point so that at point x, the density is $f(x)$ grams per centimeter. Since the density is variable, we can't use (1) directly to find the total mass of the interval. Instead, divide the interval into many small pieces (Fig. 1).

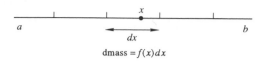

Figure 1

A typical piece has length dx and contains point x. The density within the interval varies, but if the interval is very small, its density is *almost constant* with value $f(x)$ (its value at the particular point x in the interval). Then we can use (1) to find the mass (denoted dmass) of the little piece:

$$\text{dmass} = \text{length} \times \text{almost constant mass density} = dx \times f(x) = f(x)\,dx$$

The *total* mass of the interval is found by adding the dmass's and making $dx \to 0$ (so that the "almost" goes away). That's exactly what an integral does. So

$$(2) \qquad\qquad \text{total mass} = \int_a^b \text{dmass} = \int_a^b f(x)\,dx$$

In other words, *integrate the mass density to get the total mass.*
Now we're ready to go back to probability.

Continuous Random Variables

Figure 2 shows a *discrete* random variable X which takes on values $-7, 1, \pi, 5$ with

$$(3) \qquad \begin{aligned} P(X = -7) &= .4, \quad P(X = 1) = .1, \\ P(X = \pi) &= .3, \quad P(X = 5) = .2 \end{aligned}$$

We say that X has the probability function $p(x)$, where

$$(4) \qquad \begin{aligned} p(-7) &= .4 \\ p(1) &= .1 \\ p(\pi) &= .3 \\ p(5) &= .2 \\ p(x) &= 0 \quad \text{otherwise} \end{aligned}$$

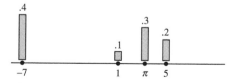

Figure 2   Probs for a discrete random variable

In general, for a *discrete* random variable taking on a few values, one unit of probability is split into chunks, which are assigned to the possible values.

If a random variable can take on a *continuum* of values, then instead of placing chunks of probability at a few points, we mash up one unit of probability and spread it out (not necessarily evenly) on the line (Fig. 3), creating a probability *density* (probability units per foot). Each individual point gets zero probability, but the amount of probability in any *interval* can be found by integrating the probability density. The random variable in this case is called *continuous*.

Figure 3   Prob density for a continuous random variable

Here's the formal definition.

Suppose X is a random variable and we have a function $f(x)$ which is integrated as follows to get probabilities of events:

$$P(a \leq X \leq b) = \int_a^b f(x)\, dx$$

$$P(X \leq b) = \int_{-\infty}^b f(x)\, dx$$

$$P(X \geq a) = \int_b^\infty f(x)\, dx$$

and so on. Then $f(x)$ is called the probability *density function* of X, and the random variable X is called *continuous*.

The set of points where $f(x) \neq 0$ is called the *universe*.

You can think of probabilities as areas under the $f(x)$ graph:

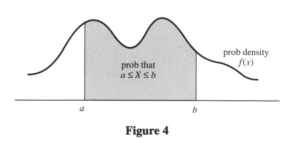

Figure 4

Properties of a Probability Density Function

Not any old function $f(x)$ can be the density of a random variable X. Probabilities are never negative and $P(-\infty < X < \infty) = 1$, so we must have

(5) $\boxed{f(x) \geq 0 \quad \text{for all } x \quad \text{(the graph never goes below the } x\text{-axis)}}$

(6) $$\boxed{\int_{-\infty}^{\infty} f(x)\, dx = 1 \quad \text{(the area under the graph is 1)}}$$

Note that (6) can't hold unless we also have

(7) $\boxed{f(\infty) = 0, \quad f(-\infty) = 0}$

Figure 5 shows three typical densities.

Figure 5   Some densities

Example 1

Let X have density

$$f(x) = \begin{cases} x^2/9 & \text{for } 0 \leq x \leq 3 \\ 0 & \text{otherwise} \end{cases} \quad \text{(Fig. 6)}$$

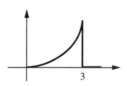

Figure 6

This is a legitimate density because $f(x) \geq 0$ and

$$\int_{-\infty}^{\infty} f(x)\, dx = \int_{0}^{3} \frac{1}{9}x^2\, dx = \left. \frac{x^3}{27} \right|_{0}^{3} = 1$$

Then

$$P(1 \le X \le 2) = \int_1^2 \frac{1}{9}x^2 \, dx = \frac{x^3}{27}\bigg|_1^2 = \frac{7}{27}$$

$$P(X \le 1) = \int_{-\infty}^1 f(x) \, dx = \int_0^1 \frac{1}{9}x^2 \, dx = \frac{x^3}{27}\bigg|_0^1 = \frac{1}{27}$$

$$P(X \ge 3) = \int_3^\infty 0 \, dx = 0$$

Notation

Instead of writing

$$f(x) = \begin{cases} x^2/9 & \text{for } 0 \le x \le 3 \\ 0 & \text{otherwise} \end{cases}$$

I'll just write

$$f(x) = \frac{x^2}{9} \quad \text{for } 0 \le x \le 3$$

and leave it understood that $f(x)$ is 0 elsewhere, that is, that the universe is [0,3].

But don't leave out the interval entirely and just write $f(x) = x^2/9$. That would be interpreted as $f(x) = x^2/9$ for *all* x, not even a legitimate density (the area under the graph is ∞, not 1).

The Event X = x

If X is a continuous random variable, then for any x,

(8)
$$\boxed{P(X = x) = 0}$$

since $\int_x^x f(x) \, dx = 0$. In other words,

even if the event $X = x$ is possible, it has probability 0.

And it follows that

$$P(a \le X \le b) = P(a < X \le b) = P(a \le X < b) = P(a < X < b)$$

In example 1,

$$P(1 \le X \le 2) = P(1 \le X < 2) = P(1 < X \le 2) = P(1 < X < 2) = \frac{7}{27}$$

Furthermore, in example 1, it makes no difference if $f(x)$ is $x^2/9$ for $0 \le x \le 3$ or for $0 \le x \le 3$ or for $0 \le x < 3$ or for $0 < x < 3$.

The Event $X \approx x$

Let X have density $f(x)$. Fix x. The probability that X *is* x is 0, not a very satisfying result. On the other hand, it *will* be useful to find the probability that X is *near* x.

We'll use this abbreviation:

> X is near x (denoted $X \approx x$) means X is in a small interval of length dx around x (Fig. 7).

The length of the interval is dx, and its density is almost constant at $f(x)$ so, by the formula density $\times$ length,

(9)
$$P(X \approx x) = f(x)\, dx$$

The event $X \approx x$ is a more realistic concept in engineering and science than is the event $X = x$. If X is the result of a physical experiment, then the values of X can be measured only to some desired degree of accuracy, say, to five decimal places without rounding off. In that case, the event $X = 3.87654$ is really $X \approx 3.87654$, in particular the event $3.87654 \le X \le 3.87655$.

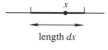

length dx

Figure 7

Warning

The values of the density function are *not probabilities*. The units on $f(x)$ are *prob per unit length*. It's $f(x)$ *times* dx that is a probability, namely,

$$f(x)\, dx = P(X \approx x)$$

Density of a Uniformly Distributed Random Variable

Suppose X is uniformly distributed on the (finite) interval $[a, b]$, meaning that

$$P(X = \text{event}) = \frac{\text{fav length}}{\text{total length}} \qquad \text{(see Section 1.5)}$$

(For all practical purposes it doesn't matter if X is picked from $[a, b]$ or (a, b) or $[a, b)$ or $(a, b]$.)

If we spread one unit of probability evenly over $b - a$ feet, there must be $1/(b - a)$ units of probability per foot, so the probability density is

(10)

$$f(x) = \frac{1}{b-a} \quad \text{for } a \le x \le b \quad \text{(Fig. 8)}$$

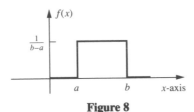

Figure 8

Another way to see this is to think of the graph of $f(x)$ in Fig. 8. On the interval $[a, b]$, $f(x)$ has a constant height (because X is uniformly distributed). The total area must be 1, the base is $b - a$, so the height must be $1/(b - a)$. For example, if X is uniformly distributed on $[-3, 7]$, then

$$f(x) = \frac{1}{10} \quad \text{for} \quad -3 \le x \le 7$$

Problems for Section 4-1

1. Let $f(x) = cx$ for $0 \le x \le 1$. Find c so that f is a probability density for a random variable X; then find $P(X < .3)$.

2. Let $f(x) = 1 - |x|$ for $-1 \le x \le 1$.

   (a) Sketch the graph of f.   (c) Find $P(-1/2 \le X \le 1/3)$.
   (b) Check that f is a density.

3. Suppose X is the number of hours a component part will last and X has density $f(x) = e^{-x}$ for $x \ge 0$.

   (a) Find the prob that a component lasts at least 6 hours.
   (b) Suppose a machine contains 3 components. Find the prob that they all last at least 6 hours, assuming that the components function independently of one another.

4. Let X have density $f(x) = \frac{1}{2}e^{-|x|}$.

   (a) Sketch the graph of f.   (c) Find $P(X^2 + X > 0)$.
   (b) Find $P(|X| < 4)$.

5. Let X be a continuous random variable. True or False?

   (a) If the prob of an event involving X is 0, then the event is impossible.
   (b) If the prob of an event involving X is 1, then the event is sure.

6. Pick a number at random between 2 and 7.
  (a) Find the probability that the number is 4.
  (b) Find the probability that the number is $\approx$ 4, meaning that the number is in an interval of length dx around 4.

SECTION 4-2   DISTRIBUTION FUNCTIONS

If you know the density function $f(x)$ of a random variable X, you can find the probability of any event involving X. You can also find these probabilities with another function called the *distribution function* of X, denoted by $F(x)$. And there are times when $F(x)$ is easier to come by or more convenient to use than $f(x)$.

The (Cumulative) Distribution Function of a Random Variable

Let X be a random variable (not necessarily continuous). Its distribution function $F(x)$ is defined by

(1)
$$F(x) = P(X \le x)$$

The distribution function gives the amount of probability that has accumulated so far. It's often called the *cumulative* distribution function and abbreviated cdf. *Every random variable (discrete, continuous, or whatever) has a distribution function.*

In the continuous case where, in addition to a distribution function, X has density $f(x)$,

$$F(x) = P(X \le x) = \int_{-\infty}^{x} f(x)dx$$

$$= \text{cumulative area under the graph of } F(x) \quad \text{(Fig. 1)}$$

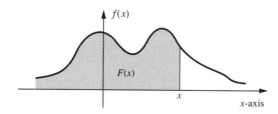

Figure 1 $F(x)$ is cumulative area under $f(x)$

The Distribution Function of a Discrete Random Variable

Suppose X takes on values 2, 3, π with probabilities

$$P(X = 2) = .5, \quad P(X = 3) = .2, \quad P(X = \pi) = .3$$

so that the probability function for X is

$$p(x) = \begin{cases} .5 & \text{if } x = 2 \\ .2 & \text{if } x = 3 \\ .3 & \text{if } x = \pi \end{cases} \qquad \text{(Fig. 2)}$$

Figure 2 Probability function for X

The probability function in Fig. 2 is sometimes referred to as the discrete density function for X. If you're familiar with *delta functions*, which allow area to be concentrated at a *point* (instead of spread out over an *interval* as with ordinary functions), then you recognize the function in Fig. 2 as

$$.5\delta(x - 2) + .2\delta(x - 3) + .3\delta(x - \pi)$$

Let $F(x)$ be the distribution function of X. Then $F(x)$ represents cumulative probability in Fig. 2. The cumulative probability is 0 until $x = 2$ when it jumps to .5. It stays .5 until $x = 3$ when it jumps to .7. It stays .7 until π when it jumps to 1 and thereafter stays 1. So

(2)
$$F(x) = \begin{cases} 0 & \text{if } x < 2 \\ .5 & \text{if } 2 \le x < 3 \\ .7 & \text{if } 3 \le x < \pi \\ 1 & \text{if } x \ge \pi \end{cases}$$

Figure 3 shows the graph of $F(x)$. The solid dots are part of the graph; the open dots are points not in the graph. At $x = 3$ the solid dot is at .7, indicating that $F(3)$ is .7, not .5. The correct value is .7 because, by the definition in (1), $F(3)$ is the probability that has accumulated in Fig. 2 up to *and including* the chunk sitting at $x = 3$.

In general, whenever there is a jump in a distribution function, the actual F value is at the top of the jump.

> The distribution function of a *discrete* random variable is a step function, rising from height 0 to height 1. There are jumps at all the possible values of X, and the size of the jump at x_0 is $P(X = x_0)$.

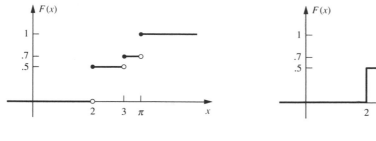

Figure 3   Figure 3'

Incorrect but Useful Graphs and Notation

Most computer graphing programs would produce the picture in Fig. 3', not Fig. 3, for $F(x)$. The vertical segments in Fig. 3' make the graph ambiguous; for example, the value of $F(3)$ looks as if it could be anything from .5 to .7. But even though Fig. 3 is the correct version, from now on in this book we're going to draw in the style of Fig. 3'. And at the risk of further enraging mathematicians, we'll refer to .7 as the upper $F(3)$ and to .5 as the lower $F(3)$. It's simply more convenient that way.

As long as we're drawing ambiguous pictures, we'll also feel free to write ambiguous formulas to match. Instead of (2), it won't hurt to write

(2')
$$F(x) = \begin{cases} 0 & \text{if } x \le 2 \\ .5 & \text{if } 2 \le x \le 3 \\ .7 & \text{if } 3 \le x \le \pi \\ 1 & \text{if } x \ge \pi \end{cases}$$

Both (2) and (2') convey a jump at $x = 3$ with upper value .7 and lower value .5.

The Distribution Function of a Uniformly Distributed Random Variable

Let X be uniformly distributed on $[a, b]$. Its density function $f(x)$ is in Fig. 4. To find $F(x)$, look at the cumulative area under $f(x)$. $F(x)$ is 0 until $x = a$. For x between a and b, $F(x)$ is the shaded area in Fig. 4, namely, $(x - a) \cdot 1/(b - a)$. For $x \ge b$, the accumulated area (probability) is 1, so $F(x) = 1$. So

$$F(x) = \begin{cases} 0 & \text{if } x \le a \\ \dfrac{x - a}{b - a} & \text{if } a \le x \le b \\ 1 & \text{if } x \ge b \qquad \text{(Fig. 5)} \end{cases}$$

In other words, $F(x)$ rises steadily from 0 to 1 on the interval $[a, b]$.

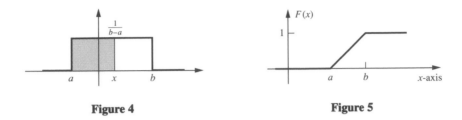

Figure 4   Figure 5

Example of a Mixed Random Variable

Toss a die. If the result is 2, then you win \$7. If the result is a non-2, then your winnings are picked at random from the interval [3,5]. Let X be your winnings. Let's find the distribution function of X.

Remember that for a *continuous* random variable, $P(X = x)$ is 0 for every x. But here, $P(X = 7) = 1/6$, so X is not continuous; that is, X doesn't have a density function. Neither is X discrete, since X can take on any of the values in the interval [3,5], as well as 7. So X doesn't have a probability function.

But you can make a pseudodensity for X. Figure 6 shows a chunk of probability of size 1/6 at $x = 7$ and the remaining 5/6 units of probability spread evenly over the interval [3,5]. The corresponding density graph is in Fig. 7: the base of the rectangle is 2, and its area must be 5/6, so the height of the rectangle is 5/12.

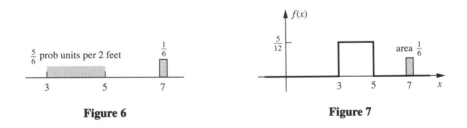

Figure 6   Figure 7

If you know about delta functions, you recognize Fig. 7 as the sum of an ordinary function and a delta function:

$$\text{Figure 7} = g(x) + \frac{1}{6}\delta(x - 7) \quad \text{where} \quad g(x) = \frac{5}{12} \quad \text{for} \quad 3 \le x \le 5$$

The distribution function $F(x)$ is the cumulative area in Fig. 7. Until $x = 3$, $F(x)$ is 0; as x goes from 3 to 5, $F(x)$ rises steadily from 0 to 5/6; it stays at height 5/6 until $x = 7$ when it jumps by 1/6 to 1 (Fig. 8).

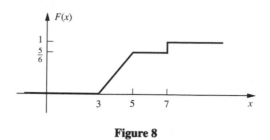

Figure 8

Properties of a Distribution Function

Let X be a random variable with distribution function $F(x)$. $F(x)$ collects cumulative probability so its graph starts at height 0 and rises to a height of 1. In other words:

(3) | F is non-decreasing (F can increase or stay level but not go down).

(4) $$F(-\infty) = 0, \quad F(\infty) = 1$$

Figure 9 shows some typical distribution functions.

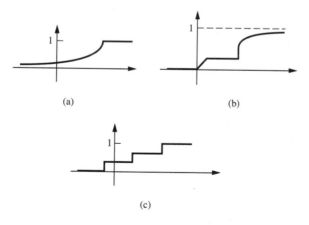

Figure 9   Typical distribution functions

Finding Probabilities from the Distribution Function

If the distribution function $F(x)$ has a jump at $x = x_0$, then it must be a jump *up* (Fig. 10), and it occurs because a chunk of probability is sitting at x_0 and just got fed in. So

$P(X = x_0) =$ size of the jump

$$P(X \le x_0) = \text{height at the top of jump, after chunk is fed in}$$
$$= \text{upper } F(x_0) \quad \text{(Fig. 10)}$$

$$P(X < x_0) = \text{height at bottom of jump, before chunk is fed in}$$
$$= \text{lower } F(x_0)$$

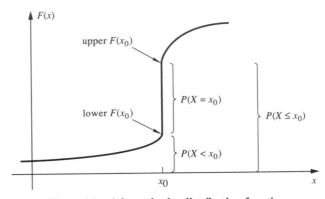

Figure 10   A jump in the distribution function

Furthermore,

$$P(X > x_0) = 1 - P(X \le x_0) = 1 - \text{ upper } F(x_0)$$

$$P(X \ge x_0) = 1 - P(X < x_0) = 1 - \text{ lower } F(x_0)$$

If F jumps at a and at b, then

$$P(a \le X \le b) = P(X \le b) - P(X < a)$$

$$= \text{upper } F(b) - \text{lower } F(a)$$
$$\text{(include the prob at } b \text{ and at } a\text{)}$$

$$P(a < X \le b) = \text{upper } F(b) - \text{ upper } F(a)$$
$$\text{(include the prob at } b \text{ and exclude the prob at } a\text{)}$$

$$P(a \le X < b) = \text{lower } F(b) - \text{ lower } F(a)$$

$$P(a < X < b) = \text{lower } F(b) - \text{upper } F(a)$$

> If there are *no jumps* in the distribution function of X, then
> $$P(X = x_0) = 0$$
> $$P(X < x_0) = P(X \le x_0) = F(x_0)$$
> $$P(X > x_0) = P(X \ge x_0) = 1 - F(x_0)$$
> $$P(a \le X \le b) = P(a < X \le b) = P(a \le X < b) = P(a < X < b)$$
> $$= F(b) - F(a)$$

When F doesn't jump, you don't have to be careful to distinguish between $\ge$ and $>$ or between $\le$ and $<$ in describing events.

   Horizontal portions of the graph of $F(x)$ (where probability is not accumulating) correspond to places where the density is 0 and events have probability 0.

Example 1

Let X have the distribution function

$$F(x) = \begin{cases} 0 & \text{if } x \le -1 \\[2mm] \dfrac{1}{8}(x+1)^2 & \text{if } -1 \le x \le 1 \\[2mm] \dfrac{3}{4} & \text{if } 1 \le x \le 2 \\[2mm] \dfrac{1}{8}x + \dfrac{1}{2} & \text{if } 2 \le x \le 4 \\[2mm] 1 & \text{if } x \ge 4 \end{cases} \qquad \text{(Fig. 11)}$$

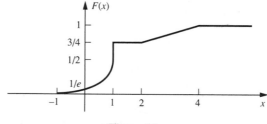

Figure 11

Note that there is a jump of 1/4 at $x = 1$ because the $F(x)$ formula for $-1 \le x \le 1$ leaves off at

$$\left. \frac{1}{8}(x+1)^2 \right|_{x=1} = \frac{1}{2}$$

but the $F(x)$ formula for $1 \le x \le 2$ picks up at 3/4.

There are no jumps anywhere else. So

$$P(X = 1) = \text{jump size } = \frac{1}{4}$$

$$P(X \le 1) = \text{upper } F(1) = \frac{3}{4}$$

$$P(X < 1) = \text{lower } F(1) = \frac{1}{2}$$

$$P(X > 1) = 1 - \text{upper } F(1) = 1 - \frac{3}{4} = \frac{1}{4}$$

$$P(X \ge 1) = 1 - \text{lower } F(1) = 1 - \frac{1}{2} = \frac{1}{2}$$

$$P(X = 0) = 0$$

$$P(X \le 0) = P(X < 0) = F(0) = \frac{1}{8}$$

$$P(X \ge 0) = P(X > 0) = 1 - F(0) = \frac{7}{8}$$

$$P(2 \le X \le 3) = P(2 < X < 3) = F(3) - F(2) = \frac{7}{8} - \frac{3}{4} = \frac{1}{8}$$

$$P(1 \le X \le 3) = F(3) - \text{lower } F(1) = \frac{7}{8} - \frac{1}{2} = \frac{3}{8}$$

$$P(1 < X \le 3) = F(3) - \text{upper } F(1) = \frac{7}{8} - \frac{3}{4} = \frac{1}{8}$$

$$P(1 < X \le 2) = 0, \quad P(X > 4) = 0$$

(the F graph is horizontal in these intervals)

The Distribution Function of a Continuous Random Variable

Every random variable has a distribution function.

Not every random variable has a density function. Those that do are called continuous.

A continuous random variable has the property that $P(X = x) = 0$ for all x, so its distribution function $F(x)$ can't jump. Conversely, it can be shown that for all practical purposes, a no-jump $F(x)$ is an exclusive property of continuous random variables, so that all in all:

> A random variable is continuous (has a density) if and only if its distribution function has no jumps.

Going from a Density to a Distribution Function

You already know how to find the distribution function given a density function:

$$F(x) = \text{cumulative area under the graph of } f(x) \quad \text{(Fig. 1)}.$$

In other words,

(5)
$$F(x) = P(X \le x) = \int_{-\infty}^{x} f(x)\, dx$$

For example, suppose X has density

$$f(x) = \begin{cases} \dfrac{1}{2} & \text{for} \quad -1 \le x \le 0 \\[2mm] \dfrac{1}{2}\, e^{-x} & \text{for} \quad x \ge 0 \qquad \text{(Fig. 12)} \end{cases}$$

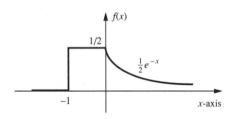

Figure 12

Then $F(x) = \displaystyle\int_{-\infty}^{x} f(x)\, dx$, but since f changes formula, you need cases to compute $F(x)$.

Case 1. $x < -1$ (Fig. 13).

$$F(x) = \int_{-\infty}^{x} 0\, dx = 0$$

Case 2. $-1 < x < 0$ (Fig. 14).

$$F(x) = \int_{-1}^{x} \frac{1}{2}\, dx = \frac{1}{2}(x+1)$$

> In case 2, use $\int_{-1}^{x}$, not $\int_{-1}^{0}$.
>
> If you use $\int_{-1}^{0}$, you are getting $F(0)$, not $F(x)$.

Case 3. $x > 0$ (Fig. 15).

$$F(x) = \int_{-1}^{0} \frac{1}{2} \, dx + \int_{0}^{x} \frac{1}{2} e^{-x} \, dx$$

$$= \frac{1}{2} - \frac{1}{2} e^{-x} \Big|_{0}^{x}$$

$$= \frac{1}{2} + \frac{1}{2}(1 - e^{-x})$$

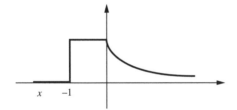

Figure 13   Case 1

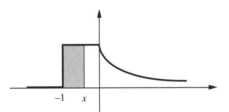

Figure 14   Case 2

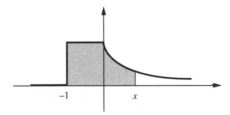

Figure 15   Case 3

So all in all (Fig. 16),

$$F(x) = \begin{cases} 0 & \text{if } x \leq -1 \\ \dfrac{1}{2}(x+1) & \text{if } -1 \leq x \leq 0 \\ 1 - \dfrac{1}{2}e^{-x} & \text{if } x \geq 0 \end{cases}$$

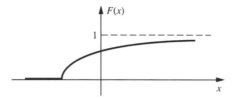

Figure 16

As a check, note that $F(-\infty) = 0$, $\quad F(\infty) = 1 - \frac{1}{2}e^{-\infty} = 1 - 0 = 1$. And as befits the distribution function of a *continuous* random variable, $F(x)$ doesn't jump: $F(x)$ comes in 3 pieces, and

the first piece leaves off at $x = -1$ with value 0

the second piece picks up at $\left. \dfrac{1}{2}(x+1) \right|_{x=-1} = 0$ (no jump at $x = -1$)

the second piece leaves off at $\left. \dfrac{1}{2}(x+1) \right|_{x=0} = \dfrac{1}{2}$

the third piece picks up at $\left. \left(1 - \dfrac{1}{2}e^{-x}\right) \right|_{x=0} = \dfrac{1}{2}$ (no jump at $x = 0$)

Warning

Here are some things that $F(x)$ is *not*:

$F(x)$ is *not* $\displaystyle\int f(x)\,dx$;

$F(x)$ is *not* $\displaystyle\int_5^7 f(x)\,dx$, even if you're in a case where $5 \leq x \leq 7$.

By (5), $F(x)$ is always $\displaystyle\int_{-\infty}^x f(x)\,dx$, which in practice becomes an integral from the *first x in the universe up to an arbitrary x*.

Going from a Distribution Function with No Jumps to a Density

It's one of the fundamental theorems of calculus[1] that if

$$F(x) = \int_a^x f(x) \, dx$$

then

$$F'(x) = f(x)$$

Applying the fundamental theorem to (5) gives

(6)
$$\boxed{F'(x) = f(x)}$$

Just *differentiate the distribution function to get the density.*
For example, let X have the distribution function

---

[1]Here's why the fundamental theorem holds. Let $F(x) = \int_a^x f(x) \, dx$. Then $F(x)$ is the cumulative area under the graph of $f(x)$ (starting from $x = a$). If x changes by dx, then the area changes by dF where (see Fig. 17)

$$dF = f(x) \, dx$$

So

$$f(x) = \frac{\text{change } dF}{\text{change } dx}$$

Take a limit as $dx \to 0$ to get

$$f(x) = F'(x)$$

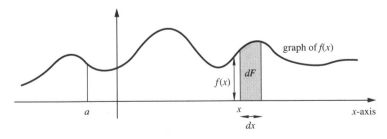

Figure 17

$$F(x) = \begin{cases} 0 & \text{if } x \le -1 \\ \dfrac{1}{8}(x+1)^2 & \text{if } -1 \le x \le 1 \\ \dfrac{1}{2} & \text{if } 1 \le x \le 4 \\ 5x - \dfrac{39}{2} & \text{if } 4 \le x \le 4.1 \\ 1 & \text{if } x \ge 4.1 \end{cases} \qquad \text{(Fig. 18)}$$

Then

$$f(x) = F'(x) = \begin{cases} 0 & \text{if } x \le -1 \\ \dfrac{1}{4}(x+1) & \text{if } -1 \le x \le 1 \\ 0 & \text{if } 1 \le x \le 4 \\ 5 & \text{if } 4 \le x \le 4.1 \\ 0 & \text{if } x \ge 4.1 \end{cases} \qquad \text{(Fig. 19)}$$

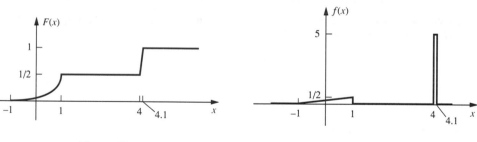

Figure 18   **Figure 19**

Note that whenever the graph of $F(x)$ is horizontal, we have $f(x) = F'(x) = 0$, indicating a whole interval of probability 0.

Summary of Discrete versus Continuous versus Mixed Random Variables

A *discrete* random variable X can assume only a finite or a countably infinite number of values.

The distribution function is a step function.

X has a probability function $p(x)$ given by $p(x) = P(X = x)$.

X doesn't have a density, although some people refer to the probability function as its discrete density.

A *continuous* random variable X takes on an uncountably infinite number of values, and probabilities are found by integrating its density $f(x)$. Even if the value x is possible, we still have $P(X = x) = 0$.

The distribution function $F(x)$ doesn't jump. (The density function *can* jump, up or down.)

A *mixed* random variable X is one that is neither discrete nor continuous.

X takes on an uncountably infinite number of values (unlike a discrete random variable).

There are x's for which $P(X = x) \neq 0$ (unlike a continuous random variable), and at each such x the distribution function has a jump whose size is $P(X = x)$.

X has a pseudodensity (part density and part probability function): Some points get chunks of probability and the rest of the probability is spread out over all or part of the universe.

Summary of the Density versus the Distribution Function

Every random variable X has a distribution function $F(x)$.

$F(x) = P(X \leq x)$.

$F(-\infty) = 0, F(\infty) = 1$.

F never decreases.

F jumps at x_0 iff $P(x = x_0) \neq 0$ and the size of the jump is $P(x = x_0)$.

$P(a \leq X \leq b) = \text{upper } F(b) - \text{lower } F(a)$

$\qquad\qquad$ (you can use plain $F(b) - F(a)$ if there are no jumps).

Not all random variables have (legal) densities (they do if you allow delta functions), but if X does have a density $f(x)$ in addition to its distribution function $F(x)$, then

$$P(a \leq X \leq b) = \int_a^b f(x) \, dx$$

$$\int_{-\infty}^{\infty} f(x) \, dx = 1$$

$$f(x) \geq 0$$

$$P(X \approx x) = f(x) \, dx$$

$$F(x) = \int_{-\infty}^{x} f(x) \, dx = \text{ cumulative area under } f(x)$$

$$f(x) = F'(x)$$

The Distribution of a Random Variable
(as Opposed to Its Distribution *Function*)

A random variable X is determined by its distribution function $F(x)$ in the sense that the probability of any event involving X can be found from $F(x)$. If X is *discrete*, then X is also determined by its probability function, and if X is *continuous*, then X is also determined by its density function.

    When we refer to the *distribution of X*, we mean that we have a way to find probabilities of X events; either we have the distribution function, or the probability function or the density function of X.

Problems for Section 4-2

1. Let X be the number of heads in 2 tosses of a fair coin.
   (a) Find the probability function $p(x)$ for X.
   (b) Find (and sketch) the distribution function $F(x)$.

2. Let X be uniformly distributed on [3,5]. Sketch the density and distribution functions.

3. Suppose X has the indicated distribution function F. Express $F(x)$ algebraically with some formula(s) and then find

   (a) $P\left(X = \dfrac{1}{2}\right)$ (f) $P(3 \le X \le 4)$

   (b) $P(X = 3)$ (g) $P(2 < X < 4)$

   (c) $P\left(X < \dfrac{1}{2}\right)$ (h) $P\left(X = 3 \text{ or } 3\tfrac{1}{2} < X < 4\right)$

   (d) $P(X < 3)$ (i) $P\left(X \ge 3 | X > \dfrac{1}{2}\right)$

   (e) $P(X \le 4)$

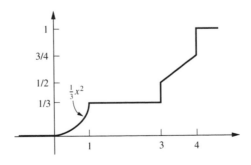

Figure P.3

4. Let X be a random variable with density $f(x)$ and distribution function $F(x)$. True or False?

(a) $f(x)$ can't be larger than 1. **(f)** $F(x)$ can't be negative.
(b) $F(x)$ can't be larger than 1. **(g)** Area under f must be 1.
(c) f can't decrease. **(h)** Area under F must be 1.
(d) F can't decrease. **(i)** f can't jump.
(e) $f(x)$ can't be negative. **(j)** F can't jump.

5. Let X have density $f(x)$ and distribution function $F(x)$. Find, if possible, $F(-\infty), F(\infty), F(0), f(-\infty), f(\infty), f(0), P(X = 7)$.

6. Suppose X has density $f(x)$ and distribution function $F(x)$. What can you conclude about $f(x)$ and $F(x)$ if

   (a) X is never between 2 and 6   **(c)** X is never ≤ 3
   (b) X is always between 2 and 6   **(d)** X is never ≥ 4

7. Define a random variable X as follows.

   > Toss a coin where $P(\text{heads}) = 1/3$.
   > If the coin is heads, then set $X = 5$.
   > If the coin is tails, then choose X at random between 3 and 6.

   X is a mixed random variable. It doesn't have a legal density; it has a pseudodensity, which is part density and part probability function.

   Find the pseudodensity of X and the (legal) distribution function $F(x)$ and draw a picture of each.

8. Let X, Y, Z be independent continuous random variables, all with the same distribution function F. Consider the minimum and maximum of X, Y, Z.
   (a) Find $P(\max \le 7)$ (in terms of F).
   (b) Find $P(\min \ge 5 \text{ and } \max \le 7)$.
   (c) Put (a) and (b) together to find $P(\min \le 5 \text{ and } \max \le 7)$.

9. For each distribution function $F(x)$, find the corresponding density if a legal density exists (otherwise, find the pseudodensity). Sketch F and f.

   (a) $F(x) = \begin{cases} \dfrac{1}{2}e^x & \text{if } x \le 0 \\[2mm] \dfrac{1}{2} & \text{if } 0 \le x \le 2 \\[2mm] \dfrac{1}{6}x + \dfrac{1}{6} & \text{if } 2 \le x \le 5 \\[2mm] 1 & \text{if } x \ge 5 \end{cases}$

   (b) $F(x) = \begin{cases} \dfrac{1}{2}e^x & \text{if } x \le 0 \\[2mm] 1 - \dfrac{1}{2}e^{-x} & \text{if } x \ge 0 \end{cases}$

(c) $F(x) = \begin{cases} \dfrac{1}{4}e^{2x} & \text{if } x \le 0 \\ 1 & \text{if } x \ge 0 \end{cases}$

10. For each density $f(x)$, find the corresponding distribution function $F(x)$.

(a) $f(x) = \dfrac{100}{x^2}$ for $x \ge 100$

(b) $f(x) = 1 - |x|$ for $|x| < 1$

(c) $f(x) = \begin{cases} 4 & \text{if } -.1 \le x \le .1 \\ \dfrac{1}{2} & \text{if } .1 \le x \le .5 \end{cases}$

11. Given this rough graph of a density $f(x)$, sketch a rough graph of the distribution function $F(x)$.

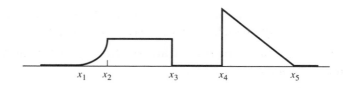

Figure P.11

SECTION 4-3   THE EXPONENTIAL DISTRIBUTION

The Exponential Density

Let $\lambda > 0$. If X has density

(1)
$$f(x) = \lambda e^{-\lambda x} \quad \text{for } x \ge 0 \quad (\text{Fig. 1})$$

then we say that X has an *exponential distribution* with parameter λ.

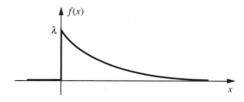

Figure 1   The exponential density

Application of the Exponential to Waiting Times

Suppose particles arrive independently.

Let λ be the arrival rate, the average number of arrivals per minute.

Let X be the waiting time for the next arrival, the waiting time between arrivals.

I'll show that X has an exponential distribution.

Let $F(x)$ be the distribution function of X. (It's easier to get $F(x)$ first and then $f(x)$.)

We know that $X \geq 0$ since the waiting time can't be negative, so

$$F(x) = 0 \quad \text{if} \quad x \leq 0.$$

For $x \geq 0$,

$$F(x) = P(X \leq x) = 1 - P(X > x)$$

Having to wait more than x minutes for the next arrival means no particles have arrived during those x minutes. So

$$F(x) = 1 - P(\text{no arrivals in } x \text{ minutes})$$

Remember that the number of arrivals in x minutes is a *discrete* random variable with a Poisson distribution, where the parameter is the average number of arrivals in x minutes, namely, λx. So

$$F(x) = 1 - \frac{e^{-\lambda x}(\lambda x)^0}{0!} = 1 - e^{-\lambda x} \quad \text{for } x \geq 0$$

Then

$$f(x) = F'(x) = \lambda e^{-\lambda x} \quad \text{for } x \geq 0$$

So X has an exponential distribution, QED.

The parameter λ is the arrival rate, but it can also be expressed in terms of the *average waiting time* because the arrival rate and the average waiting time are reciprocals. (If the average waiting time between arrivals is, say, 2 minutes, then the particles arrive at the average rate of $\frac{1}{2}$ particle per minute.) All in all:

(2)

> If particles arrive independently and X is the waiting time for the next arrival, then X has an exponential distribution with
>
> $$\lambda = \text{arrival rate} = \frac{1}{\text{average waiting time}}$$

Example 1

Telephone calls arrive independently at a switchboard at the average rate of 2 per hour. To find the probability that it will be at least 3 hours until the next call, let X be the waiting time between calls. The best model for X is the exponential distribution with $\lambda = 2$. So

$$P(X \geq 3) = \int_3^\infty 2e^{-2x}\, dx = -e^{-2x}\Big|_3^\infty = e^{-6}$$

Application of the Exponential to Lifetimes

Suppose the age of a wineglass has no effect on how likely it is to break. Let X be the lifetime of a wineglass. Here's why X has an exponential distribution.

Imagine a wineglass that eventually breaks and is replaced by another wineglass that eventually breaks and is replaced, and so on. The *lifetime* of a wineglass is the *waiting time for a breakage to occur*. The hypothesis that a 10-year-old wineglass is as likely to break as a new wineglass means that the odds on when the next breakage arrives are the same whether the last breakage was 10 years ago or an instant ago; that is, the breakage-arrivals are independent. Since X is the waiting time for breakage-arrivals, where arrivals are independent, X has an exponential distribution with

$$\lambda = \frac{1}{\text{average waiting time between breakage-arrivals}}$$

$$= \frac{1}{\text{average lifetime of a wineglass}}$$

So, as a corollary of (2), and expressed in terms of particles rather than wineglasses, we have this model:

Let X be the lifetime of a particle.

Assume that the particle's chances of dying don't depend on its age (an old particle has as much life to look forward to as a new particle).

Then X has an exponential distribution where

$$\lambda = \frac{1}{\text{average lifetime}}$$

Memoryless Feature of the Exponential

Telephone calls arrive at a switchboard independently at the rate of 2 per hour. It is now 4:00 P.M. The telephone rang at 2:30 P.M. and has not rung since then. Find the probability that the phone will still not have rung by 7:00 P.M.

We want the probability of having to wait a total of at least $4\frac{1}{2}$ hours between calls *having already waited* $1\frac{1}{2}$ *hours*; that is, the prob of having to wait at least *another* 3 hours *having already waited* $1\frac{1}{2}$ *hours*. Look at it like this.

If you toss a coin, the probability of a tail on the 9th toss, *having gotten heads on the first 8 tosses*, is still 1/2. The tosses are independent, so the odds on the 9th toss don't change, no matter what happened previously.

Similarly, the calls arrive independently, so the fact that you have waited $1\frac{1}{2}$ hours doesn't make the next call more (or less) likely to arrive:

$P\left(\text{total wait between calls is } \geq 4\frac{1}{2} \text{ hours} \mid \text{waited } 1\frac{1}{2} \text{ hours already}\right)$

$= P\left(\text{have to wait at least another 3 hours} \mid \text{waited } 1\frac{1}{2} \text{ hours already}\right)$

$= P(\text{waiting time for a call is } \geq 3 \text{ hours from scratch})$

$= \int_{3}^{\infty} 2e^{-2x}\, dx$ (waiting time is exponentially distributed with $\lambda = 2$)

$= e^{-6}$ (same answer as example 1)

In other words, if X has an exponential distribution, then

$$P\left(X \geq 4\tfrac{1}{2} \mid X \geq 1\tfrac{1}{2} \right) = P(X \geq 3)$$

and, more generally,

(3)
$$P\left(X \geq x + t \mid X \geq t\right) = P(X \geq x)$$

> If the waiting time for an arrival has an exponential distribution, then the *conditional* probability that you must wait at least x *more* hours after having already waited t hours is the same as the *unconditional* probability of having to wait at least x hours initially.
> And if the lifetime of a particle has an exponential distribution, then the probability that the particle will last at least x *more* hours is the same for an old particle as for a new particle.
> The exponential distribution is said to be *memoryless* because of (3).[2]

---

[2]Furthermore, it can be shown (but it's harder) that the exponential is the *only* memoryless continuous distribution.

We already justified (3) by thinking of the physical model of *independent* arrivals. Here's a formal proof that depends only on the formula for the exponential density.

First,

$$P(X \geq x) = \int_x^\infty \lambda e^{-\lambda x} \, dx = e^{-\lambda x}$$

On the other hand,

$$P(X \geq x + t | X \geq t) = \frac{P(X \geq x + t \text{ and } X \geq t)}{P(X \geq t)}$$

$$= \frac{P(X \geq x + t)}{P(X \geq t)}$$

$$= \frac{\int_{x+t}^\infty \lambda e^{-\lambda x} \, dx}{\int_t^\infty \lambda e^{-\lambda x} \, dx}$$

$$= \frac{e^{-\lambda(x+t)}}{e^{-\lambda t}}$$

$$= e^{-\lambda x}$$

So $P(X \geq x + t | X \geq t) = P(X \geq x)$, QED.

Warning (the Poisson versus the Exponential)

Suppose particles arrive independently at the average rate of 2 per minute.

The *number of particles arriving in any 1-minute period* is a *discrete* random variable and has a *Poisson* distribution with $\lambda = 2$. The number arriving in, say, any 7-minute period has a Poisson distribution with $\lambda = 14$. For example,

$$P(10 \text{ particles arrive in any particular 7-minute period}) = \frac{e^{-14} \, 14^{10}}{10!}$$

On the other hand, the *waiting time for the next arrival* is a *continuous* random variable and has an *exponential* distribution with $\lambda = 2$. For example,

$$P(\text{it takes 3 to 8 minutes for the next particle to arrive}) = \int_3^8 2e^{-2x} \, dx$$

The Gamma Distribution

Suppose particles arrive independently at the average rate of λ per second. Let X be the waiting time for the nth arrival. Then the density of X is

(4) $$f(x) = \frac{\lambda^n e^{-\lambda x} x^{n-1}}{(n-1)!} \quad \text{for } x \geq 0$$

called the Gamma distribution.

The Gamma distribution has two parameters, n and λ. The exponential distribution is the special case of the Gamma distribution when $n = 1$.

Here's a derivation of the Gamma distribution. We'll find $F(x)$ first and then $f(x)$.

Since waiting times can't be negative,

$$F(x) = 0 \text{ if } x \le 0$$

For $x \ge 0$,

$$
\begin{aligned}
F(x) &= P(X \le x) \\
&= 1 - P(X \ge x) \\
&= 1 - P(\text{waiting time for } n\text{th particle is } \ge x) \\
&= 1 - P(\text{number of arrivals in the first } x \text{ seconds is} \\
&\qquad\qquad\qquad\qquad\qquad 0 \text{ or } 1 \text{ or } \ldots \text{ or } n - 1) \\
&= 1 - [P(0 \text{ arrivals in } x \text{ secs}) + P(1 \text{ arrival in } x \text{ secs}) \\
&\qquad + P(2 \text{ arrivals in } x \text{ secs}) + \cdots + P(n - 1 \text{ arrivals in } x \text{ secs})]
\end{aligned}
$$

The number of arrivals in x seconds has a Poisson distribution with parameter λx. So

(5) $$F(x) = 1 - e^{-\lambda x}\left[1 + \lambda x + \frac{(\lambda x)^2}{2!} + \cdots + \frac{(\lambda x)^{n-1}}{(n-1)!}\right]$$

Then, for $x \ge 0$, use the derivative product rule to get

$$f(x) = F'(x) = -e^{-\lambda x}\left[\lambda + \frac{2\lambda^2 x}{2!} + \frac{3\lambda^3 x^2}{3!} + \cdots + \frac{(n-1)\lambda^{n-1}\, x^{n-2}}{(n-1)!}\right]$$

$$+ \lambda e^{-\lambda x}\left[1 + \lambda x + \frac{(\lambda x)^2}{2!} + \cdots + \frac{(\lambda x)^{n-1}}{(n-1)!}\right]$$

Look carefully to see that everything except the last term in the second bracket cancels out, leaving the formula in (4).

Problems for Section 4-3

(These problems can be done in more than one way. So when you check your answers, you may find that we have the same final answer but different methods.)

1. Check that the exponential density in (1) is a legal density.

2. Particles arrive (independently) at a detector at the average rate of 3 per second.
   (a) Find the prob that you have to wait no more than 2 seconds for an arrival.
   (b) Find the prob that you have to wait between 2 and 5 seconds for an arrival.
   (c) Nothing has arrived for the past 6 seconds. Find the prob that nothing will arrive in the next 6 seconds. (Do this one twice, once with the exponential and again with the Poisson to illustrate that they both work.)
   (d) You are timing how long it takes from the last arrival to the next arrival. The reading on the stopwatch is past 7 seconds already. Find the prob the final reading will be under 20 seconds.
   (e) Find the prob that between 1 and 5 particles arrive in the next second.

3. Telephone calls to a radio talk show last 3 minutes on the average. Assume that the exponential distribution applies.
   (a) Find the probability that a call will last less than 2 minutes.
   (b) Find the probability that a call will last longer than average.
   (c) Suppose a call has already lasted for 4 minutes. Find the probability that it will last less than 2 more minutes.
   (d) Suppose a call has lasted 2 minutes already. Find the probability that it will last longer than average.
   (e) What assumptions must be made about the telephone calls so that the exponential distribution really does apply?

4. A gidget on your car lasts 50,000 miles on the average. An old gidget has as much life in it as a new gidget. Find the probability that the gidget dies sometime during your 5000-mile vacation trip.

5. Let X be the duration in minutes of a volley in tennis. Assume X is exponentially distributed with parameter λ.
   (a) Find λ if on the average there are 2 volleys per minute.
   (b) Find λ if on the average there are 20 volleys per hour.
   (c) Find λ if on the average a volley lasts for 2 minutes.
   (d) What assumptions must be made about volleys to make the exponential distribution the right model for the duration of a volley?

6. On the average, it takes 5 minutes for a bank teller to process a customer. Assume that a customer in the midst of transacting business is just as likely to finish up in t more minutes as a customer who has just reached the teller.
   (a) A customer has just stepped up to the window, and you are now at the head of the line. Find the prob that you will have to wait between 5 and 10 minutes before being helped.
   (b) You have been at the head of the line since 11:03. It is now 11:10. Find the prob that you will be taken by 11:30.

(c) If there are 2 customers ahead of you on the line, find the prob that you have to wait at least 10 minutes to get taken.

(d) Find the prob that no more than 2 customers will be taken in the next 3 minutes.

7. It is now 4:20. Particles arrive independently at the rate of 2 per minute. Find the probability that

(a) 4 arrive between 4:30 and 4:40

(b) 4 arrive between 4:40 and 4:50

(c) the next arrival is between 4:30 and 4:40

(d) the next arrival is between 4:40 and 4:50

(e) it's at least 5 minutes before the next particle arrives

(f) 4 particles arrive between 6:40 and 6:50 given that none arrived between 6:30 and 6:40

(g) no particles arrive between 6:30 and 6:45 and then 3 arrive between 6:45 and 6:50

(h) a particle finally comes within the next 3 minutes given that we have already waited 2 days

8. You have 6 wineglasses bought from a company whose glasses last 3 years on the average. You only use 1 glass at a time so you won't use the second glass until the first breaks, and so on. Once all 6 are broken, your wine drinking is over. Let X be the number of your wine drinking years. Find the distribution of X.

SECTION 4-4   THE NORMAL (GAUSSIAN) DISTRIBUTION

The Normal Density

Let $\sigma > 0$. If X has density

(1)
$$f(x) = \frac{1}{\sigma\sqrt{2\pi}} \, e^{-(x-\mu)^2/2\sigma^2}$$

we say that X has a *normal* distribution with parameters μ and σ^2. The normal distribution is also called the *Gaussian* distribution.[3]

When we define the mean (expected value) and variance of a *continuous* random variable in Chapter 7, it will turn out that the parameter μ is the mean of X and the parameter σ^2 is the variance of X (so that σ is the standard deviation).

Figure 1 shows the graph of $f(x)$ with large versus small σ^2. Remember that the variance is a measure of spread around the mean, so in Fig. 1 the area (probability) is more concentrated near the mean μ when σ^2 is small. In any case, the density is symmetric around the mean μ and is often referred to as a bell-shaped curve.

---

[3]The distribution is legal because $f(x) \geq 0$ for all x, and if you spend a lot of time on $\int_{-\infty}^{\infty} f(x)\, dx$ you will see that it is 1.

Experimental data shows that random variables such as height, weight, test scores, error in measurement, and many, many more seem to have a normal distribution.

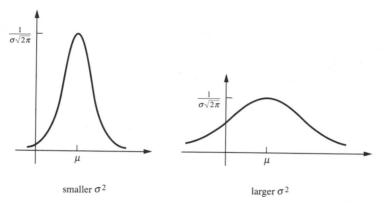

smaller σ^2 larger σ^2

Figure 1

The Unit Normal

The normal distribution with $\mu = 0$, $\sigma^2 = 1$ is called the unit or standard normal. I'll use the notation X^*, f^*, F^* for the random variable, the density, and the distribution function. (The distribution function of the unit normal is often denoted by $\Phi(x)$.)

Figure 2 shows the graph of $f^*(x)$, symmetric with respect to the y-axis.

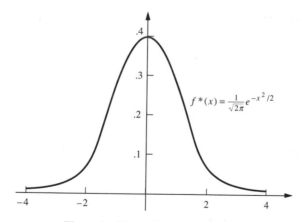

$$f^*(x) = \frac{1}{\sqrt{2\pi}} e^{-x^2/2}$$

Figure 2   The unit normal density

Table 1 gives some values of $F^*(x)$ (a good calculator can go much farther and find $F(x)$ values given *any* μ and σ^2) that can be used to find various probabilities. For example,

$$P(X^* \geq .3) = 1 - P(X^* \leq .3) = 1 - F^*(.3) = 1 - .618 = .382$$

Alternatively, using symmetry,

$$P(X^* \geq .3) = P(X^* \leq -.3) = F^*(-.3) = .382$$

As another example,

$$P(|X^*| \leq 2) = P(-2 \leq X^* \leq 2) = F^*(2) - F^*(-2) = .954$$

Table 1
The unit normal distribution function

x	$F^*(x)$	x	$F^*(x)$	x	$F^*(x)$	x	$F^*(x)$
-3.0	.001	-1.5	.067	.1	.540	1.64	.950
-2.9	.002	-1.4	.081	.2	.579	1.7	.955
-2.8	.003	-1.3	.087	.3	.618	1.8	.964
-2.7	.004	-1.28	.100	.4	.655	1.9	.971
-2.6	.005	-1.2	.115	.5	.691	1.96	.975
-2.5	.006	-1.1	.136	.6	.726	2.0	.977
-2.4	.008	-1.0	.159	.7	.758	2.1	.982
-2.3	.011	$-.9$	.184	.8	.788	2.2	.986
-2.2	.014	$-.8$	.212	.9	.816	2.3	.989
-2.1	.018	$-.7$	.242	1.0	.841	2.33	.990
-2.0	.023	$-.6$	.274	1.1	.864	2.4	.992
-1.96	.025	$-.5$	.309	1.2	.885	2.5	.994
-1.9	.029	$-.4$	.345	1.28	.900	2.6	.995
-1.8	.036	$-.3$	.382	1.3	.903	2.7	.996
-1.7	.045	$-.2$	.421	1.4	.919	2.8	.997
-1.64	.050	$-.1$	.460	1.5	.933	2.9	.998
-1.6	.055	0	.500	1.6	.945	3.0	.999

The New Random Variable aX + b

If

$$X \text{ is normal with mean } \mu \text{ and variance } \sigma^2$$

then

$$X + b \text{ is normal with mean } \mu + b \text{ and variance } \sigma^2$$

(2) $$aX \text{ is normal with mean } a\mu \text{ and variance } a^2\sigma^2$$

$$aX + b \text{ is normal with mean } a\mu + b \text{ and variance } a^2\sigma^2$$

Proof

Suppose X (the old random variable) is normal with mean μ and variance σ^2.

Part I.   First I'll work on $X + b$ and try an informal approach. For convenience, let's try $X + 7$ specifically so that

$$\text{new random variable} = \text{old} + 7$$

I want the density of the new.

Figure 3 shows the density of X. The thin area at $x = 5$ represents the probability that $X \approx 5$.

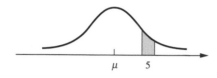

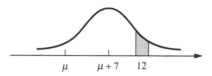

Figure 3   Density of X **Figure 4**   Density of $X + 7$

For the density of the new random variable in Fig. 4, the same amount of probability should be located at $x = 12$ because $P(\text{old} \approx 5) = P(\text{new} \approx 12)$.

In general, the thin strip at $x = x_0$ under the old density should be the same as the thin strip at $x = x_0 + 7$ under the new density. So the new density can be gotten by shifting the old density to the right by 7. The new density has exactly the same shape as the old but is centered around $\mu + 5$ instead of μ. In other words, the new density is normal with mean $\mu + 7$ and the same variance as the old.

Part II.   Now let's try a more rigorous method for the new random variable $Y = aX$. I'll do the case where $a > 0$ (a similar argument holds if $a < 0$). Then

$$F_Y(y) = P(Y \le y) = P(aX \le y) = P\left(X \le \frac{y}{a}\right) \text{ (because } a > 0)$$

$$= \int_{-\infty}^{y/a} \frac{1}{\sigma\sqrt{2\pi}} e^{-(x-\mu)^2/2\sigma^2} \, dx$$

Let $t = ax$. Then $x = t/a$, $dt = a\,dx$. If $x = y/a$, then $t = y$; if $x = -\infty$, then $t = -\infty$. So

$$F_Y(y) = \int_{t=-\infty}^{y} \frac{1}{\sigma\sqrt{2\pi}} e^{-(t/a-\mu)^2/2\sigma^2} \, \frac{dt}{a}$$

$$= \int_{t=-\infty}^{y} \frac{1}{a\sigma\sqrt{2\pi}} e^{-(t-a\mu)^2/2a^2\sigma^2} \, dt$$

The integrand has the form of the normal density with $a\mu$ sitting in the μ spot and $a^2\sigma^2$ playing the role of σ^2. So Y is normally distributed with parameters $a\mu$, $a^2\sigma^2$.

Put Parts I and II together to see that $aX + b$ is normal with mean $a\mu + b$ and variance $a^2\sigma^2$.

Switching from a Non-Unit Normal to the Unit Normal

(3) | Let X be normal with parameters μ and σ^2. Then $(X - \mu)/\sigma$ is normal with mean 0 and variance 1; that is, $(X - \mu)/\sigma$ is the unit normal X^*.

This works because $(X - \mu)/\sigma$ is the special case of $aX + b$ where $a = 1/\sigma$, $b = -\mu/\sigma$.

So $(X - \mu)/\sigma$ is normal with

$$\text{mean} = a\mu + b = \frac{1}{\sigma}\mu - \frac{\mu}{\sigma} = 0$$

and

$$\text{variance} = a^2\sigma^2 = \left(\frac{1}{\sigma}\right)^2 \sigma^2 = 1$$

Example 1

Suppose that height X is normal with $\mu = 66$ inches and $\sigma^2 = 9$. You can find the probability that a person picked at random is under 6 feet, using only the table for the *unit* normal.

You want the probability that $X < 72$.

Subtract μ and divide by σ on both sides to rewrite $X < 72$ as

$$\frac{X - \mu}{\sigma} < \frac{72 - \mu}{\sigma}$$

Then

$$P(X < 72) = P\left(\frac{X - 66}{3} < \frac{72 - 66}{3}\right)$$
$$= P(X^* < 2) \quad \text{(by (3))}$$
$$= F^*(2)$$
$$= .977$$

Example 1 continued

Find b so that you can be 95% sure that a person picked at random will be within b of the mean.

You want b so that

$$P(66 - b \leq X \leq 66 + b) = .95$$

$$P\left(-\frac{b}{3} \leq \frac{X - 66}{3} \leq \frac{b}{3}\right) = .95 \quad \text{(subtract 66 and divide by 3)}$$

$$P\left(-\frac{b}{3} \leq X^* \leq \frac{b}{3}\right) \quad \text{(by (3))}$$

$$P\left(X^* \leq \frac{b}{3}\right) = .975 \quad \text{(by symmetry—see Fig. 5)}$$

$$F^*\left(\frac{b}{3}\right) = .975$$

$$\frac{b}{3} = 1.96 \quad \text{(Table 1)}$$

$$b = 5.88$$

So 95% of the population is between 60.12 and 71.88 inches tall.

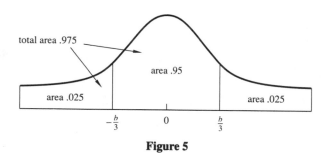

Figure 5

The Probability That a Normal Random Variable Is within *k* Standard Deviations of the Mean

Let X be normal with parameters μ, σ^2. Remember that σ is called the standard deviation. Let's look at the probability that when you perform the experiment, X is within k standard deviations of the mean. We have

$$P(\mu - k\sigma \leq X \leq \mu + k\sigma) = P\left(-k \leq \frac{X - \mu}{\sigma} \leq k\right) \quad \text{(by algebra)}$$

so

(4)
$$\boxed{P(\mu - k\sigma \leq X \leq \mu + k\sigma) = P(-k \leq X^* \leq k)}$$

In other words:

> (5) | The probability that X is within k standard deviations of its mean is the *same* for all normal distributions. In particular, it is the same as the probability that the unit normal X^* is between $-k$ and k (Fig. 6).

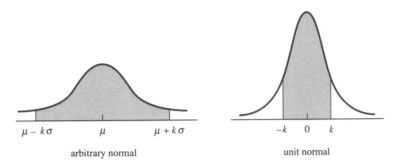

Figure 6   Equal areas

Because of (5), in problems involving a normal random variable, a standard deviation is often used as the unit of measure for distance, and distance is often measured from the mean μ rather than from 0. That way you can draw conclusions that hold for *all* normals.

For example, for the *unit* normal

$$P(-1.96 \leq X^* \leq 1.96) = F^*(1.96) - F^*(-1.96) = .95$$

So 95% of the area under the *unit* normal density lies between -1.96 and 1.96 (Fig. 7).

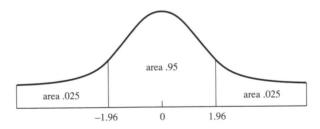

Figure 7   Unit normal

So, for the normal density with parameters μ and σ^2, 95% of the area lies between $\mu - 1.96\sigma$ and $\mu + 1.96\sigma$ (Fig. 8); that is, for *any* normal random variable X,

(6) $$P(\mu_X - 1.96\sigma_X \leq X \leq \mu_X + 1.96\sigma_X) = .95$$

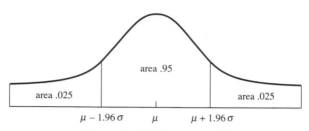

Figure 8   Arbitrary normal

Similarly,

$$P(-2.6 \le X^* \le 2.6) = F^*(2.6) - F^*(-2.6) = .99$$

so for *any* normal,

(7) $$P(\mu_X - 2.6\sigma_X \le X \le \mu_X + 2.6\sigma_X) = .99$$

That is, the probability is .99 that any normally distributed random variable lies within 2.6 standard deviations of its mean.

The Normal Approximation to the Binomial

Let X be the number of successes in n Bernoulli trials where $P(\text{success}) = p$. Then X is a discrete random variable with a binomial distribution, but there is a connection between X and the normal distribution. First we'll get some pictures that show the resemblance between the probability function of X and the bell-shaped normal density.

Suppose $n = 100$ and $p = 1/2$. Then X can take on any of the values $0, 1, 2, \ldots, 100$ where

$$P(X = 50) = \binom{100}{50} \left(\frac{1}{2}\right)^{100} = .0796$$

$$P(X = 49) = P(X = 51) = \binom{100}{49} \left(\frac{1}{2}\right)^{100} = .078$$

$$P(X = 48) = P(X = 52) = \binom{100}{48} \left(\frac{1}{2}\right)^{100} = .074$$

$$\vdots$$

$$P(X = 0) = P(X = 100) = \left(\frac{1}{2}\right)^{100} = 7.9 \times 10^{-31}$$

Figure 9 shows the probability function. The mean of a binomial random variable is np, 50 in this case, and the chunks of probability are symmetric about the mean; the envelope looks like a normal.

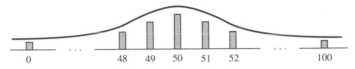

Figure 9   Binomial distribution, $n = 100, p = 1/2$

At another extreme, suppose $n = 100$ and $p = 1$. Then all the probability is concentrated at the value $X = 100$ (Fig. 10) and again the distribution looks like a normal.

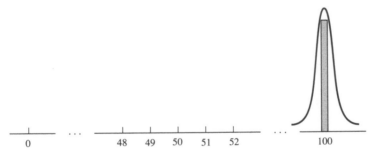

Figure 10   Binomial distribution, $n = 100, p = 1$

When we get back to variance in Chapter 7 I'll show that the variance of a binomial random variable is npq. In Fig. 9 the variance is 25, and you see a large spread around the mean. The variance in Fig. 10 is 0 and you see no spread around the mean (the mean is $np = 100$ and all the probability is concentrated there).

Here's the connection between the binomial and the normal (the proof is very difficult).

The normal distribution can be used to approximate the binomial. If X has a binomial distribution with parameters n, p (so that X has mean np and variance npq) where n is large, then X is approximately normal with

$$\mu = np \quad \text{and} \quad \sigma^2 = npq$$

Many books claim that the approximation seems to be good if $npq \geq 10$.

Example 2

A coin has $P(\text{heads}) = .3$. Toss the coin 1000 times so that the expected number of heads is 300. Find the probability that the number of heads is 400 or more.

Let X be the number of heads. We'll find $P(X \geq 400)$.

Method 1. X has a binomial distribution with parameters $n = 1000$ and $p = .3$. So

$$P(X \geq 400) = P(X = 400) + P(X = 401) + \cdots + P(X = 1000)$$

$$= \binom{1000}{400}(.3)^{400}(.7)^{600} + \binom{1000}{401}(.3)^{401}(.7)^{599} + \cdots + (.3)^{1000}$$

Too much arithmetic!

Method 2. X is approximately normal with $\mu = np = 300, \sigma^2 = npq = 210$. Relying on Table 1, we have

$$P(X \geq 400) = P\left(\frac{X - 300}{\sqrt{210}} \geq \frac{400 - 300}{\sqrt{210}}\right) \quad \text{(algebra)}$$

$$= P(X^* \geq 6.9) \quad \left(\frac{X - \mu}{\sigma} \text{ is the unit normal}\right)$$

$$= 1 - \underbrace{P(X^* \leq 6.9)}_{\text{practically 1}}$$

$$= 0$$

Method 3. X is approximately normal with mean $\mu = 300, \sigma^2 = 210$, $\sigma \sim 14.5$. The value 400 differs from the mean 300 by 100, which is about 6.9 standard deviations. From (7) we know that there is only a 1% chance for X to be as far from the mean as 2.6 standard deviations. So the probability is virtually 0 that X is as far as 6.9 standard deviations from the mean.

Problems for Section 4-4

(The solutions use Table 1 without bothering to interpolate so the answers are only approximations. You can probably get answers faster using a calculator.)

1. Find the probability that
   (a) $X^* \leq .5$
   (b) $X^* = .5$
   (c) $X^* < .5$
   (d) $X^* \geq .5$
   (e) $|X^*| \leq .5$
   (f) $-2 \leq X^* \leq 3$
   (g) $|X^*| > 1$

2. Find b so that
   (a) $P(X^* \leq b) = .75$
   (b) $P(X^* \geq b) = .3$
   (c) $P(-b \leq X^* \leq b) = .8$

3. Suppose you lose the first half of Table 1 and only have $F^*(x)$ for positive x. How would you find $F^*(-.3)$?

4. If X is normal with $\mu = 10, \sigma^2 = 9$, find

   (a) $P(X \geq 10)$ (b) $P(|X - 10| \geq 1)$

5. A company wants to buy cylinders with diameter 2 and is willing to accept diameters that are off by as much as .05. The factory produces a diameter that is normally distributed with $\mu = 2$.
   (a) If $\sigma = .08$, what percentage is rejected by the company?
   (b) If the rejection rate is 20%, find σ.

6. Everyone in a school is weighed. Half weigh over 110 pounds and 15% weigh over 150 pounds. Assume that weight is normally distributed.
   (a) Find the best choice for μ and σ.
   (b) The heavyweight club enrolls everyone over 150 pounds. Find the probability that a heavyweight weighs over 170 pounds.

7. Test scores are normally distributed with parameters μ, σ^2.
   (a) Find the cutoff for the top quartile (how high you have to score to be in the top 25%).
   (b) Find the cutoffs for the next two quartiles.

8. If X is normal with mean μ and variance σ^2, find $P(|X - \mu| \leq .5\sigma)$.

9. Suppose test scores are normally distributed with parameters μ, σ^2. Letter grades are assigned to scores as follows.

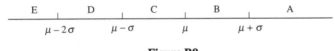

Figure P.9

What percentage of the class gets A's? B's? C's? D's? E's?

10. Assuming that test scores are normally distributed, what is the best choice for μ and σ if the average score is 60 and 30% of the scores are over 73?

11. Find the probability of getting at most 510 heads if you toss a coin 1000 times.

12. The total rainfall in a summer is normally distributed with $\mu = 10$ and $\sigma^2 = 16$. If the summer's rainfall is under 4 inches, farmers declare a drought. Find the probability that there will be no drought in the next 10 years. (Assume weather conditions are independent from one summer to the next.)

13. Only 45% of the voters support candidate Z. If a poll samples 400 voters, find the probability that more than half the sample is for Z.

14. On the average, 20% of airplane ticket holders are no-shows. If a plane has 300 seats, how many seats can the airline overbook and still be 95% sure that it won't have to bump anyone?

15. Let $P(\text{heads}) = .3$. Toss 1000 times. Find b so that you can be 90% sure of getting at least b heads.

16. A thousand people arrive independently and at random between 12:00 and 5:00. Find the probability that at least 175 will arrive in the first hour.

17. Suppose X is normal with parameters μ_X, σ_X^2. Find the distribution of

    (a) $5X - 2$ **(b)** $X - \mu_X$

SECTION 4-5   THE ELECTION PROBLEM

This section contains one lovely application of probability. If you skip over to the next section you won't lose the thread of the course, but you will miss one of the most famous problems in statistics.

   The television anchor reports that according to the latest poll of 2000 people, 52% of the voters are for candidate A give or take an error of 3%. The implication is that between 49% and 55% of the voters in the country are for A.

   How much faith should you have in this?
   How did they decide on how large a sample to take in the first place?
   (And what crucial idea has the anchor left out?)

Let

$$p = \text{percentage of A voters } \textit{in the country}$$

In other words,

$$p = \text{probability that a voter picked at random is for A}$$

Let n be the size of the polling sample. And let

$$X = \text{the number of A voters in the sample}$$

so that

$$\frac{X}{n} = \text{percentage of A voters } \textit{in the sample}$$

The pollsters want to use the sample percentage X/n to estimate the overall percentage p. The problem is to decide on what size n to use so that you have confidence in the estimate.

   If you want to be 100% sure that p actually *equals* X/n, you'd better poll the entire population.

   You can poll less than the entire population if you're willing to settle for an error, say, $\pm 3\%$, and conclude only that

$$\frac{X}{n} - .03 \leq p \leq \frac{X}{n} + .03$$

But you can't get even this much *for sure*. The best you can do is choose a less-than-sure probability you can live with, say, 95%, and find n so that

(1) $$P\left(\frac{X}{n} - .03 \leq p \leq \frac{X}{n} + .03\right) \geq .95$$

In other words, you want to find n so that you can be at least 95% sure that the actual percentage of A voters is within 3% of the sample percentage.

The interval

$$\left[\frac{X}{n} - .03, \quad \frac{X}{n} + .03\right]$$

is called the *confidence interval*. And the .95 is called the *confidence level*. The smaller the confidence interval and the higher the confidence level desired, the larger you'll have to make n.

I'll show you how to find n for the interval and level in (1). Similar calculations can be done for any interval and level you like.

Rewrite (1) as the inequality

(2) $$P(np - .03n \leq X \leq np + .03n) \geq .95$$

I'll actually solve the *equation*

(3) $$P(np - .03n \leq X \leq np + .03n) = .95$$

for n. Then any larger n will satisfy (2). (If a sample of size n makes you 95% sure, then a larger sample can only serve to increase your sureness.)

Remember that X is the number of A voters in a sample of size n where the probability of an A voter is p. Drawings with*out* replacement from a *large* population can be considered to be independent trials, so X has a binomial distribution with parameters n and p. The key idea is that if n is large, the binomial can be approximated by the normal so that X is approximately normal with

$$\mu = np, \quad \sigma^2 = npq$$

Subtract μ and divide by σ throughout the parenthesized inequality in (3) to get

$$P\left(-\frac{.03n}{\sqrt{npq}} \leq \frac{X - \mu}{\sigma} \leq \frac{.03n}{\sqrt{npq}}\right) = .95$$

$$P\left(-\frac{.03\sqrt{n}}{\sqrt{pq}} \leq X^* \leq \frac{.03\sqrt{n}}{\sqrt{pq}}\right) = .95$$

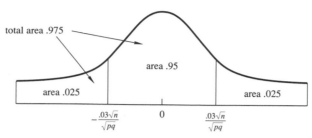

Figure 1

If 95% of the area under the unit normal is to lie between $-.03\sqrt{n}/\sqrt{pq}$ and $.03\sqrt{n}/\sqrt{pq}$ (Fig. 1), then 97.5% of the area lies to the left of $.03\sqrt{n}/\sqrt{pq}$. The normal tables list

$$F^*(1.96) = .975$$

so

$$\frac{.03\sqrt{n}}{\sqrt{pq}} = 1.96$$

$$\sqrt{n} = \frac{1.96\sqrt{pq}}{.03}$$

and the solution to (3) is

$$n = 4268pq$$

The solution to the original inequality in (2) is

$$n \geq 4268pq$$

The catch is that we don't know p or q. The whole point of taking the sample of size n in the first place is to be able to predict p.

But the worst (biggest) that pq can be is 1/4 (when $p = 1/2$ and $q = 1/2$), so it's always safe to choose

$$n \geq 4268 \times \frac{1}{4} = 1067$$

With this n you can be 95% sure that the p you're looking for is within 3% of the sample percentage X/n (give or take the error introduced, because we approximated the binomial with the normal).

The TV anchors faithfully report the error margin, that is, the confidence interval, say, from 49% to 55%. But they make it sound like a sure thing that the number of A voters lies in this interval. They never mention the confidence *level* which warns that this is, say, only 95% likely.

SECTION 4-6   FUNCTIONS OF A RANDOM VARIABLE

Suppose you know the distribution of the radius R of a circle and want the distribution of its area πR^2. (For example, if R is picked at random between 0 and 5, then the area must be between 0 and 25π, but is it uniformly distributed?)

Or suppose you know the distribution of a signal X. Before the signal reaches you, it's squared and shifted so that what you actually receive is $X^2 + 7$. We want the distribution of the received signal.

In general, given the distribution of X, we'll find the distribution of some function Y of X. There are two methods.

1. Find the distribution function $F(y)$ first and then find the density $f(y)$.

2. Find the density $f(y)$ directly.

Throughout, X and Y are random variables, while x and y are values of the random variables.

The Distribution Function Method

Let X have an exponential distribution with $\lambda = 1$ so that

$$f_X(x) = e^{-x} \quad \text{for } x \geq 0$$

Let

$$Y = 2 - X^3$$

We'll find the density of Y by first finding the distribution function of Y. We know that X is always ≥ 0. By inspection, this makes Y always ≤ 2 and so

$$F_Y(y) = 1 \text{ if } y \geq 2 \quad \text{(all the probability has accumulated by } y = 2\text{)}$$

Consider $y \leq 2$. We have

$$F_Y(y) = P(Y \leq y) = P(2 - X^3 \leq y)$$

In this example you can easily use algebra to solve the inequality

$$2 - X^3 \leq y$$

and get

$$X \geq \sqrt[3]{2 - y}$$

But, in general, inequalities can be tricky to solve (remember that when you multiply or divide by a negative number it reverses the inequality). So we'll solve the inequality graphically as follows.

The top picture in Fig. 1 is the graph of $Y = 2 - X^3$ (and below it, for reference, is the density of X). On the Y-axis is a typical y less than 2 and the corresponding value $\sqrt[3]{2 - y}$ on the X-axis.

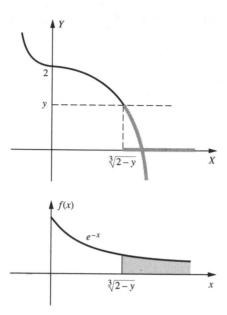

Figure 1

Figure 1, top diagram, shows that Y is *below* the y-mark iff X is to the *right* of the $\sqrt[3]{2-y}$-mark. So

$$F_Y(y) = P(Y \leq y)$$

$$= P\left(X \geq \sqrt[3]{2-y}\right)$$

$$= \text{area under } f_X(x) \text{ after } \sqrt[3]{2-y} \qquad \text{(lower picture in Fig. 1)}$$

$$= \int_{\sqrt[3]{2-y}}^{\infty} e^{-x}\, dx = -e^{-x}\bigg|_{\sqrt[3]{2-y}}^{\infty} = e^{-\sqrt[3]{2-y}}$$

All in all,

$$F_Y(y) = \begin{cases} e^{-\sqrt[3]{2-y}} & \text{if } y \leq 2, \\[2mm] 1 & \text{if } y \geq 2 \quad \text{(Fig. 2)} \end{cases}$$

and

$$f_Y(y) = F_Y'(y) = \begin{cases} \dfrac{e^{-\sqrt[3]{2-y}}}{3(2-y)^{2/3}} & \text{if } y \leq 2 \\[4mm] 0 & \text{if } y \geq 2 \quad \text{(Fig. 3)} \end{cases}$$

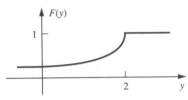

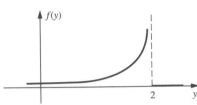

Figure 2   **Figure 3**

Notation

Instead of writing $f_X(x)$ and $f_Y(y)$ for the densities of X and Y, respectively, I'll often leave out the subscripts and write $f(x)$ and $f(y)$. It should be clear from the letter used (x or y) which density I'm referring to.

Similarly, I'll write $F(x)$ instead of $F_X(x)$ and $F(y)$ instead of $F_Y(y)$.

Values of *F(y)* and *f(y)* by Inspection

In the last example, we got a headstart on the distribution function of Y by noticing that Y is always ≤ 2. Here's a summary of conclusions you may be able to draw easily before you jump into the hard stuff.

Suppose you can tell that Y is always ≥ 5. Then

$$F(y) = 0 \quad \text{for } y \leq 5 \quad \text{(no probability has accumulated yet)}$$

$$f(y) = 0 \quad \text{for } y \leq 5 \quad \text{(impossible values of Y)}$$

Suppose Y is always ≤ 5. Then

$$F(y) = 1 \quad \text{for } y \geq 5 \quad \text{(all the probability has accumulated by } y = 5\text{)}$$

$$f(y) = 0 \quad \text{for } y \geq 5 \quad \text{(impossible values of } Y\text{)}$$

Suppose Y is always between 3 and 7. Then

$$F(y) = \begin{cases} 0 & \text{for } y \leq 3 \\ \text{don't know yet} & \text{for } 3 \leq y \leq 7 \\ 1 & \text{for } y \geq 7 \end{cases}$$

$$f(y) = \begin{cases} \text{don't know yet} & \text{for } 3 \leq y \leq 7 \\ 0 & \text{otherwise} \end{cases}$$

Example 1

Let X have density

$$f(x) = \begin{cases} \dfrac{1}{2} & \text{if } 0 \le x \le 1 \\[2mm] -\dfrac{1}{4}x + \dfrac{3}{4} & \text{if } 1 \le x \le 3 \end{cases}$$

Let

$$Y = X^3$$

Use the distribution function method to find the density of Y.

By inspection, $0 \le X \le 3$. This keeps Y between 0 and 27, so

$$F(y) = \begin{cases} 0 & \text{if } y \le 0 \\ 1 & \text{if } y \ge 27 \end{cases}$$

Now consider $0 \le y \le 27$. We have

$$F(y) = P(Y \le y)$$

Figure 4 shows the graph of $Y = X^3$ with a typical y-value between 0 and 27 on the Y-axis and the corresponding value $\sqrt[3]{y}$ on the X-axis.

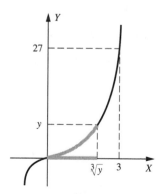

Figure 4

Look at the graph to see that Y is *below* the y-mark iff X is to the *left* of the $\sqrt[3]{y}$-mark. So

$$F(y) = P(Y \le y) = P\left(X \le \sqrt[3]{y}\right) = \int_{x=-\infty}^{\sqrt[3]{y}} f(x)\, dx$$

But $f(x)$ changes formula at $x = 1$, so the integral depends on where $\sqrt[3]{y}$ is, which in turn depends on where y is. So we need cases.

Case 1. $0 \le y \le 1$ so that $0 \le \sqrt[3]{y} \le 1$ (Fig. 5)

$$F(y) = \int_{x=-\infty}^{\sqrt[3]{y}} f(x)\, dx = \text{ indicated area in Fig. 5 } = \frac{1}{2}\sqrt[3]{y}$$

Case 2. $1 \le y \le 27$ so that $1 \le \sqrt[3]{y} \le 3$ (Fig. 6)

$$F(y) = \int_{x=-\infty}^{\sqrt[3]{y}} f(x)\, dx = \text{ indicated area in Fig. 6 }$$

$$= \frac{1}{2} + \int_{x=1}^{\sqrt[3]{y}} \left(-\frac{1}{4}x + \frac{3}{4} \right) dx = \frac{3}{4}\sqrt[3]{y} - \frac{1}{8}y^{2/3} - \frac{1}{8}$$

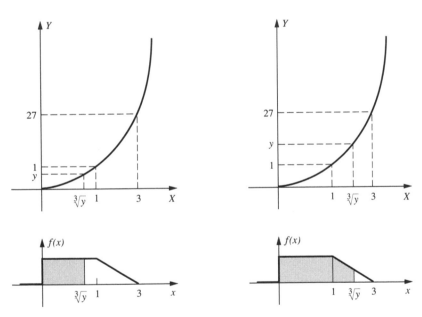

Figure 5 $0 \le y \le 1$ **Figure 6** $1 \le y \le 27$

If you didn't see by inspection that $F(y) = 1$ for $y \ge 27$, then you would go on to the next case (Fig. 7) where $y \ge 27$: If $y \ge 27$, then $\sqrt[3]{y} \ge 3$, so

$$F(y) = \int_{x=-\infty}^{\sqrt[3]{y}} f(x)\, dx = \text{ total area under the } f(x) \text{ graph}$$

$$= 1 \text{ (as predicted)}$$

All in all,

$$F(y) = \begin{cases} 0 & \text{if } y \le 0 \\[2mm] \dfrac{1}{2}\sqrt[3]{y} & \text{if } 0 \le y \le 1 \\[3mm] \dfrac{3}{4}\sqrt[3]{y} - \dfrac{1}{8}y^{2/3} - \dfrac{1}{8} & \text{if } 1 \le y \le 27 \\[3mm] 1 & \text{if } y \ge 27 \quad \text{(Fig. 8)} \end{cases}$$

and

$$f(y) = F'(y) = \begin{cases} \dfrac{1}{6y^{2/3}} & \text{if } 0 \le y \le 1 \\[4mm] \dfrac{1}{4y^{2/3}} - \dfrac{1}{12y^{1/3}} & \text{if } 1 \le y \le 27 \quad \text{(Fig. 9)} \end{cases}$$

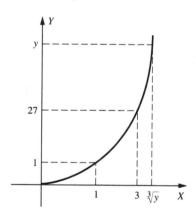

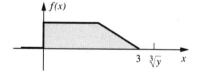

Figure 7 $y \ge 27$

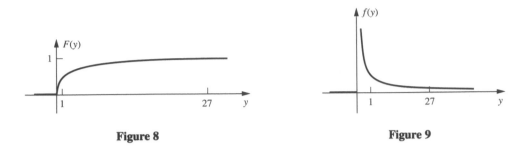

Figure 8   **Figure 9**

Density Function Method

Suppose X has density $f(x)$. Let Y be a function of X (Fig. 10). I'll get a formula for the density $f(y)$ so that you can find it directly without having to first find the distribution function.

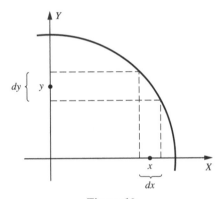

Figure 10

In Fig. 10, Y is within dy of the indicated y iff X is within dx of the corresponding x so

$$P(Y \approx y) = P(X \approx x)$$

Computing these probabilities we have

(1)
$$\boxed{f(y)\,dy = f(x)\,dx}$$

So

(2)
$$f(y) = f(x)\,\frac{\text{length } dx}{\text{length } dy}$$

Both dx and dy in Fig. 10 and in (1) are *lengths*; neither was meant to be negative. If $dy \to 0$, then

$$\frac{\text{length } dx}{\text{length } dy} \rightarrow \text{absolute value of the derivative } \frac{dx}{dy}$$

and (2) becomes

(3)
$$\boxed{f_Y(y) = f_X(x) \quad \left|\frac{dx}{dy}\right|}$$

The formula in (3) lets you find $f(y)$, knowing $f(x)$, provided that you use the connection between x and y (which is the same as the connection between X and Y) to compute the derivative dx/dy and eventually to *replace all x's on the right-hand side of (3) by y's.*

I'll do some examples so you can see how to use (3). (I'll take the examples I just did with the distribution function method and do them again with the density function method.)

Example 2

Let X have an exponential distribution with $\lambda = 1$; that is,

$$f(x) = e^{-x} \quad \text{for } x \geq 0$$

Let

$$Y = 2 - X^3$$

We want to find the density $f(y)$ of Y. Figure 11 shows Y versus X in the top picture and the density $f(x)$ in the bottom picture.

By inspection, $X > 0$, which makes $Y < 2$, so

$$f(y) = 0 \quad \text{if } y \geq 2$$

Let $y \leq 2$. Then

(4)
$$f(y) = f(x)\left|\frac{dx}{dy}\right|$$

In Fig. 11, if $y \leq 2$, then $x \geq 0$ and

(5)
$$f(x) = e^{-x}$$

And since $Y = 2 - X^3$, the same relationship holds for the values x and y in Fig. 11. So

$$y = 2 - x^3$$

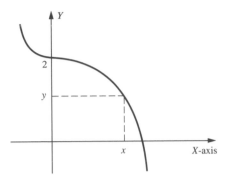

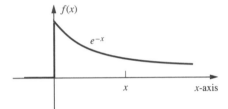

Figure 11

(6)
$$x = \sqrt[3]{2 - y}$$

(7)
$$\frac{dx}{dy} = -\frac{1}{3(2 - y)^{2/3}}$$

Substitute (5) and (7) into (4) to get

$$f(y) = e^{-x}\, \frac{1}{3(2 - y)^{2/3}}$$

and then get rid of any x's still in the answer by using (6). The final answer is

$$f(y) = \frac{e^{-\sqrt[3]{2-y}}}{3(2 - y)^{2/3}} \quad \text{for } y \leq 2$$

Warning

1. Remember to take the *absolute value* of dx/dy when you use (3).
2. Don't forget to switch from x's to y's. There should be no x's in $f(y)$.

Example 3

Let X have density

$$f(x) = \begin{cases} \dfrac{1}{2} & \text{if } 0 \leq x \leq 1 \\[2mm] -\dfrac{1}{4}x + \dfrac{3}{4} & \text{if } 1 \leq x \leq 3 \end{cases}$$

Let

$$Y = X^3$$

Use the density function method to find the density of Y.

Figures 12 and 13 each show the graph of Y versus X, and below it, the graph of $f(x)$ versus x.

We know $0 \leq X \leq 3$, so Y is always between 0 and 27 and

$$f(y) = 0 \quad \text{if } y \leq 0 \text{ or } y \geq 27$$

Now look at $0 \leq y \leq 27$. The answer will be

$$f(y) = f(x) \left| \frac{dx}{dy} \right|$$

where $y = x^3$, so we plug in

$$x = \sqrt[3]{y}, \quad \frac{dx}{dy} = \frac{1}{3y^{2/3}}$$

But we can't plug in the $f(x)$ formula immediately because $f(x)$ has two formulas. Which one we use depends on where x is, which in turn depends on where y is, so we need cases.

Case 1. $0 \leq y \leq 1$ (Fig. 12)

Then $0 \leq x \leq 1$ and $f(x) = 1/2$, so

$$f(y) = \frac{1}{2} \frac{1}{3y^{2/3}}$$

Case 2. $1 \leq y \leq 27$ (Fig. 13)

Then $1 \leq x \leq 3$ and $f(x) = -\frac{1}{4}x + \frac{3}{4}$ so

$$f(y) = \left(-\frac{1}{4}x + \frac{3}{4} \right) \frac{1}{3y^{2/3}}$$

$$= \left(-\frac{1}{4}\sqrt[3]{y} + \frac{3}{4} \right) \frac{1}{3y^{2/3}} \quad \text{(don't forget to switch from x's to y's)}$$

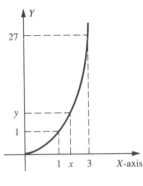

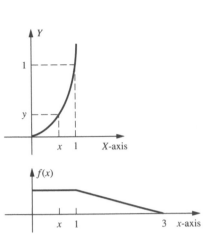

Figure 12   **Figure 13**

All in all,

$$
f(y) = \begin{cases} \dfrac{1}{6y^{2/3}} & \text{if } 0 \le y \le 1 \\[4mm] \dfrac{1}{4y^{2/3}} - \dfrac{1}{12y^{1/3}} & \text{if } 1 \le y \le 27 \end{cases}
$$

More on the Density Function Method

Let X have density $f(x)$. Let Y be a non-one-to-one function of X (Fig. 14), so that for some value(s) y of Y there is more than one corresponding value of X (see x_1 and x_2 in Fig. 14).

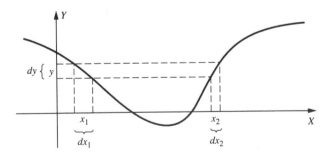

Figure 14

In Fig. 14,

Y is within dy of indicated y iff X is in the x_1-interval *or* x_2-interval.

So

$$P(Y \approx y) = P(X \approx x_1) + P(X \approx x_2)$$

which gives

(8)
$$\boxed{f(y)\, dy = f(x_1)dx_1 + f(x_2)dx_2}$$

So

(9)
$$f(y) = f(x_1)\frac{\text{length } dx_1}{\text{length } dy} + f(x_2)\frac{\text{length } dx_2}{\text{length } dy}$$

As in the earlier argument, dx_1 and dx_2 are lengths; they are never negative. So as $dy \to 0$, the quotients in (9) become absolute values of derivatives:

(10)
$$\boxed{f(y) = f(x_1)\left|\frac{dx_1}{dy}\right| + f(x_2)\left|\frac{dx_2}{dy}\right|}$$

The formula in (10) lets you find $f(y)$ if you know $f(x)$, provided you use the connection between x_1 and y and between x_2 and y to compute the derivatives and to eventually replace all x's on the right-hand side of (10) by y's.

Here's an example to show how (10) works.

Example 4

The head of a lawn sprinkler revolves back and forth so that drops of water shoot out at angles between 0° and 90°. The gardener wants to know if her lawn is being uniformly watered and, if not, where the soggy and dry spots are.

Let θ be the angle at which the drop shoots out and let x be how far from the sprinkler the drop lands (Fig. 15).

Physics and calculus books show that if a particle flies off from the origin at time $t = 0$ at angle θ, then its position at time t is given by the parametric equations

$$x = v_0\, t \cos\theta, \qquad y = v_0\, t \sin\theta - gt$$

where g is the gravitational constant and v_0 is the initial speed of the drop (imparted to it by the sprinkler). Set $y = 0$ and solve for t to find the time that the particle lands. Plug that t into the first equation to find x. It turns out that

$$x = k \sin 2\theta \qquad \text{where } k = \frac{v_0^2}{g}$$

For example, the 45° drop lands the farthest (Fig. 16); the 15° drop and 75° drop land in the same place.

To find out if the lawn is uniformly watered, we have this problem.

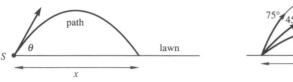

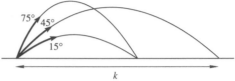

Figure 15   Path of a drop   **Figure 16**   Several drops

Let Θ be uniform on $[0, \pi/2]$.
Let $X = k \sin 2\Theta$, where k is a fixed positive constant.
Find the distribution of X.

Let's use the density function method.

We know that X is between 0 and k, so

$$f(x) = 0 \quad \text{for } x \leq 0 \text{ or } x \geq k$$

Now consider $0 \leq x \leq k$.

A horizontal line intersects the graph of $X = k \sin 2\Theta$ twice (Fig. 17), so

(11) $$f(x) = f(\theta_1)\left|\frac{d\theta_1}{dx}\right| + f(\theta_2)\left|\frac{d\theta_2}{dx}\right|$$

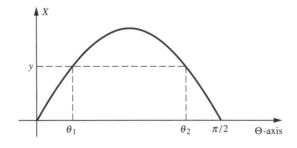

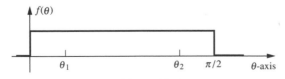

Figure 17

In order to use (11) we have to express θ_1 and θ_2 in terms of x. For the θ_1 solution we have

$$\sin 2\theta_1 = \frac{x}{k}$$

where $2\theta_1$ is between 0 and $\pi/2$, so

$$2\theta_1 = \arcsin \frac{x}{k} \qquad \text{(see arcsin review box, next page)}$$

$$\theta_1 = \frac{1}{2} \arcsin \frac{x}{k}$$

By symmetry, the second solution is

$$(12) \qquad \qquad \theta_2 = \frac{\pi}{2} - \frac{1}{2} \arcsin \frac{x}{k}$$

So

$$(13) \qquad \frac{d\theta_1}{dx} = \frac{1}{2} \frac{1}{k\sqrt{1-(x/k)^2}}, \qquad \frac{d\theta_2}{dx} = -\frac{1}{2} \frac{1}{k\sqrt{1-(x/k)^2}}$$

Since Θ is uniform on $[0, \pi/2]$,

$$(14) \qquad \qquad f(\theta_1) = f(\theta_2) = \frac{2}{\pi}$$

Plug (13) and (14) into (11):

$$f(x) = \frac{2}{\pi} \frac{1}{2} \frac{1}{k\sqrt{1-(x/k)^2}} + \frac{2}{\pi} \frac{1}{2} \frac{1}{k\sqrt{1-(x/k)^2}}$$

So the X density is[4]

$$f(x) = \frac{2}{\pi k\sqrt{1-x^2/k^2}} \qquad \text{for} \;\; 0 \leq x \leq k$$

It blows up at $x = k$ (Fig. 18). So the outer edges of the lawn get very soggy; drops are much more likely to land there.

---

[4]Because of the symmetry of $\sin 2\Theta$ and because Θ is uniform on $[0, \pi/2]$, we have

$$P(\Theta \approx \theta_1) = P(\Theta \approx \theta_2)$$

So (11) becomes

$$f(x) = 2 \, f(\theta_1) \left| \frac{d\theta_1}{dx} \right|$$

which is a shortcut in this particular example.

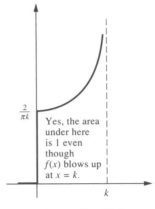

Figure 18 $f(x)$

Warning

Watch for multiple intersections when you use the density method.

Review of $\sin^{-1} x$ and $\tan^{-1} x$ (arcsin x and arctan x)

If $y = \sin x$, then it is *not* necessarily true that $x = \sin^{-1} y$. What *is* true is that if $y = \sin x$ and $-\frac{1}{2}\pi \le x \le \frac{1}{2}\pi$, then $x = \sin^{-1} y$.

In other words, for any y between -1 and 1, there are many angles whose sin is y; $\sin^{-1} y$ is meant to be the particular angle between $-\frac{1}{2}\pi$ and $\frac{1}{2}\pi$. The other angles can be expressed in terms of $\sin^{-1} y$ (as we did in (12)), but only one of the angles is $\sin^{-1} y$.

Similarly, if $y = \tan x$, then it is *not* necessarily true that $x = \tan^{-1} y$. What *is* true is that if $y = \tan x$ and $-\frac{1}{2}\pi < x < \frac{1}{2}\pi$, then $x = \tan^{-1} y$.

In other words, for any given y there are many angles whose tan is y; $\tan^{-1} y$ is meant to be the particular angle between $-\frac{1}{2}\pi$ and $\frac{1}{2}\pi$.

Warnings

1. If the "answer" involves a formula *on an interval* and you supply the formula but not the interval, then your answer is incomplete.

2. Suppose X is a random variable and Y is a function of X.

   Then $f_X(x)$ is a *density* and the graph of $f_X(x)$ versus x must *look* like a density (nonnegative, area under graph is 1, $f_X(-\infty) = 0$, $f_X(\infty) = 0$) and similarly for $f_Y(y)$ versus y.

And $F_X(x)$ is a *distribution function* and the graph of $F_X(x)$ versus x must *look* like a distribution function ($F_X(-\infty) = 0, F_X(\infty) = 1$, F never decreases) and similarly for $F_Y(y)$ versus y.

But the function Y of X has no restrictions. The graph of Y versus X can be anything.

3. Don't confuse x with X, y with Y, or f with F either in your mind or in your handwriting.

> X is a *random variable*—it's a *function* that takes on a different value for each outcome of an experiment.
> x is a value that X can assume, a *real number*.
> f is a *density* function; F is a *distribution* function.

The graph of the density $f_X(x)$ is drawn in a coordinate system with a horizontal x-axis (not an X-axis) and a vertical f_X axis (and similarly for the graph of F_X).

4. *Draw pictures.* You'll need the Y versus X graph and also the $f(x)$ graph. Solving inequalities and finding cases is much easier using the pictures.

Problems for Section 4-6

(Each solution includes the distribution function method and the density method.)

1. Pick X at random between 0 and 5. Let Y be the volume of a sphere with radius X. Then $Y = \frac{4}{3}\pi X^3$, but for convenience let's take $Y = X^3$. Find the density of Y.

2. Let X have density

$$f(x) = \frac{2x}{R^2} \quad \text{for} \ \ 0 \le x \le R$$

Let

$$Y = \frac{X}{R}$$

Find the density of Y.

3. Let X have density

$$f(x) = \frac{1}{x^2} \quad \text{for} \ \ x \ge 1$$

Let

$$Y = \begin{cases} 2X & \text{if } X \le 2 \\ X^2 & \text{if } X \ge 2 \end{cases}$$

Find the density of Y.

4. Find the density of Y if $Y = 1/X$ and X is uniform on $[-1, 1]$.

5. Let X have density

$$f(x) = \begin{cases} \dfrac{1}{2} & \text{if } -1 \le x \le 0 \\ \dfrac{1}{2}e^{-x} & \text{if } x \ge 0 \end{cases}$$

Let $Y = |X|$. Find the density of Y.

6. Let X be exponential with $\lambda = 1$. Find the density of Y if

$$Y = \begin{cases} X & \text{if } X \le 1 \\ \dfrac{1}{X} & \text{if } X \ge 1 \end{cases}$$

7. Let

$$f(x) = \frac{1}{x^2} \quad \text{for } x \ge 1$$

Let

$$Y = \begin{cases} 2X & \text{if } X \le 3 \\ 6 & \text{if } X \ge 3 \end{cases}$$

(a) Find $P(Y = 6)$ and show that Y is a mixed random variable.
(b) Find $F(y)$.
(c) Find the pseudodensity of Y, which is partly a density and partly a probability function.

8. Find the density of Y if $Y = X^2$ where X has density

$$f(x) = \begin{cases} \dfrac{1}{2} & \text{if } -1 \le x \le 0 \\ \dfrac{1}{2}e^{-x} & \text{if } x \ge 0 \end{cases}$$

SECTION 4-7   SIMULATING A RANDOM VARIABLE

Simulating a Continuous Random Variable

A computer with a random number generator can pick a number at random in [0,1]. In other words, the computer can be used to simulate a random variable X uniformly distributed on [0,1]. Here's how you can take advantage of this to simulate *any* continuous random variable.

Let X be uniformly distributed on [0,1]. We can use X as follows to create a new random variable Y with a prescribed distribution function $F(y)$ (Fig. 1).

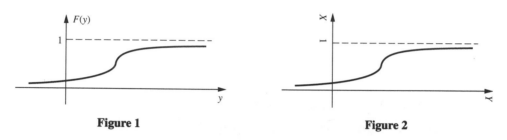

Figure 1   **Figure 2**

Look at Fig. 2, which is just Fig. 1 from a new point of view and with the axes relabeled. Figure 2 defines Y as a function of X, and we can show that Y does have the distribution function $F(y)$ from Fig. 1.

> Let X be uniform on [0,1]. To get Y with distribution function $F(y)$, let $X = F(Y)$ and solve for Y.

Here's why it works. With the Y created this way, Fig. 3 shows that Y is below the y-mark iff X is below the $F(y)$-mark, so

$$F_Y(y) = P(Y \leq y) = P(X \leq F(y))$$

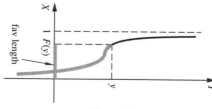

Figure 3

Since X is uniform on [0,1],

$$P(X \leq F(y)) = \frac{\text{fav length}}{\text{total length}} = \frac{F(y)}{1} = F(y)$$

So $F_Y(y)$ does equal the desired $F(y)$, QED.

An example will help make the process clear.

Suppose you want Y to have the distribution function $F(y)$ in Fig. 4.

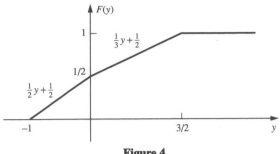

Figure 4

Begin with X uniform on $[0,1]$. Look at Fig. 5, which shows $X = F(Y)$ (this is Fig. 4 from a new point of view). To get the explicit formula for Y, you need two cases.

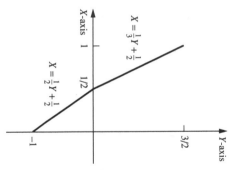

Figure 5

If $0 \le X \le \frac{1}{2}$, solve $X = \frac{1}{2}Y + \frac{1}{2}$ to get $Y = 2X - 1$.

If $\frac{1}{2} \le X \le 1$, solve $X = \frac{1}{3}Y + \frac{1}{2}$ to get $Y = 3X - \frac{3}{2}$.

So

$$
Y = \begin{cases} 2X - 1 & \text{if } 0 \le X \le \dfrac{1}{2} \\[2ex] 3X - \dfrac{3}{2} & \text{if } \dfrac{1}{2} \le X \le 1 \end{cases}
$$

In other words, the following experiment produces the random variable Y with the distribution function $F(y)$ in Fig. 4.

Have your computer pick a number X at random between 0 and 1.

If X is between 0 and 1/2, take Y to be $2X - 1$.

If X is between 1/2 and 1, take Y to be $3X - 3/2$.

Figure 5 shows Y as a function of X with off-beat axes; Fig. 6 shows Y as a function of X with the usual axes.

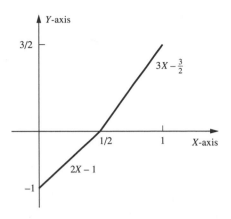

Figure 6

Simulating a Discrete Random Variable

Let X be uniform on $[0,1]$. To get Y so that

$$P(Y = -5) = .7$$
$$P(Y = \pi) = .2$$
$$P(Y = 6) = .1$$

let

$$Y = \begin{cases} -5 & \text{if } 0 \le X \le .7 \\ \pi & \text{if } .7 \le X \le .9 \\ 6 & \text{if } .9 \le X \le 1 \end{cases}$$

There are many other possibilities, for example,

$$Y = \begin{cases} \pi & \text{if } .5 \le X \le .7 \\ -5 & \text{if } .1 \le X \le .5 \text{ or } .7 \le X \le 1 \\ 6 & \text{if } 0 \le X \le .1 \end{cases}$$

Any scheme works as long as the length of the X-interval assigned to $Y = \pi$ matches the desired probability, namely, .2, for the event $Y = \pi$, and so on.

Simulating a Mixed Random Variable

Let X be uniform on $[0,1]$. Suppose you want Y to have the distribution function in Fig. 7.

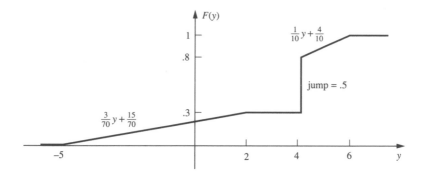

Figure 7

Case 1.   If $0 \le X \le .3$, let

$$X = \frac{3}{70}Y + \frac{15}{70}, \quad Y = \frac{70X - 15}{3}$$

Case 2.   Let $.3 \le X \le .8$. Figure 7 shows a jump from .3 to .8 at $y = 4$ so $P(Y = 4) = .5$.

To get this, let $Y = 4$ for $.3 \le X \le .8$.

Case 3.   If $.8 \le X \le 1$, let

$$X = \frac{1}{10}Y + \frac{4}{10}, \quad Y = 10X - 4$$

All in all,

$$Y = \begin{cases} \dfrac{70X - 15}{3} & \text{if } 0 \le X \le .3 \\[2mm] 4 & \text{if } .3 \le X \le .8 \\[2mm] 10X - 4 & \text{if } .8 \le X \le 1 \quad \text{(Fig. 8)} \end{cases}$$

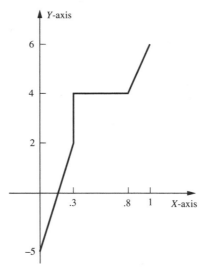

Figure 8

Problems for Section 4-7

Let X be uniformly distributed on $[0,1]$.

1. Express Y in terms of X so that Y has an exponential distribution with parameter λ.

2. (a) Express Y in terms of X so that Y has the distribution function in the diagram.

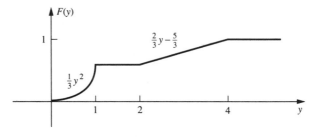

Figure P.2(a)

   (b) As a check, start with the Y you created in part (a) and find its distribution function (using the dist function method from the preceding section) to see if it does come out to be the F in the picture.

3. Find Y in terms of X so that

$$F(y) = \begin{cases} 0 & \text{if } y \leq 2 \\ 1 - \dfrac{2}{y} & \text{if } 2 \leq y \leq 4 \\ 1 - \dfrac{1}{\sqrt{y}} & \text{if } y \geq 4 \end{cases}$$

4. Find Y in terms of X so that
   (a) $P(Y = 1) = .1, P(Y = 2) = .4, P(Y = e) = .3, P(Y = -3) = .2$
   (b) $P(Y = 1) = 1/2, P(Y = 2) = 1/4, P(Y = 3) = 1/8$, and, in general

$$P(Y = n) = \frac{1}{2^n} \quad \text{for } n = 1, 2, 3, \ldots$$

   (c) Y has a Poisson distribution

5. Find Y in terms of X.
   (a) (watch for jumps)

$$F(y) = \begin{cases} 0 & \text{if } y \leq -10 \\ \dfrac{1}{y^2} & \text{if } -10 \leq y \leq -2 \\ \dfrac{1}{4} & \text{if } -2 \leq y \leq 3 \\ \dfrac{1}{4}y - \dfrac{1}{2} & \text{if } 3 \leq y \leq 4 \\ 1 & \text{if } y \geq 4 \end{cases}$$

   (b)

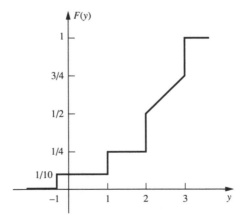

Figure P.5(b)

6. Find Y in terms of X so that Y is uniform on [3,7].

REVIEW FOR THE NEXT CHAPTER

Some More Inequalities

The review in Section 1.5 graphed a few inequalities. Here are some trickier ones.

The inequality $y/x > 3$ is equivalent to

$$y > 3x \quad \text{if} \quad x > 0$$

$$y < 3x \quad \text{if} \quad x < 0$$

So the graph is *above* line $y = 3x$ on the right side and *below* $y = 3x$ on the left side (Fig. 1).

The inequality $\max(x, y) < 4$ means $x < 4$ *and* $y < 4$ so the graph is to the left of line $x = 4$ and below line $y = 4$ (Fig. 2).

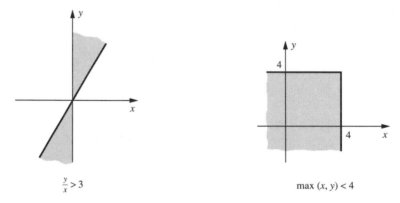

$\frac{y}{x} > 3$

Figure 1

max $(x, y) < 4$

Figure 2

The inequality $\max(x, y) > 4$ means $x > 4$ *or* $y > 4$ so the graph includes all points that are either to the right of line $x = 4$ or above line $y = 4$ (Fig. 3).

The inequality $\min(x, y) < 4$ means $x < 4$ *or* $y < 4$ so the graph includes all points that are either to the left of line $x = 4$ or below line $y = 4$ (Fig. 4).

min $(x, y) < 4$

Figure 3

max $(x, y) > 4$

Figure 4

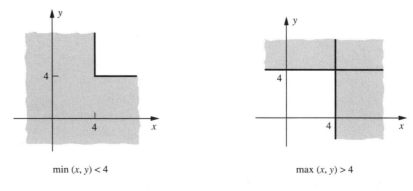

Double Integrals

In Chapter 5 we'll start using double integrals. In case you don't remember them too well, here's a quick rundown on how to set them up as successive single integrals. If you want more details, see *The Calculus Tutoring Book* (Ash and Ash, IEEE Press, 1986).

There are two ways to set up $\int f(x, y)\, dA$ over a region R.

$$\int_R f(x,y)\, dA = \int_{\text{lowest } y \text{ in } R}^{\text{highest } y \text{ in } R} \int_{x \text{ on left bdry}}^{x \text{ on right bdry}} f(x,y)\, dx\, dy \text{ (Fig. 5)}$$

$$\int_R f(x,y)\, dA = \int_{\text{leftmost } x \text{ in } R}^{\text{rightmost } x \text{ in } R} \int_{y \text{ on lower bdry}}^{y \text{ on upper bdry}} f(x,y)\, dx\, dy \text{ (Fig. 6)}$$

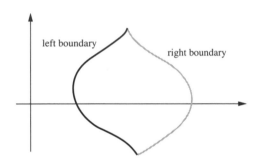

Figure 5

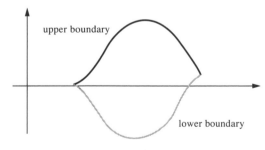

Figure 6

Example 1

I'll set up

$$\int_R f(x,y)\, dA$$

where R is the triangular region in Fig. 7.

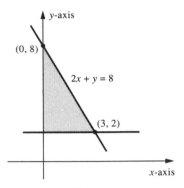

Figure 7

Method 1 (Fig. 8).

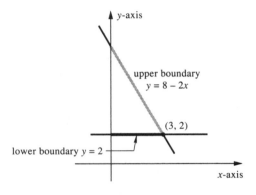

Figure 8

$$\int_R f(x, y) \, dA = \int_{x=0}^{3} \int_{y=2}^{y=8-2x} f(x, y) \, dy \, dx$$

Method 2 (Fig. 9).

$$\int_R f(x, y) \, dA = \int_{y=2}^{8} \int_{x=0}^{4-y/2} f(x, y) \, dx \, dy$$

Warning

Inner limits are *boundary* values; outer limits are *extreme* values.

The inner limits will *contain the other variable* unless the boundary in question is a vertical or horizontal line with an equation as simple as $x = 0$ or $y = 2$.

The limits on the outer integral are *always constants*.

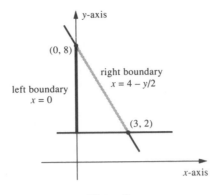

Figure 9

Integrating on a Region with a Two-Curve Boundary

Consider $\int_R f(x, y)\, dA$, where R is the region bounded by the line $x + y = 6$ and the parabola $x = y^2$ (Fig. 10).

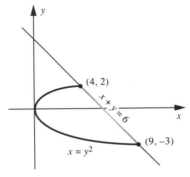

Figure 10

Method 1. (Fig. 11).

$$\int_R f(x, y)\, dA = \int_{y=-3}^{2} \int_{x=y^2}^{x=6-y} f(x, y)\, dx\, dy$$

Method 2. (Fig. 12).

The lower boundary is the parabola $x = y^2$; solve the equation for y to get $y = -\sqrt{x}$ (choose the negative square root because y is negative on the lower part of the parabola).

But *the upper boundary consists of two curves*, the parabola and the line. To continue with this order of integration, divide the region into the two indicated parts, I and II:

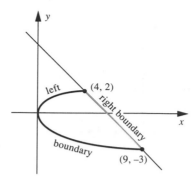

Figure 11

$$\int_R f(x,y)\, dA = \int_I f(x,y)\, dA + \int_{II} f(x,y)\, dA$$

$$= \int_{x=0}^4 \int_{y=-\sqrt{x}}^{\sqrt{x}} f(x,y)\, dy\, dx + \int_{x=4}^9 \int_{y=-\sqrt{x}}^{y=6-x} f(x,y)\, dy\, dx$$

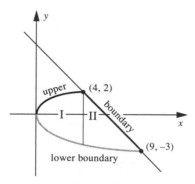

Figure 12

Warning

Examine the boundaries carefully to catch the ones consisting of more than one curve.

Examples

Here are two regions of integration with limits put in both ways.

For the unbounded region in Fig. 13, the limits are

$$\int_{y=-\infty}^0 \int_{x=1/y}^\infty + \int_{y=0}^\infty \int_{x=-\infty}^{1/y}$$

and also

$$\int_{x=-\infty}^{0} \int_{y=1/x}^{\infty} + \int_{x=0}^{\infty} \int_{y=-\infty}^{1/x}$$

For the region in Fig. 14, the limits are

$$\int_{y=0}^{2} \int_{x=y}^{x=y+3}$$

and also

$$\int_{x=0}^{2} \int_{y=0}^{x} + \int_{x=2}^{3} \int_{y=0}^{2} + \int_{x=3}^{5} \int_{y=x-3}^{2}$$

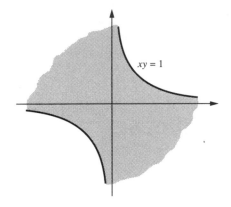

Figure 13

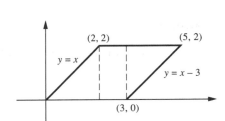

Figure 14

Double Integrating in Polar Coordinates

To find $\int_R f(X, y) \, dA$, let

$$x = r \cos \theta$$
$$y = r \sin \theta$$
$$r^2 = x^2 + y^2$$
$$dA = r \, dr \, d\theta$$

and use these limits of integration (Fig. 15):

$$\int_{\text{smallest } \theta}^{\text{largest } \theta} \int_{r \text{ on inner bdry}}^{r \text{ on outer bdry}}$$

For example,

$$\int x^2\, dA \quad \text{on the semicircular region in Fig. 16}$$

$$= \int_{\theta=\pi}^{2\pi} \int_{r=0}^{4} (r\cos\theta)^2\, r\sin\theta\, r\, dr\, d\theta$$

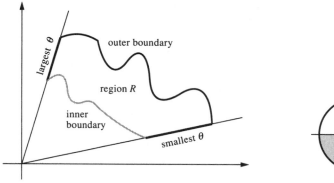

Figure 15 **Figure 16**

Warning

Don't forget that it's $dA = \boxed{r}\, dr\, d\theta$. Remember the extra r.

CHAPTER 5

Jointly Distributed Random Variables

SECTION 5-1   JOINT DENSITIES

Before we start on the 2-dimensional continuous case (two random variables with a joint density function), we'll summarize the 1-dimensional continuous case (one random variable with a density function).

Let X be a random variable—for each outcome of an experiment, X takes on a value. Suppose X has density $f(x)$. You can think of the experiment as throwing a dart at a line where those pieces of the line with larger f values are more likely to be hit.

The universe is the set of points where $f(x) \neq 0$.

Probabilities are found by integrating $f(x)$:

$$P(a \leq X \leq b) = \int_a^b f(x) \, dx$$

If we think of the interval $[a, b]$ as the region in the universe *favorable* to the event $a \leq X \leq b$, then we can express the rule as

$$P(X\text{-event}) = \int_{\text{fav region}} f(x) \, dx$$

You'll see that similar ideas hold for the 2-dimensional case.

Jointly Continuous Random Variables

Suppose the result of an experiment is a pair of numbers X, Y (imagine tossing a dart at the x, y plane). Each event corresponds to a region in the plane.

For example, consider the event $Y > X$. After the underlying experiment is performed, X takes on value x, Y takes on value y, and the favorable region is the sets of points in the plane where $y > x$ (Fig. 1).

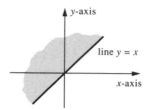

Figure 1   Favorable region for the event $Y > X$

Suppose we have a function $f(x, y)$ that can be integrated as follows to get probabilities of events:

$$(1) \qquad P(\text{event involving } X \text{ and } Y) = \int_{\text{fav region}} f(x, y)\, dA$$

Then $f(x, y)$ is called the *joint density function* of X and Y, and X and Y are said to be jointly continuous.

   The set of points in the plane where $f(x, y) \neq 0$ is called the *universe*. In other words, in the 2-dimensional continuous case, to find the probability of an event, integrate the joint density over the favorable region in the universe.

Think of the density $f(x, y)$ as the probability per unit area. Then $f(x, y)\, dA$ is the probability in a little region of area dA, and $\int_{\text{fav}} f(x, y)\, dA$ adds up the little probabilities to produce the total probability in the favorable region.

Properties of a Joint Density

Not any old function $f(x, y)$ can be a joint density. Probabilities are never negative and $P(\text{universe}) = 1$, so we must have

$$(2) \qquad\qquad f(x, y) \geq 0 \quad \text{for all } x, y$$

$$(3) \qquad\qquad \int_{\text{universe}} f(x, y)\, dA = 1$$

Example 1

Let X and Y have joint density

$$f(x, y) = e^{-(x+y)} \quad \text{for } x \geq 0, y \geq 0$$

We'll find

$$P(X \geq Y \geq 2)$$

Figure 2 shows the universe, the set of points (x, y) where $f(x, y) \neq 0$. Figure 3 shows the favorable region, the set of points in the universe where $x \geq y \geq 2$. (The favorable region is always the graph of an inequality. Check with the review section back in Section 1.5 and the one before this section for a summary of graphing inequalities.)

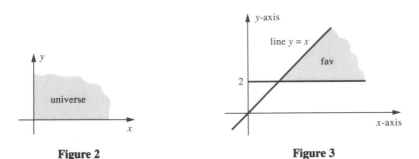

| **Figure 2** | **Figure 3** |

To find the probability that $X \geq Y \geq 2$, *integrate the density over the favorable region* (check with the review section preceding this chapter if you don't remember how to do double integrals):

$$P(X \geq Y \geq 2) = \int_{\text{fav}} e^{-(x+y)} \, dA = \int_{x=2}^{\infty} \int_{y=2}^{x} e^{-(x+y)} dy \, dx$$

$$\text{inner integral} = -e^{-(x+y)} \Big|_{y=2}^{x} = -e^{-2x} + e^{-(2+x)}$$

$$\text{outer} = \left(\frac{1}{2} e^{-2x} - e^{-(2+x)} \right) \Big|_{x=2}^{\infty} = \frac{1}{2} e^{-4}$$

Warning

For correct mathematical style, pictures of the universe and the favorable region should have axes labeled x and y, not X and Y. Integrals should involve x and y, not X and Y.

Example 2

Let X and Y have joint density

$$f(x, y) = \begin{cases} .1 & \text{in region I} \\ .2 & \text{in region II} \\ .3 & \text{in region III} \\ .4 & \text{in region IV} \quad \text{(Fig. 4)} \end{cases}$$

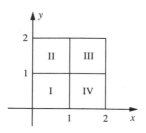

Figure 4   Universe

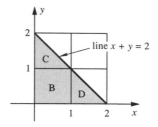

Figure 5   Fav region

Find $P(X + Y < 2)$.

The favorable region is shown in Fig. 5.

$$P(X + Y < 2) = \int_{\text{fav}} f(x,y) \, dA$$

$$= \int_B .1 \, dA + \int_C .2 \, dA + \int_D .4 \, dA$$

$$= .1 \int_B dA + .2 \int_C dA + .4 \int_D dA$$

$$= .1 \times \text{ area } B + .2 \times \text{ area } C + .4 \times \text{ area } D$$

$$= .1 \times 1 + .2 \times \frac{1}{2} + .4 \times \frac{1}{2}$$

$$= .4$$

The Event $X \approx x$ and $Y \approx y$

Let X and Y have joint density $f(x,y)$. As in the 1-dimensional case, any individual point has probability 0; that is, for any x and y,

$$P(X = x, Y = y) = 0$$

(In the 2-dimensional continuous case, not only does a point-event have probability 0, so does any event where the corresponding favorable region has zero area. For example, the event $Y = X^2$ has probability 0, since the favorable region is just a curve—too skinny to carry any probability.)

     On the other hand, it will be useful to find the probability that X is *near* x and (simultaneously) Y is *near* y.

     If $X \approx x$ *and* $Y \approx y$, then point (X, Y) is in the small box with dimensions dx by dy around point (x, y) (Fig. 6).

     The area of the box is $dx \, dy$, and its probability density (which gives the prob per unit area) is almost constant at $f(x,y)$. So

$$\text{box prob} = \text{prob density} \times \text{area}$$

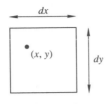

Figure 6

That is,

(4)
$$P(X \approx x \text{ and } Y \approx y) = f(x,y)\, dx\, dy$$

Independent Random Variables

We have already defined *events* A and B as independent if

$$P(A \text{ and } B) = P(A)\, P(B)$$

We call *random variables* X and Y independent if

$$P(X\text{-event and } Y\text{-event}) = P(X\text{-event})P(Y\text{-event})$$

for all X-events and Y-events.

The Joint Density of Independent Random Variables

John and Mary arrive independently . John's arrival time is X, where

$$f(x) = e^{-x} \quad \text{for } x \geq 0 \qquad \text{(Fig. 7)}$$

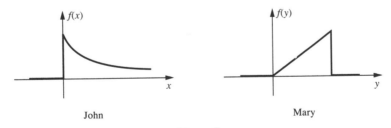

Figure 7

(He'll arrive some time after time 0 and is more likely to arrive early than late.) Mary's arrival time is Y, where

$$f(y) = \frac{2}{9}y \quad \text{for } 0 \leq y \leq 3 \qquad \text{(Fig. 7)}$$

(She'll arrive between times 0 and 3 and is more likely to arrive late than early.) After Mary arrives she'll wait up to an hour for John (but not vice versa). We'll find the probability that they meet.

The universe for the experiment is the region in 2-space where $x \geq 0$, $0 \leq y \leq 3$ (Fig. 8). John and Mary will meet if she arrives first (i.e., X is larger than Y) and then he arrives within an hour. So

$$P(\text{they meet}) = P(0 \leq X - Y \leq 1)$$

The favorable region is the graph of the inequality $0 \leq x - y \leq 1$ within the universe (Fig. 9).

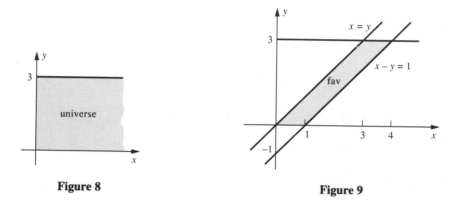

Figure 8   **Figure 9**

To find the probability that they meet, we want to integrate the joint density over the favorable region. But we don't have the joint density. All we have are two individual densities. So, first we need the following rule for getting the joint density.

Suppose

X takes on values in $[a, b]$ with density $f_X(x)$,
Y takes on values in $[c, d]$ with density $f_Y(y)$.

(The intervals can be finite or infinite.)

If X and Y are independent, then X and Y have joint density

(5) $f(x, y) = f_X(x) f_Y(y)$ for the rectangle $a \leq x \leq b, c \leq y \leq d$

In other words, *for independent random variables, just multiply the individual densities to get the joint density.*

Here's why the rule works.

$$f(x,y)\ dx\ dy = P(X \approx x \text{ and } Y \approx y)$$
$$= P(X \approx x)\,P(Y \approx y) \qquad \text{(by independence)}$$
$$= f_X(x)\ dx\ f_Y(y)\ dy$$

Cancel dx and dy to get $f(x,y) = f_X(x)f_Y(y)$.
   Now back to John and Mary. The joint density of X and Y is

$$f(x,y) = \frac{2}{9}y\,e^{-x} \quad \text{for } 0 \leq y \leq 3, x \geq 0$$

and

$$P(\text{they meet}) = \int_{\text{fav}} \frac{2}{9}y\,e^{-x}dA$$

$$= \int_{y=0}^{3} \int_{x=y}^{x=1+y} \frac{2}{9}y\,e^{-x}\ dx\ dy$$

$$= \int_{0}^{3} \frac{2}{9}y\left(e^{-y} - e^{-(1+y)}\right) dy$$

$$= .112$$

Uniform Joint Densities

Look at the 1-dimensional case first. If X is picked at random in an interval, then

$$P(\text{event}) = \frac{\text{favorable length}}{\text{total length}}$$

and X is said to be uniformly distributed in the interval. The density of X is

$$f(x) = \frac{1}{\text{length of interval}} \quad \text{for } x \text{ in the interval}$$

For the 2-dimensional version, pick point (X,Y) at random from a region so that

$$P(\text{event}) = \frac{\text{favorable area}}{\text{total area}}$$

Then we say that X and Y have a *joint uniform distribution*. Analogous to the 1-dimensional case:

If X and Y are jointly uniform on a region, the joint density is

(6) $f(x, y) = \dfrac{1}{\text{area of region}}$ for (x, y) in the region

---

For example, if (X, Y) is picked at random from a circle of radius 3, then the random variables X and Y have the uniform joint density

$$f(x, y) = \frac{1}{9\pi} \quad \text{for } (x, y) \text{ in the circle}$$

The Joint Density of Independent *Uniform* Random Variables

---

(7)   Let X be uniformly distributed in $[a, b]$.
Let Y be uniformly distributed in $[c, d]$.
Let X and Y be independent.
Then X and Y are jointly uniform in the rectangle
$a \le x \le b, \; c \le y \le d$, and

$$P(\text{event}) = \frac{\text{fav area}}{\text{total area}}$$

This holds intuitively because picking X at random in $[a, b]$ and independently picking Y at random in $[c, d]$ is the same as picking the point (X, Y) at random in the rectangle $a \le x \le b, \; c \le y \le d$. (We took this for granted in Section 1.5.)

It follows formally as a special case of (5) since

$$f(x, y) = f_X(x) \, f_Y(y) = \frac{1}{b - a} \frac{1}{d - c} = \frac{1}{\text{area of rectangle}}$$

for $0 \le x \le b, \; 0 \le y \le c$, which makes $f(x, y)$ a uniform density.

Example 3

Pick X at random between 0 and 2. Independently, pick Y at random between 0 and 3. Find the probability that $X + Y < 1$.

By (7), X and Y are jointly uniform in the rectangle in Fig. 10, so

$$P(X + Y < 1) = \frac{\text{favorable area}}{\text{total area}} = \frac{1/2}{6} = \frac{1}{12}$$

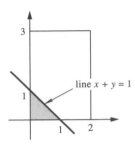

Figure 10

If you integrate the joint density over the favorable region, you'll get the same answer, but it takes longer. The joint uniform density is

$$f(x,y) = \frac{1}{\text{total area}} = \frac{1}{6} \quad \text{for } 0 \le x \le 2, \quad 0 \le y \le 3$$

so

$$P(X + Y < 1) = \int_{\text{fav}} \frac{1}{6}\, dA = \frac{1}{6} \int_{\text{fav}} dA = \frac{1}{6} \times \text{fav area } = \frac{1}{6} \times \frac{1}{2} = \frac{1}{12}$$

You can be even more inefficient by doing it like this:

$$P(X + Y < 1) = \int_{\text{fav}} \frac{1}{6}\, dA$$

$$= \int_{x=0}^{1} \int_{y=0}^{1-x} \frac{1}{6}\, dy\, dx$$

$$= \frac{1}{6} \int_0^1 (1 - x)\, dx$$

$$= \frac{1}{6} \left(x - \frac{x^2}{2} \right) \Bigg|_0^1$$

$$= \frac{1}{12}$$

Warning

In the *joint uniform* case, you can use $\int_{\text{fav}} f(x,y)\, dA$. But if areas are easy to find, as they often are, fav area/total area is much easier.

On the other hand, if X and Y are *not* jointly uniform (as in the John and Mary problem), then you *can't* use fav/total and *must* use $\int_{\text{fav}} f(x,y)\, dA$.

$y\backslash x$	1	2	3	4	5	6
2	$\frac{1}{36}$	0	0	0	0	0
3	0	$\frac{2}{36}$	0	0	0	0
4	0	$\frac{1}{36}$	$\frac{2}{36}$	0	0	0
5	0	0	$\frac{2}{36}$	$\frac{2}{36}$	0	0
6	0	0	$\frac{1}{36}$	$\frac{2}{36}$	$\frac{2}{36}$	0
7	0	0	0	$\frac{2}{36}$	$\frac{2}{36}$	$\frac{2}{36}$
8	0	0	0	$\frac{1}{36}$	$\frac{2}{36}$	$\frac{2}{36}$
9	0	0	0	0	$\frac{2}{36}$	$\frac{2}{36}$
10	0	0	0	0	$\frac{1}{36}$	$\frac{2}{36}$
11	0	0	0	0	0	$\frac{2}{36}$
12	0	0	0	0	0	$\frac{1}{36}$

Figure 11   Joint probability function $p(x, y)$

The Joint Distribution of *Discrete* Random Variables

If X and Y are *discrete* random variables, then their *joint probability function* $p(x, y)$ is given by

$$p(x, y) = P(X = x, Y = y)$$

Here's an example. Toss two dice, D_1 and D_2.

Let X be the maximum of the two faces and let Y be the sum of the two faces.

Then X and Y have joint probability function $p(x, y)$ where

$p(1, 2) = P(X = 1, Y = 2)$

$\qquad = P(\text{max is } 1, \text{sum is } 2) = P(D_1 = 1, D_2 = 1) = \dfrac{1}{36}$

$p(1, 3) = P(X = 1, Y = 2) = P(\text{max is } 1, \text{sum is } 3) = 0$

$p(3, 5) = P(X = 3, Y = 5)$

$\qquad = P(\text{max is } 3, \text{sum is } 5) = P(D_1 = 3, D_2 = 2 \text{ or vice versa}) = \dfrac{2}{36}$

etc.

The complete probability function is given in Fig. 11.

Problems for Section 5-1

1. Let X and Y have joint density

$$f(x, y) = Ke^{-(x+y)} \quad \text{for } x \geq 0, y \geq x$$

   (a) Find K to make the density legitimate.
   (b) Find $P(Y < 2X)$.
   (c) Find the probability that the maximum of X and Y is ≤ 4.

2. Let X and Y have joint density

$$f(x, y) = \frac{6}{7}x \quad \text{for } 1 \leq x + y \leq 2, x \geq 0, y \geq 0$$

   Express with integrals but stop before the actual computation (just put in the limits).
   (a) $P(Y > X^2)$
   (b) $P(\text{max} > 1)$ (max means the maximum of X and Y)

3. Let X and Y have joint density

$$f(x, y) = \begin{cases} 3/2 & \text{in region A} \\ 1/2 & \text{in region B} \end{cases} \quad \text{(see the diagram)}$$

   Find
   (a) $P(X < 1/2, Y > 1/2)$
   (b) $P(X < 1/2 | Y > 1/2)$
   (c) $P(\text{min} < 2/3)$ (min means the minimum of X and Y)

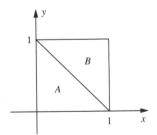

Figure P.3

4. Let X and Y have a joint uniform distribution in the region in the diagram (Fig. P.4, next page).
   (a) Find $P(Y/X < -3/2)$. (Be careful with the graph of $y/x < -3/2$.)
   (b) Find the joint density of X and Y.

5. Let X be uniformly distributed in [0,2] and let Y have an exponential distribution with $\lambda = 1$. If X and Y are independent, find $P(Y - X \geq 3)$.

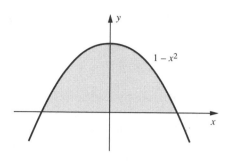

Figure P.4

6. Let X and Y be independent and each uniformly distributed on $[-1,1]$. Find
   (a) $P(XY \leq 1/4)$
   (b) $P(Y/X \leq 1/2)$
   (c) $P(|X| + |Y| \leq 1)$
   (d) the joint density $f(x,y)$

7. Mary will arrive at random between 3:15 and 3:45 and John will arrive at random between 3:00 and 4:00 (independently of Mary). Each agrees to wait up to 5 minutes for the other before leaving.
   (a) Find the probability that they meet.
   (b) Find the probability that John arrives first.

8. Toss two dice, D_1 and D_2. Let X be the face value of D_2, and let Y be the maximum of D_1 and D_2.
   (a) Make a table showing the joint probability function of X and Y.
   (b) Let $F(x,y) = P(X \leq x, Y \leq y)$, the joint distribution function. Find $F(2,5)$.

9. Suppose $X_1, \ldots, X_4$ are independent and identically distributed (iid). Find their joint density $f(x_1, x_2, x_3, x_4)$ if
   (a) each is exponential with $\lambda = 1$
   (b) each has density $f(x) = \frac{1}{4}x$ for $1 \leq x \leq 3$
   (c) each has density g

SECTION 5-2   MARGINAL DENSITIES

Joint Versus Marginal Densities

If X and Y have a joint density $f(x,y)$, then their individual densities $f_X(x)$ and $f_Y(y)$ are sometimes called *marginal densities*.

The probability of an event involving just X alone can be found by (double) integrating the joint density *or* by (single) integrating its marginal density.

The only way to find the probability of an event involving X *and* Y is to integrate the joint density.

Going from the Marginals to the Joint

We showed in the last section that *if X and Y are independent,* then the joint density can be found from the marginals:

(1) $$f(x, y) = f_X(x) f_Y(y)$$

If X and Y are not independent, then the joint density is not determined by the marginals; in this case you can't get the joint from the marginals.

Going from the Joint to the Marginals

Let X and Y have joint density $f(x, y)$. Here's how to get the marginals.

> (2) $$f_X(x) = \int_{y=-\infty}^{\infty} f(x, y)\, dy$$
> (integrate out the y, the *other* variable)
>
> (3) $$f_Y(y) = \int_{y=-\infty}^{\infty} f(x, y)\, dx$$
> (integrate out the x, the *other* variable)

And here's why. We have

$$f_X(x)\, dx = P(X \approx x)$$
$$= \text{probability that } (X, Y) \text{ is in the thin strip in Fig. 1}$$

The strip can be divided into little boxes (Fig. 2), so the strip prob is a sum of box probs. A typical box prob is $f(x, y)\, dx\, dy$. So

$$f_X(x)\, dx = \text{sum of box probs from Fig. 2}$$
$$= \int_{y=-\infty}^{\infty} f(x, y)\, dx\, dy$$

Cancel the dx's to get (2), QED.

Example 1

Pick point (X, Y) at random from the region in Fig. 3. I'll find the marginal densities of X and Y.

This particular universe has area 1, so

$$f(x, y) = \frac{1}{\text{total area}} = 1 \quad \text{for } (x, y) \text{ in the universe}$$

To find $f_X(x)$, *integrate up a y-line* located at position x (Fig. 4).

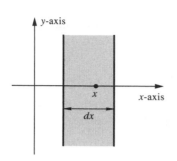

Figure 1

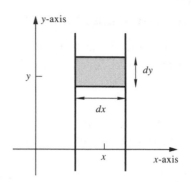

Figure 2   Box prob is $f(x,y)dx\,dy$

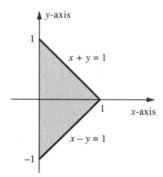

Figure 3

For $0 \le x \le 1$,

$$f_X(x) = \int_{y=-\infty}^{\infty} f(x,y)\,dy = \int_{y=x-1}^{y=1-x} 1\,dy = 2 - 2x$$

To find $f_Y(y)$, *integrate across an x-line* located at height y. But this time the integration depends on where y is, so there are cases (Fig. 5).

Case 1. $0 \le y \le 1$

Integrate across the *upper* line in Fig. 5:

$$f_Y(y) = \int_{x=-\infty}^{\infty} f(x,y)\,dx = \int_{x=0}^{x=1-y} 1\,dx = 1 - y$$

Case 2. $-1 \le y \le 0$

Integrate across the *lower* line in Fig. 5:

$$f_Y(y) = \int_{x=0}^{1+y} 1\,dx = 1 + y$$

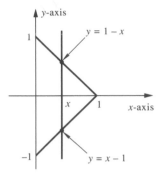

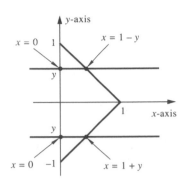

Figure 4   **Figure 5**

So all in all,

$$f_X(x) = 2 - 2x \quad \text{for } 0 \leq x \leq 1 \quad \text{(Fig. 6)}$$

and

$$f_Y(y) = \begin{cases} 1 + y & \text{if } -1 \leq y \leq 0 \\ \\ 1 - y & \text{if } 0 \leq y \leq 1 \end{cases} \quad \text{(Fig. 7)}$$

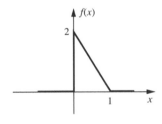

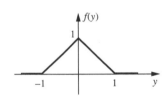

Figure 6   **Figure 7**

Warning

Look at the joint universe of X and Y in Figs. 3 and 4.
If you look at the upper and lower boundaries, you see $x - 1 \leq y \leq 1 - x$.
If you look at the extreme upper and lower values, you see $-1 \leq y \leq 1$.
When you integrate out the y to find the marginal density of X, you should
use $x - 1 \leq y \leq 1 - x$ as the limits of integration.
On the other hand, the Y universe (to go with the marginal density of Y) is
$-1 \leq y \leq 1$. The universes for X and Y individually always comes from the
extreme values of x and y in the joint universe.

Special Case of a Separable Joint Density on a Rectangle

There's a special case where you can get the marginals almost by inspection.

Suppose X and Y have joint density

(4) $$f(x,y) = \frac{1}{2}x^3 y \quad \text{for } 0 \leq x \leq 2, 0 \leq y \leq 1$$

and we want the marginals $f(x)$ and $f(y)$.

The joint density in (4) is special because it consists of a function of x (namely, $\frac{1}{2}x^3$) times a function of y (namely, y).

The key idea is that we'll be able to prove that X and Y are independent. Then, by (1),

$$f(x)f(y) = \frac{1}{2}x^3 y$$

Since $f(x)$ contains only x's and $f(y)$ contains only y's, the only way this can happen is for

(5) $$f(x) = K_1 x^3 \quad \text{for } 0 \leq x \leq 2$$

(6) $$f(y) = K_2 y \quad \text{for } 0 \leq y \leq 1$$

where

(7) $$K_1 K_2 = \frac{1}{2}$$

So, as soon as we decide how to split up the 1/2 into factors K_1 and K_2, we'll have the marginals.
To find K_1, make $K_1 x^3$ a legal density on [0,2]:

$$\int_0^2 K_1 x^3 \, dx = 1$$

$$K_1 = \frac{1}{\int_0^2 x^3 \, dx} = \frac{1}{4}$$

Then by (7),

$$K_2 = \frac{1/2}{K_1} = 2$$

This value of K_2 will automatically make $K_2 y$ legitimate on [0,1] so that

$$f(x) = \frac{1}{4}x^3 \quad \text{for } 0 \leq x \leq 2 \qquad \text{and} \qquad f(y) = 2y \quad \text{for } 0 \leq y \leq 1$$

Here's the general result:

> Suppose the joint density of X and Y has the form
>
> (8) $\quad f(x,y) = \text{constant} \times x\text{-factor} \times y\text{-factor}$
> for the rectangle (finite or infinite) $a \leq x \leq b$, $c \leq y \leq d$
>
> Then X and Y are independent.
> And the x-factor is $f(x)$ and the y-factor is $f(y)$, provided that you juggle the constants (as illustrated above) so that $f(x)$ and $f(y)$ are legal densities.
> Furthermore, if the joint density does *not* have the form in (8) (either the $f(x,y)$ formula doesn't consist of separate x and y factors or the universe is not a rectangle), then X and Y are not independent.

Here's the proof that X and Y with the joint density in (4) *are* independent, which justifies the method. I'll show that knowing X doesn't change the odds on Y.

Look at

(9) $\quad P(Y \approx y | X \approx x) = \dfrac{P(Y \approx y \text{ and } X \approx x)}{P(X \approx x)} = \dfrac{f(x,y)\,dx\,dy}{f(x)\,dx} = \dfrac{\frac{1}{2}x^3 y\,dy}{f(x)}$

Get the marginal $f(x)$ from $f(x,y)$ by integrating out the y:

$$f(x) = \int_0^1 \frac{1}{2}x^3 y\,dy = \frac{1}{2}x^3 \int_0^1 y\,dy = \frac{1}{4}x^3$$

Plug into (9) to get

$$P(Y \approx y | X \approx x) = \frac{\frac{1}{2}x^3 y\,dy}{\frac{1}{4}x^3} = 2y\,dy$$

So $P(Y \approx y | X \approx x)$ has no x's in it; that is, its value is independent of x. So Y and X are independent.

Warning

1. A non-rectangular universe, regardless of the $f(x,y)$ formula, automatically makes X and Y *not* independent.

2. A non-separable formula for $f(x,y)$, such as $3x+5y$, regardless of the universe, automatically makes X and Y *not* independent.

Joint Uniform versus Marginal Uniforms

Here's a result from the preceding section.

> If X and Y are independent and each is uniformly distributed in an interval, then their joint distribution is uniform on a (finite) rectangle.

Now we have the converse.

(10)

> Suppose X and Y have a joint uniform distribution on the (finite) rectangle $a \leq x \leq b, c \leq y \leq d$.
> Then
>
> X and Y are independent.
> X is uniform on $[a, b]$.
> Y is uniform on $[c, d]$.

This holds intuitively because picking a point (X, Y) at random in the rectangle is the same as picking X at random in $[a, b]$ and independently picking Y at random in $[c, d]$.

It follows formally since a uniform joint density on a rectangle has the form in (8); in particular, $f(x, y)$ is a constant. So X and Y are independent. And $f(x)$ and $f(y)$, the x-factor and y-factor, are themselves constant, which makes them uniform.

Warning

If X and Y have a joint uniform distribution but in a *non*-rectangle, then X and Y are *not* independent and are *not* individually uniform.

For example, if X and Y are uniform on the circular region in Fig. 8, then X and Y are *not* independent (knowing one of them changes the odds on the other) and are *not* individually uniform (each is more likely to be near 0 than near 3 or -3).

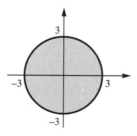

Figure 8

Discrete Marginals

Suppose X and Y have the joint probability function $p(x, y)$ in Fig. 9. Then X takes on values 2, 3, 7, and Y takes on values 1, 3, π, 6, and

$$P(X = 2, Y = 1) = p(2, 1) = \frac{1}{20}$$

$$P(X = 2, Y = 3) = p(2, 3) = \frac{3}{20}$$

etc.

Let $p_X(x)$ be the marginal probability function of X. Then

$$\begin{aligned}
p_X(2) &= P(X = 2) \\
&= P(X = 2, Y = 1) + P(X = 2, Y = 3) + P(X = 2, Y = \pi) \\
&\quad + P(X = 2, Y = 6) \\
&= \text{sum down the column where } x \text{ is 2} \\
&= \frac{9}{20}
\end{aligned}$$

$y \backslash x$	2	3	7	$p_Y(y)$
1	$\frac{1}{20}$	$\frac{2}{20}$	0	$\frac{3}{20}$
3	$\frac{3}{20}$	0	$\frac{1}{20}$	$\frac{4}{20}$
π	$\frac{5}{20}$	$\frac{1}{20}$	$\frac{2}{20}$	$\frac{8}{20}$
6	0	$\frac{4}{20}$	$\frac{1}{20}$	$\frac{5}{20}$
$p_X(x)$	$\frac{9}{20}$	$\frac{7}{20}$	$\frac{4}{20}$	

Figure 9   Joint probability function $p(x, y)$ marginals $p_X(x)$ and $p_Y(y)$

Similarly,

$$p_X(3) = \text{sum down the column where } x \text{ is 3} = \frac{7}{20}$$

$$p_X(7) = \text{sum down the column where } x \text{ is 7} = \frac{4}{20}$$

In general, to find the marginal probability functions, sum out the other variable:

$$p_X(x) = \sum_y p(x, y) \qquad \text{and} \qquad p_Y(y) = \sum_x p(x, y)$$

The values of the marginal $p_X(x)$ appear in the lower margin in Fig. 9, and the values of the marginal $p_Y(y)$ appear in the right-hand margin.

Problems for Section 5-2

1. If X and Y have a joint uniform distribution in the indicated universe, find the marginals.

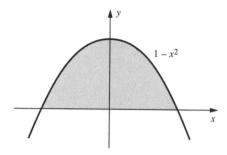

Figure P.1

2. Let X and Y have the joint density

$$f(x, y) = \frac{6}{7}x \quad \text{for } x \geq 0, y \geq 0, 1 \leq x + y \leq 2$$

Find (a) $f(x)$ and (b) $f(y)$.

3. Let X and Y have the joint density

$$f(x, y) = \begin{cases} \dfrac{3}{2} & \text{in region I} \quad \text{(see the diagram)} \\[2mm] \dfrac{1}{2} & \text{in region II} \end{cases}$$

Find $f_X(x)$.

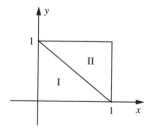

Figure P.3

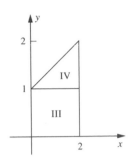

Figure P.4

4. Let X and Y have the joint density

$$f(x,y) = \begin{cases} \dfrac{1}{4} & \text{in triangle IV} \quad \text{(see the diagram)} \\[2mm] \dfrac{x+y}{4} & \text{in rectangle III} \end{cases}$$

Find (a) $f_X(x)$ and (b) $f_Y(y)$.

5. If X and Y have the following joint distribution, are they independent? If so, find the marginals $f(x)$ and $f(y)$.
   (a) $f(x,y) = \frac{6}{7}x$ for $x \geq 0, y \geq 0, 1 \leq x + y \leq 2$
   (b) $f(x,y) = xye^{-y^2/4}$ for $0 \leq x \leq 1, y \geq 0$
   (c) X and Y jointly uniform on the elliptical region $x^2 + 3y^2 \leq 4$
   (d) $f(x,y) = 2(x+y)$ for $1 \leq x \leq 2, 0 \leq y \leq 1$
   (e) X and Y jointly uniform in the region $1 \leq x \leq 2, 0 \leq y \leq 3$

6. If X and Y have joint density

$$f(x,y) = \frac{1}{2}e^{-2y} \text{ for } 0 \leq x \leq 4, y \geq 0$$

find the marginals

(a) by inspection   (b) with integration (for practice)

7. Toss two dice, D_1 and D_2. Let X be the min and Y the max. Make a table showing the joint prob function $p(x,y)$ and the marginal prob functions $p_X(x)$ and $p_Y(y)$. Are X and Y independent?

8. Pick a point (X,Y) at random in a circle with center at the origin and radius 2. Find $f_X(x)$.

9. True or False?
   (a) If the X universe is $0 \leq x \leq 1$ and the Y universe is $0 \leq y \leq 1$, then X and Y jointly have the rectangular universe $0 \leq x \leq 1$, $0 \leq y \leq 1$.
   (b) If X and Y are each uniform, then X, Y are jointly uniform.

(c) If X, Y are jointly uniform, then X and Y are each uniform.

(d) If $f(x, y) = \sin(x + y)$, then X and Y can't be independent.

(e) If $f(x, y) = Kx^3 y^2$, then X and Y are independent.

(f) If X, Y are jointly uniform on $0 \le x \le 2, 5 \le y \le 6$, then X and Y are independent.

(g) If X, Y are jointly uniform on a circular universe, then X and Y are independent.

SECTION 5-3 FUNCTIONS OF SEVERAL RANDOM VARIABLES

Section 4.5 showed how to find the distribution of a function Y of X (e.g., $Y = X^2$) given the distribution of X itself. There were two methods, one that found the distribution function first and another that found the density first.

Now we'll find the distribution function $F(z)$ of a function Z of X and Y (e.g., $Z = X^2 Y^3$) given the joint density of X and Y. (We won't have a method for jumping directly to the density of Z.)

Distribution of a Function of X and Y

To illustrate the idea, suppose X and Y are independent and each has an exponential distribution with parameter λ. Consider the new random variable

$$Z = Y - X$$

We'll find the distribution of Z, first its distribution function and then its density function.

To begin, we need the joint density of X and Y. Since X and Y are independent and

$$f(x) = \lambda e^{-\lambda x} \quad \text{for } x \ge 0$$

$$f(y) = \lambda e^{-\lambda} \quad \text{for } y \ge 0$$

we have

$$f(x, y) = f(x)f(y) = \lambda^2 e^{-\lambda(x+y)} \quad \text{for } x \ge 0, y \ge 0 \quad \text{(Fig. 1)}$$

Then

$$F_Z(z) = P(Z \le z) = P(Y - X \le z) = \int_{\text{fav}} f(x, y) \, dA$$

The favorable region is the set of points (x, y) in the universe such that

$$y - x \le z$$

where z is thought of as a constant. But the favorable region looks different for different z's, so we need cases.

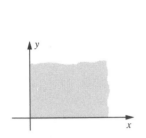

Figure 1   Universe

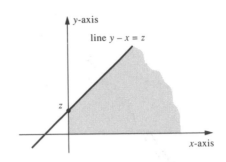

Figure 2 $z \geq 0$

Case 1. $z \geq 0$ (Fig. 2)

$$F_Z(z) = \int_{\text{fav}} f(x,y)\, dA = \int_{x=0}^{\infty} \int_{y=0}^{x+z} \lambda^2 e^{-\lambda(x+y)}\, dy\, dx$$

$$\text{inner} = -\lambda e^{-\lambda(x+y)}\Big|_{y=0}^{x+z} = -\lambda e^{-\lambda(2x+z)} + \lambda e^{-\lambda x}$$

$$\text{outer} = \left(\frac{1}{2}e^{-\lambda(2x+z)} - e^{-\lambda x}\right)\Big|_{x=0}^{\infty} = 1 - \frac{1}{2}e^{-\lambda z}$$

Case 2. $z \leq 0$ (Fig. 3)

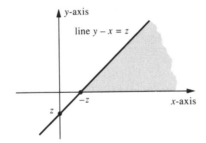

Figure 3 $z \leq 0$

$$F_Z(z) = \int_{\text{fav}} f(x,y)\, dA = \int_{x=-z}^{\infty} \int_{y=0}^{x+z} \lambda^2 e^{-\lambda x}\, dy\, dx$$

$$\text{inner} = \text{same as case 1}$$

$$\text{outer} = \left(\frac{1}{2}e^{-\lambda(2x+z)} - e^{-\lambda x}\right)\Big|_{x=-z}^{\infty} = \frac{1}{2}e^{-\lambda z}$$

So

$$F_Z(z) = \begin{cases} \dfrac{1}{2}e^{\lambda z} & \text{if } z \leq 0 \\[2mm] 1 - \dfrac{1}{2}e^{-\lambda z} & \text{if } z \geq 0 \quad \text{(Fig. 4)} \end{cases}$$

and

$$f_Z(z) = F_Z'(z) = \begin{cases} \dfrac{1}{2}\lambda e^{\lambda z} & \text{if } z \le 0 \\[2ex] \dfrac{1}{2}\lambda e^{-\lambda z} & \text{if } z \ge 0 \quad \text{(Fig. 5)} \end{cases}$$

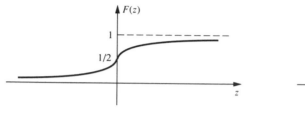

Figure 4

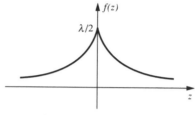

Figure 5

Distribution of the Maximum of X and Y

Pick point (X, Y) at random in the unit circle, centered at the origin. I'll find the distribution function of the maximum of X and Y (but leave the integrals unevaluated).

We have $-1/\sqrt{2} \le \max \le 1$, so

$$F_{\max}(z) = \begin{cases} 0 & \text{if } z \le -1/\sqrt{2} \\[1ex] 1 & \text{if } z \ge 1 \end{cases}$$

For $-1/\sqrt{2} \le z \le 1$,

$$F_{\max}(z) = P(\max \le z) = \frac{\text{fav area}}{\text{total area}}$$

The total area is π. Now let's get the favorable area.

The graph of $\max(x, y) \le z$ is shown in Fig. 6 (for positive z), but we need cases to accommodate the different ways it can overlap the universe.

Case 1. $-1/\sqrt{2} \le z \le 0$ (Fig. 7)

$$\text{fav area} = \int_{-\sqrt{1-z^2}}^{z} \left(z + \sqrt{1 - x^2} \right) dx$$

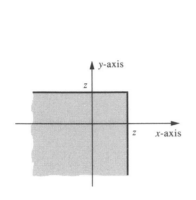

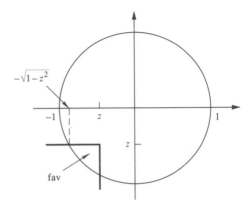

Figure 6 $\max(x, y) \leq z$

Figure 7 $-1/\sqrt{2} \leq z \leq 0$

Case 2. $0 \leq z \leq 1/\sqrt{2}$ (Fig. 8)

$$\text{fav area} = \text{quarter circle} + \text{square} + 2\,\text{rectangles} + 2\,\text{blips}$$

$$= \frac{1}{4}\pi + z^2 + 2z\sqrt{1 - z^2} + 2\int_{-1}^{-\sqrt{1-z^2}} \sqrt{1 - x^2}\,dx$$

Case 3. $1/\sqrt{2} \leq z \leq 1$ (Fig. 9)

$$\text{fav} = \text{total} - 4\,\text{blips} = \pi - 4\int_{z}^{1} \sqrt{1 - x^2}\,dx$$

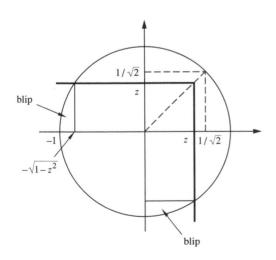

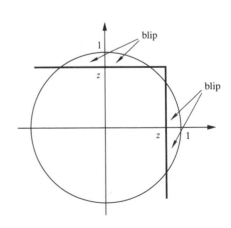

Figure 8 $0 \leq z \leq 1/\sqrt{2}$

Figure 9 $1/\sqrt{2} \leq z \leq 1$

Problems for Section 5-3

1. Suppose

   X and Y are independent,
   X is uniform on [0,1],
   Y is exponential with $\lambda = 1$.

   Let $Z = X + Y$. Find the density of Z.

2. Let I have density

$$f_I(x) = 6x(1 - x) \quad \text{for } 0 \le x \le 1$$

   Let R have density

$$f_R(y) = 2y \quad \text{for } 0 \le y \le 1$$

   If I and R are independent and $Z = I^2 R$, find the density of Z.

3. Let X and Y be independent, each uniform on [0,1].
   (a) Find the density of the maximum of X and Y.
   (b) Find the density of XY.

4. Pick a point (X, Y) at random in the indicated triangle.   Let $Z = |Y - X|$. Find the distribution function of Z.

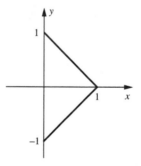

Figure P.4

5. Let X and Y be independent, each with a unit normal distribution. Find the density and distribution function of

$$T = \frac{1}{\sqrt{X^2 + Y^2}}$$

6. Let X and Y be independent with densities f_X and f_Y, respectively. If $Z = Y/X$, find the distribution function F_Z in terms of f_X and f_Y.

7. Let X and Y have joint density

$$f(x, y) = 2e^{-(x+y)} \quad \text{for } y \geq x \geq 0$$

Find the density of (a) the maximum and (b) the minimum of X and Y.

Review Problems for Chapters 4 and 5

1. Suppose X never takes on values between 2 and 8 (inclusive). What can you conclude about $f_X(x)$ and $F_X(x)$?

2. The median of a continuous random variable X is the number m such that $P(X \leq m) = 1/2$ and $P(X \geq m) = 1/2$. Find the median if X is
   (a) uniformly distributed on $[a, b]$
   (b) normal with parameters μ and σ^2
   (c) exponentially distributed with parameter λ

3. If the density of X is

$$f(x) = \begin{cases} x & \text{if } 0 \leq x \leq 1 \\ 2 - x & \text{if } 1 \leq x \leq 2 \end{cases}$$

and

$$Y = \begin{cases} 2X & \text{if } X \leq 1 \\ \dfrac{2}{X^2} & \text{if } X \leq 1 \end{cases}$$

find the density of Y.

4. Pick a point (X, Y) at random from the region where $0 \leq y \leq 1 - x^2$.
   (a) Find $P(X + Y \geq 1)$.
   (b) Find the joint density of X and Y.

5. Out of a large batch of gizmos, a sample of 100 is tested. The whole batch will be rejected if 15% or more of the sample is defective. Find the probability that the batch is rejected if it actually is 20% defective.

6. Pick a point (X, Y) at random in the indicated triangle. Find the joint density $f(x, y)$ and the marginals $f(x)$ and $f(y)$.

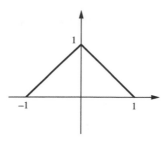

Figure P.6

7. The input X into a system is normal with $\mu = 1, \sigma^2 = 4$. The output of the system is Y, where Y is the function of X given in the diagram. Find the probability function $p(y)$ of Y.

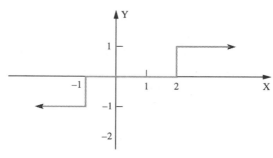

Figure P.7

8. Let X and Y be non-negative, independent and identically distributed with common density g. Find the distribution of $Y - X^2$ (either its distribution function or its density function) in terms of g and integrals.

9. The temperature in a gizmo is a random variable X with density

$$f(x) = 11(1 - x)^{10} \quad \text{for } 0 \le x \le 1$$

The gizmo has a cutoff number so that it turns off automatically when the temperature goes above the cutoff. If it is observed that the gizmo turns off with probability 10^{-22}, find its cutoff number.

10. Let X be uniform on [0,2] and let Y be uniform on [0,3]. If X and Y are independent, find

(a) $P(Y \ge 2X)$ (b) the joint density of X and Y

11. Choose a point at random in a sphere with radius a. Let R be the distance from the point to the center of the sphere. Find the distribution function $F_R(r)$ and the density $f_R(r)$.

12. Let X have distribution function

$$
F(x) = \begin{cases}
0 & \text{if } x \leq 0 \\[2mm]
\dfrac{1}{8} + \dfrac{1}{8}x & \text{if } 0 \leq x \leq 1 \\[2mm]
\dfrac{1}{2} & \text{if } 1 \leq x \leq 2 \\[2mm]
\dfrac{1}{2} + \dfrac{1}{8}x & \text{if } 2 \leq x \leq 4 \\[2mm]
1 & \text{if } x \geq 4
\end{cases}
$$

Find

(a) $P(X \text{ is } 2)$
(b) $P(X \text{ is more than } 2)$
(c) $P(X \text{ is less than } 2)$
(d) $P(X \text{ is } 3)$
(e) $P(X \text{ is more than } 1 \text{ but less than } 3)$
(f) $P(X \text{ is } 2 \text{ or more} \mid X \text{ is more than } 0)$

13. Let X have density

$$
f(x) = \begin{cases}
\dfrac{1}{4} & \text{if } -1 \leq x \leq 1 \\[3mm]
\dfrac{1}{2x^2} & \text{if } x \geq 1
\end{cases}
$$

(a) Check that $f(x)$ is a legal density.
(b) Find $P(0 \leq X \leq 2)$.
(c) Find $F(x)$.

14. Let X be a random variable with distribution function $F(x)$. Express the distribution function of X^3 in terms of $F(x)$.

15. Circuit extraphasors on the average last for 10 round trips. Assume that their lifetimes have an exponential distribution.
(a) Find the prob that after lasting 5 round trips, an extraphasor will last at least another 5 round trips.
(b) Find the prob that you'll have a total of 3 breakdowns in your next 2 trips. (Once a breakdown occurs, the extraphasor is replaced and the trip continues.)
(c) Find the prob that you'll have 1 breakdown in each of your next 2 trips.

16. Suppose X has density Ke^{-x^2-3x}, where K is a constant. Show that X has a normal distribution (complete the square), and find (easily) μ, σ^2, and K.

17. Think of particles arriving and show (without any calculations) that if X and Y are independent and each is exponentially distributed with parameter λ (the same λ), then the minimum of X and Y is also exponentially distributed. And find the parameter.

18. Two light bulbs have respective lifetimes X and Y (measured in months), where X and Y are independent and each is exponential with $\lambda = 1$.

    You use the bulbs to light your workplace. Let Z be the lifetime of your workplace, that is, the length of time your workplace is sufficiently lit. Find $f_Z(z)$ if
    (a) you need the light from both bulbs in order to work
    (b) you can manage with the light from one bulb, so you don't use the second bulb until the first one wears out
    (c) you can manage with one bulb, but you wastefully turn both on and keep them on until both are dead

CHAPTER 6

Jointly Distributed Random Variables Continued

SECTION 6-1 SUMS OF INDEPENDENT RANDOM VARIABLES

In Section 5.3 you saw how to find the distribution function of any combination of X and Y (e.g., XY^3, XY, X/Y, max) given the joint density of X and Y. In this section we'll look at the particular combination $X + Y$ in the special case that X and Y are independent.

First, we'll get the distribution of the sum of some famous random variables; then, we'll get a formula for the distribution of the sum in general.

The Sum of Independent Binomials with a Common p

Suppose

X_1 has a binomial distribution with parameters n_1, p,
X_2 has a binomial distribution with parameters n_2, p (*same p*),
X_1 and X_2 are independent.

Then $X_1 + X_2$ has a binomial distribution with parameters $n_1 + n_2, p$.

Proof

Think of X_1 as the number of successes when a coin is tossed n_1 times where $P(\text{success}) = p$.

Think of X_2 as the number of successes when the *same* coin is tossed n_2 times, independent of the other n_1 tosses. Then $X + Y$ is the number of successes in $n_1 + n_2$ tosses, so $X + Y$ has a binomial distribution with parameters $n_1 + n_2$, p.

The Sum of Independent Poissons

Suppose

X_1 has a Poisson distribution with parameter λ_1,
X_2 has a Poisson distribution with parameter λ_2,
X_1 and X_2 are independent.

Then $X + Y$ has a Poisson distribution with parameter $\lambda_1 + \lambda_2$.

First Proof

Think of X_1 as the number of blue particles arriving in a minute, where the blues arrive independently at the average rate of λ_1 per minute.

Think of X_2 as the number of red particles arriving in a minute, where the reds arrive independently at the average rate of λ_2 per minute.

Since X_1 and X_2 are independent, the blues arrive independently of the reds.

Then $X_1 + X_2$ is the number of arrivals of particles (red and blue) in a minute, where on the average, $\lambda_1 + \lambda_2$ particles arrive per minute.

So $X_1 + X_2$ has a Poisson distribution with parameter $\lambda_1 + \lambda_2$.

Second Proof   (which doesn't depend on any physical model)

We'll begin by finding $P(X_1 + X_2 = 7)$ to illustrate the pattern.

$$
\begin{aligned}
P(X_1 + X_2 = 7) \\
= P(X_1 = 0, X_2 = 7 \text{ or } X_1 = 1, X_2 = 6 \\
\text{ or } X_1 = 2, X_2 = 5 \text{ or } \ldots \text{ or } X_1 = 7, X_2 = 0) \\
= P(X_1 = 0, X_2 = 7) + P(X_1 = 1, X_2 = 6) \\
+ \cdots + P(X_1 = 7, X_2 = 0) \text{ (OR rule)} \\
= P(X_1 = 0)\,P(X_2 = 7) + P(X_1 = 1)\,P(X_2 = 6) \\
+ \cdots + P(X_1 = 7)\,P(X_2 = 0) \\
\text{(AND rule for independent events)}
\end{aligned}
$$

$$= \frac{e^{-\lambda_1}\lambda_1^0}{0!} \frac{e^{-\lambda_2}\lambda_2^7}{7!} + \frac{e^{-\lambda_1}\lambda_1^1}{1!} \frac{e^{-\lambda_2}\lambda_2^6}{6!}$$

$$+ \cdots + \frac{e^{-\lambda_1}\lambda_1^7}{7!} \frac{e^{-\lambda_2}\lambda_2^0}{0!}$$

$$= e^{-(\lambda_1+\lambda_2)} \left[\frac{\lambda_1^0 \lambda_2^7}{0!\,7!} + \frac{\lambda_1^1 \lambda_2^6}{1!\,6!} + \frac{\lambda_1^2 \lambda_2^5}{2!\,5!} + \cdots + \frac{\lambda_1^7 \lambda_2^0}{0!\,7!} \right]$$

To identify the sum in the brackets, remember the formula

$$(x+y)^7 = \binom{7}{0}x^0 y^7 + \binom{7}{1}x^1 y^6 + \cdots + \binom{7}{7}x^7 y^0$$

So

$$\text{sum in brackets} \times 7! = (\lambda_1 + \lambda_2)^7$$

and

$$P(X_1 + X_2 = 7) = e^{-(\lambda_1+\lambda_2)} \frac{(\lambda_1 + \lambda_2)^7}{7!}$$

Similarly, we can show that, in general,

$$P(X_1 + X_2 = k) = e^{-(\lambda_1+\lambda_2)} \frac{(\lambda_1 + \lambda_2)^k}{k!}$$

This is the probability function for a Poisson random variable, so $X_1 + X_2$ has a Poisson distribution with parameter $\lambda_1 + \lambda_2$.

The Sum of Independent Exponentials with a Common λ

Let $X_1, \ldots, X_n$ be independent, each exponentially distributed with parameter λ. Then $X_1 + \cdots + X_n$ has a gamma distribution with parameters n, λ.

Proof

Consider the physical model where particles arrive independently at the average rate of λ per minute.

Then X_1 can be thought of as the waiting time for the first arrival, X_2 can be thought of as the waiting time *from* the first arrival *to* the second arrival, and in general X_i can be thought of as the waiting time between the $(i-1)$st and ith arrivals.

So $X_1 + \cdots + X_n$ is the (total) waiting time for the nth arrival. But that means $X_1 + \cdots + X_n$ has a gamma distribution with parameters n, λ (see Section 4.3).

The Density of the Sum of Two Arbitrary Independent Random Variables

Let X and Y be any independent random variables with densities $f_X(x)$ and $f_Y(y)$, respectively. Here's a formula for the density of $X + Y$:

$$(1) \quad f_{X+Y}(z) = \int_{x=-\infty}^{\infty} f_X(x) f_Y(z-x)\, dx = \int_{y=-\infty}^{\infty} f_X(z-y) f_Y(y)\, dy$$

Each of these integrals is called the *convolution* of f_X and f_Y and is denoted by $f_X * f_Y$.

> The density of the sum of two independent random variables is the convolution of the individual densities.

Here's an informal argument to show why (1) holds. Start with

$$f_{X+Y}(z)\, dz = P(X + Y \approx z) = P(z \leq X + Y \leq z + dz)$$

To find this probability, look at all the ways in which the event can happen as X takes on all its possible values:

$$P(z \leq X + Y \leq z + dz) = \sum_x P(X \approx x \text{ and then } z - x \leq Y \leq z - x + dz)$$

$$= \sum_x P(X \approx x) P(z - x \leq Y \leq z - x + dz)$$
$$\text{(since } X \text{ and } Y \text{ are independent)}$$

$$= \sum_x f_X(x)\, dx f_Y(z-x)\, dz$$

So

$$f_{X+Y}(z)\, dz = \sum_x f_X(x)\, dx f_Y(z-x)\, dz$$

Cancel the dz's and use an integral to do the adding, and you'll get the first convolution formula in (1).

If you reverse the roles of X and Y and use

$$P(z \leq X + Y \leq z + dz) = P(Y \approx y \text{ and then } z - y \leq X \leq z - y + dz)$$

you end up with the second convolution integral in (1), QED.

The catch is that convolutions can be tricky to compute. If you've seen them before, you know that it's not usually a simple matter of plugging into the formula in (1). So we'll use the formula only once, to prove that the sum of independent normals is normal.

The Sum of Independent Normals

Suppose

X_1 is normal with mean μ_1 and variance σ_1^2,

X_2 is normal with mean μ_2 and variance σ_2^2,

X_1 and X_2 are independent.

Then $X_1 + X_2$ is normal with mean $\mu_1 + \mu_2$ and variance $\sigma_1^2 + \sigma_2^2$.

Proof

When we get to continuous expectation and variance in Chapter 7, it will be easy to show that for any X and Y,

$$E(X + Y) = E(X) + E(Y)$$

and if X and Y are independent, then

$$\text{Var}(X + Y) = \text{Var}(X) + \text{Var}(Y)$$

Once we do that, it follows here that $X_1 + X_2$ has mean $\mu_1 + \mu_2$ and variance $\sigma_1^2 + \sigma_2^2$. The hard part is to show that $X_1 + X_2$ is *normal*.

First, we'll prove it in the special case that $\mu_1 = \mu_2 = 0$. Let $Z = X_1 + X_2$. Using convolution, we have

$$f(z) = \int_{x=-\infty}^{\infty} f_1(z - x) f_2(x)\, dx$$

$$= \int_{x=-\infty}^{\infty} \frac{1}{\sigma_1 \sqrt{2\pi}}\, e^{-(z-x)^2/2\sigma_1^2}\, \frac{1}{\sigma_2 \sqrt{2\pi}}\, e^{-x^2/2\sigma_2^2}\, dx$$

After a lot of algebra and square completing, we get

$$f(z) = \frac{1}{\sigma_1 \sigma_2\, 2\pi}\, e^{-z^2/2(\sigma_1^2 + \sigma_2^2)} \int_{x=-\infty}^{\infty} e^{-b(x-a)^2}\, dx$$

where

$$a = \frac{z\sigma_1^2}{\sigma_1^2 + \sigma_1^2} \qquad \text{and} \qquad b = \frac{\sigma_1^2 + \sigma_2^2}{2\sigma_1^2 \sigma_2^2}$$

The integral can be evaluated using the substitution $t = x - a$ and integral tables. Its value is

$$\sqrt{2\pi} \; \frac{\sigma_1 \, \sigma_2}{\sqrt{\sigma_1^2 + \sigma_2^2}}$$

so that eventually we get

$$f(z) = \frac{1}{\sqrt{\sigma_1^2 + \sigma_2^2} \sqrt{2\pi}} \; e^{-z^2/2(\sigma_1^2 + \sigma_2^2)}$$

This is the normal density with 0 in the μ spot and $\sigma_1^2 + \sigma_2^2$ in the σ^2 spot. So $X_1 + X_2$ is normal with mean 0 and variance $\sigma_1^2 + \sigma_2^2$.

Now consider the general case where

X_1 has mean μ_1 and variance σ_1^2,
X_2 has mean μ_2 and variance σ_2^2.

We'll use the fact that $X_1 + X_2$ can be written as

$$(X_1 - \mu_1) + (X_2 - \mu_2) + (\mu_1 + \mu_2)$$

Remember from Section 4.4, (2), that if X is normal with parameters μ, σ^2, then $X + b$ is normal with parameters $\mu + b, \sigma^2$. So

$X_1 - \mu_1$ is normal with mean 0, variance σ_1^2,
$X_2 - \mu_2$ is normal with mean 0, variance σ_2^2.

Also $X_1 - \mu_1$ and $X_2 - \mu_2$ are independent because X_1 and X_2 are independent. So by the special case already established,

$$(X - \mu_1) + (X_2 - \mu_2) \text{ is normal with mean 0 and variance } \sigma_1^2 + \sigma_2^2$$

Tack on $\mu_1 + \mu_2$ and use the $X + b$ rule to see that

$$(X - \mu_1) + (X_2 - \mu_2) + (\mu_1 + \mu_2) \text{ is normal}$$
$$\text{with mean } 0 + (\mu_1 + \mu_2) \text{ and variance } \sigma_1^2 + \sigma_2^2$$

But this says that $X_1 + X_2$ is normal with mean $\mu_1 + \mu_2$ and variance $\sigma_1^2 + \sigma_2^2$, QED.

Problems for Section 6-1

1. Suppose X and Y are independent with the indicated distributions. Find the distribution of $X + Y$ if it can be done by inspection.
   (a) X is binomial with $n = 10$, $p = 1/3$; Y is binomial with $n = 20$, $p = 1/3$.
   (b) X is binomial with $n = 10$, $p = 1/3$; Y is binomial with $n = 10$, $p = 1/2$.
   (c) X is Poisson with $\lambda = 3$; Y is Poisson with $\lambda = 3$.
   (d) X is Poisson with $\lambda = 3$; Y is Poisson with $\lambda = 4$.
   (e) X is exponential with $\lambda = 3$; Y is exponential with $\lambda = 3$.
   (f) X is exponential with $\lambda = 2$; Y is exponential with $\lambda = 3$.
   (g) X is normal with $\mu = 2$, $\sigma^2 = 5$; Y is normal with $\mu = 3$, $\sigma^2 = 6$.

2. If $X_1, \ldots, X_{10}$ are independent and each has a geometric distribution with parameter p (Section 2.2), find the distribution of $X_1 + \cdots + X_{10}$ by thinking about a physical model.

3. Suppose

   X has a gamma distribution with parameters $n = 5$, $\lambda = 2$,
   Y has a gamma distribution with parameters $n = 7$, $\lambda = 2$,
   Z has a gamma distribution with parameters $n = 8$, $\lambda = 2$.

   If X, Y, and Z are independent, find the distribution of $X + Y + Z$ by thinking about a physical model.

4. A machine has one component and is packaged with 9 spare components, all identical. When the original and the 9 spares are eventually used up, the machine dies.

   If the lifetimes of the components are iid, all exponential with parameter λ, find the distribution of the *machine's* lifetime.

SECTION 6-2   ORDER STATISTICS

Section 5.3 showed how to find the distribution of a function of X and Y given the distributions of X and Y. In Section 6.1 we looked in detail at the special function $X_1 + \cdots + X_n$, where the X_i's are independent. In this section we'll look at some more special functions.

Order Statistics

I'll illustrate the new idea and its significance with an example.

A machine has 10 components with respective lifetimes $X_1, \ldots, X_{10}$ (measured in hours). For the machine to work properly, at least 3 of its 10 components must work (the machine can manage with 7 dead components but not with 8).

Let's figure out the connection between $X_1, \ldots, X_{10}$ and the lifetime of the machine itself.

Suppose the experiment is performed (i.e., run all the components simultaneously until they all die) and the outcome is the set of component lifetimes listed in Fig. 1. Then the machine kept working even though X_2 died immediately, it kept working when X_{10} died a half hour later, and it kept working after the deaths of $X_5, X_6, X_9, X_1,$ and X_3. But when X_4 died (after 9.2 hours), there were only 2 components left so the machine died. So for this outcome, the machine's lifetime is 9.2.

For the outcome in Fig. 2, the machine quit when X_{10} died, leaving only X_6 and X_8 alive. So for this outcome, the machine's lifetime is 21.

Overall, the machine's lifetime is the 3rd largest of $X_1, \ldots, X_{10}$.

In general, if a machine needs k of its n components, then the machine's lifetime is the kth largest of the n component lifetimes.

$X_1 = 4.1$	$X_1 = 5$
$X_2 = 0$	$X_2 = 1$
$X_3 = 8$	$X_3 = 3.2$
$X_4 = 9.2$	$X_4 = \pi$
$X_5 = 2$	$X_5 = 7.1$
$X_6 = 2$	$X_6 = 32$
$X_7 = 20$	$X_7 = 3$
$X_8 = 10$	$X_8 = 24$
$X_9 = 3$	$X_9 = 2$
$X_{10} = .5$	$X_{10} = 21$

Figure 1   Outcome 1   **Figure 2**   Outcome 2

Given random variables $X_1, \ldots, X_n$, the new random variables

maximum of $X_1, \ldots, X_n$
2nd largest of $X_1, \ldots, X_n$
3rd largest of $X_1, \ldots, X_n$

$\vdots$

$(n-1)$st largest of $X_1, \ldots, X_n$
minimum of $X_1, \ldots, X_n$

are called *order statistics*.

I'll show you how to find the distribution function and density function of the order statistics, provided that $X_1, \ldots, X_n$ are *iid*, meaning *independent* and *identically distributed* (so that the machine's components are identical and work independently of one another).

The Distribution Function of the *k*th Largest of iid Random Variables

Let $X_1, \ldots, X_{10}$ be iid with common distribution function $F(x)$ and common density $f(x)$. Here's how to find the distribution function $F_{3rd}(x)$ of the 3rd largest of $X_1, \ldots, X_{10}$.

Start with

$$F_{3rd}(x) = P(\text{3rd largest of the } X_i\text{'s is } \leq x)$$

$$= P(\text{at most 2 are } \geq x) \quad \text{(Fig. 3)}$$

$$= P\left(\begin{matrix} \text{all are} \leq x \\ \text{none is} \geq x \end{matrix} \quad \text{or} \quad \begin{matrix} 9 \text{ are } \leq x \\ 1 \text{ is } \geq x \end{matrix} \quad \text{or} \quad \begin{matrix} 8 \text{ are } \leq x \\ 2 \text{ are } \geq x \end{matrix}\right)$$

$$= P\left(\begin{matrix} \text{all are} \leq x \\ \text{none is} \geq x \end{matrix}\right) + P\left(\begin{matrix} 9 \text{ are } \leq x \\ 1 \text{ is } \geq x \end{matrix}\right) + P\left(\begin{matrix} 9 \text{ are } \leq x \\ 2 \text{ are } \geq x \end{matrix}\right)$$

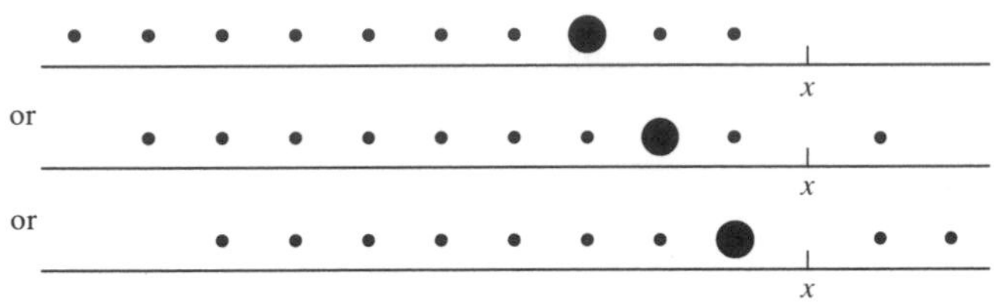

Figure 3   3rd largest is $\leq x$

Because the X_i's are iid, we can think of them as the results of 10 Bernoulli trials (i.e., independent trials of the same experiment, like coin tosses) where in any one trial,

$$P(\text{success}) = P(X_i \leq x) = F(x)$$
$$P(\text{failure}) = P(X_i \geq x) = 1 - F(x)$$

So

(1) $F_{3rd}(x) = P(10S) + P(9S) + P(8S)$

$$= [F(x)]^{10} + \binom{10}{9} [F(x)]^9 [1 - F(x)] + \binom{10}{8} [F(x)]^8 [1 - F(x)]^2$$

You can get the distribution function of the *k*th largest for any *k* in the same way.

The Density Function of the *k*th Largest of iid Random Variables

Let $X_1, \ldots, X_{10}$ be iid with common distribution function $F(x)$ and common density $f(x)$. Here's how to find the density function $f_{3rd}(x)$ of the 3rd largest of $X_1, \ldots, X_{10}$.

Start with

$$
\text{(2)} \quad
\begin{aligned}
f_{3\text{rd}}(x)\, dx &= P(3\text{rd largest is} \approx x) \\
&= P(\text{one of the } X_i\text{'s is} \approx x,\ \text{two are} \geq x,\ \text{seven are} \leq x)
\end{aligned}
$$

Because the X_i's are iid, you can think of them as the results of 10 tosses of a 3-sided die, where

$$
\begin{aligned}
P(\text{die comes up} \approx x) &= P(X_i \approx x) = f(x)\, dx \\
P(\text{die comes up} \leq x) &= P(X_i \leq x) = F(x) \\
P(\text{die comes up} \geq x) &= P(X_i \geq x) = 1 - F(x)
\end{aligned}
$$

The result of the 10 tosses has a multinomial distribution, so

$$
\text{(3)} \quad
\begin{aligned}
f_{3\text{rd}}(x)\, dx &= P(3\text{rd largest is} \approx x) \\
&= P(\text{one is} \approx x,\ \text{two are} \geq x,\ \text{seven are} \leq x) \\
&= \frac{10!}{1!\,2!\,7!}\, [f(x)\, dx]^1\, [1 - F(x)]^2\, [F(x)]^7
\end{aligned}
$$

Cancel the dx's to get

$$
\text{(4)} \quad f_{3\text{rd}}(x) = \frac{10!}{1!\,2!\,7!}\, f(x)\, [1 - F(x)]^2\, [F(x)]^7
$$

Another way to get $f_{3\text{rd}}(x)$ is first to find $F_{3\text{rd}}(x)$ (which we did in (1)) and then differentiate, using product and chain rules:

$$
\begin{aligned}
f_{3\text{rd}}(x) &= F'_{3\text{rd}}(x) \\
&= 10F^9 \cdot F' + \binom{10}{9}\, [F^9 \cdot -F' + (1 - F) \cdot 9F^8 \cdot F'] \\
&\quad + \binom{10}{8}\, [F^8 \cdot 2(1 - F) \cdot -F' + (1 - F)^2 \cdot 8F^7 \cdot F']
\end{aligned}
$$

This comes out the same as (4) if you use some algebra and remember that $F'(x) = f(x)$.

Problems for Section 6-2

1. Suppose $X_1, \ldots, X_{17}$ are iid with common density $f(x)$ and common distribution function $F(x)$. Find the distribution function and density function of the

(a) 5th largest

(b) 10th largest

(c) 13th largest

(d) minimum

2. Suppose $X_1, \ldots, X_{32}$ are iid with common density $f(x)$ and common distribution function $F(x)$. Find the distribution function and density function of the
   (a) 4th largest (and find the probability that it's between 7 and 8)
   (b) maximum
   (c) 27th largest
   (d) 3rd smallest

3. Let $X_1, \ldots, X_9$ be iid, each uniform on [1,12].
   (a) Find the probability that the next to smallest is ≥ 4.
   (b) Find the distribution function of the 7th largest.
   (c) Find the density function of the 7th largest.

4. Let $X_1, \ldots, X_n$ be iid with common density $f(x)$ and common distribution function $F(x)$.
   (a) Find the probability that the maximum is ≤ 9.
   (b) Find the distribution function of the maximum.
   (c) Find the density of the maximum
      (i) by differentiating the distribution function from part (b)
      (ii) directly.
   (d) Find the probability that the minimum is ≥ 2.
   (e) Find the distribution function of the minimum.
   (f) Find the density function of the minimum.

5. A machine has 7 identical components operating independently with respective lifetimes $X_1, \ldots, X_7$. Let $f(x)$ be their common density and $F(x)$ their common distribution function. Find the probability that the machine lasts at most 5 hours if
   (a) it keeps going until all its components are dead
   (b) it dies as soon as one of its components dies
   (c) it dies when it has only one component left

6. A machine needs 4 out of its 6 identical independent components to operate. Let $X_1, \ldots, X_6$ be the lifetimes of the respective components. Suppose each is exponentially distributed with parameter λ.

   Find (a) the distribution function and (b) the density function of the machine's lifetime.

7. Let $X_1, \ldots, X_7$ be iid with common dist function $F(x) = x/(x+1)$ for $x \geq 0$. Find the probability that each of the intervals $[0, 1], [1, 2], [2, 3], \ldots, [6, 7]$ contains exactly one of the X_i's.

   In other words, find the probability that after the 7 X_i's are picked, 1 of them is between 0 and 1, another is between 1 and 2, $\ldots$, and another is between 6 and 7.

8. If $X_1, \ldots, X_7$ are iid with common density $f(x)$, find (easily) the expected number of X_i's ≥ 5.

9. Let $X_1, \ldots, X_7$ be iid with common density $f(x)$ and common distribution function $F(x)$. Find the probability that
    (a) the maximum is ≥ 9 and the next largest is ≤ 5
    (b) the maximum is ≤ 9 and the next largest is ≤ 5

10. Let $X_i, \ldots, X_8$ be iid, each uniform on $[0,10]$. Find the probability of a gap of at least 6 between the maximum and the next largest.

    Suggestion: First find the joint density $f(x, y)$ of the max and the next largest.

11. This section showed how to find the distribution of the kth largest of iid random variables $X_1, \ldots, X_n$. How does the method break down if $X_1, \ldots, X_n$
    (a) are not independent
    (b) are independent but not identically distributed

12. Let $X_1, \ldots, X_7$ be iid with common density function $f(x)$.
    (a) Find the joint density $f(x_1, \ldots, x_7)$ of $X_1, \ldots, X_7$.
    (b) Find the joint density $g(x_1, \ldots, x_7)$ of the 7 order statistics (the new random variables, largest, 2nd largest, 3rd largest, $\ldots$, 7th largest).

CHAPTER 7

Expectation Again

SECTION 7-1   EXPECTATION OF A RANDOM VARIABLE

Definition of Expected Value

Remember that if X is discrete, then the expectation $E(X)$, also denoted EX, is given by

$$E(X) = \sum_x x\, P(X = x)$$

The expectation of X is a weighted average of the values of X where each value is weighted by the probability that it occurs.

In the continuous case where X has density $f(x)$, the analogous sum is

$$\sum_x x \underbrace{P(X \approx x)}_{f(x)dx}$$

which suggests this definition:

If X is continuous with density $f(x)$ then the *expectation* or *expected value* or *mean* of X is

(1) $$E(X) = \int_{-\infty}^{\infty} x\, f(x)\, dx$$

The mean of X is often called μ_X.

Mean of an Exponentially Distributed Random Variable

If X has an exponential distribution with parameter λ, then

$$EX = 1/\lambda$$

Here's why. The density of X is $\lambda e^{\lambda x}$ for $\lambda \geq 0$, so

$$EX = \int_{-\infty}^{\infty} x\, f(x)\, dx = \int_{0}^{\infty} x\, \lambda\, e^{-\lambda x} dx = \lambda \cdot \frac{1}{\lambda^2} \text{ (see the integral tables)}$$

$$= \frac{1}{\lambda}$$

When we used the exponential in Section 4.3 as a model for a lifetime, we chose

$$\lambda = \frac{1}{\text{average lifetime}}$$

So

$$\text{average lifetime} = \frac{1}{\lambda}$$

which agrees with the mathematical conclusion that the mean of the exponential is $1/\lambda$.

Mean of a Uniformly Distributed Random Variable

If X is uniform on $[a,b]$, then

$$EX = \frac{a+b}{2} = \text{ midpoint of the interval}$$

This is intuitively what you would expect. And here's the official proof.
    The density of X is

$$f(x) = \frac{1}{b-a} \text{ for } a \leq x \leq b$$

so

$$EX = \int_{-\infty}^{\infty} x \, f(x) \, dx$$

$$= \int_{a}^{b} x \, \frac{1}{b-a} \, dx$$

$$= \frac{1}{b-a} \, \frac{x^2}{2} \Big|_{a}^{b}$$

$$= \frac{1}{b-a} \, \frac{b^2 - a^2}{2}$$

$$= \frac{1}{b-a} \, \frac{(b-a)(b+a)}{2}$$

$$= \frac{a+b}{2}$$

Law of the Unconscious Statistician

Suppose X is uniformly distributed on [3,5] and you want the expectation not of X (which is 4, by inspection) but of X^3. You have to see the long way of doing it before you can appreciate the short way.

Let $Y = X^3$. To find EY using the definition in (1) (the *long* way), first find $f(y)$ by the methods of Section 4.5:

$$f(y) = f(x) \left| \frac{dx}{dy} \right| = \frac{1}{2} \cdot \frac{1}{3} y^{-2/3} \quad \text{for } 27 \le y \le 125$$

Then

$$EY = \int_{y=-\infty}^{\infty} y \, f(y) \, dy = \int_{27}^{125} y \, \frac{1}{2} \cdot \frac{1}{3} \, y^{-2/3} \, dy = 68$$

I'll derive a method that avoids having to find $f(y)$ explicitly. Start with

$$E(X^3) = EY = \int_{27}^{125} y \, f_Y(y) \, dy$$

From (1) in Section 4.5,

$$f_Y(y) \, dy = f_X(x) \, dx \quad \text{(Fig. 1)}$$

So instead of adding $y f(y) \, dy$'s on the y-interval [27,125] in Fig. 1, you can add $x^3 f(x) \, dx$'s on the x-interval [3,5]. So

$$E(X^3) = \int_{3}^{5} x^3 f(x) \, dx = \int_{3}^{5} x^3 \, \frac{1}{2} \, dx = 68$$

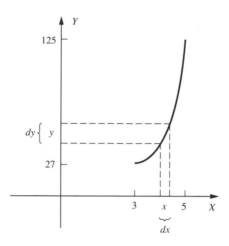

Figure 1

Here's the general rule in the 1-dimensional case.

Suppose you know the distribution of X and want the expected value of $g(X)$, some function of X.

If X is discrete, then

$$E(g(X)) = \sum_x g(x)\, P(X = x)$$

and if X is continuous, then

$$E(g(X)) = \int_{x=-\infty}^{\infty} g(x)\, f_X(x)\, dx$$

For example, if X is discrete, then

$$E(X^2) = \sum_x x^2 P(X = x)$$

and if X is continuous, then

$$E(\sin X) = \int_{x=-\infty}^{\infty} \sin x\, f_X(x)\, dx$$

The same shortcut holds in the 2-variable case. Suppose you know the joint density of X and Y and want the expected value of $X^2 Y^3$. The *long* way is to let $Z = X^2 Y^3$, find $f_Z(z)$ with the method of Section 5.3, and use

$$EZ = \int_{z=-\infty}^{\infty} z \, f_Z(z) \, dz$$

This adds all the possible values of $X^2 Y^3$, each one weighted by the probability of getting that value.

The *short* way is to find

$$\sum_x \sum_y x^2 y^3 \, P(X \approx x, Y \approx y)$$

that is,

$$\int_{x,y\text{plane}} x^2 y^3 \, f(x,y) \, dA$$

This adds an $x^2 y^3$ from each little region in the universe, each weighted by the probability of hitting that little region.

Here's the general rule in the 2-dimensional continuous case.

Suppose you know the joint density $f(x,y)$ of X and Y and want the expected value of $g(X,Y)$, some function of X and Y. Then

$$E(g(X,Y)) = \int_{\text{plane}} g(x,y) \, f(x,y) \, dA$$

For example,

$$E\left(\frac{Y}{X}\right) = \int_{\text{plane}} \frac{y}{x} \, f(x,y) \, dA$$

As a special case you can get the individual expected values of X and Y from the joint density:

$$EY = \int_{\text{plane}} y \, f(x,y) \, dA, \quad EX = \int_{\text{plane}} x \, f(x,y) \, dA$$

Example 1

Let X have density

$$f(x) = e^{-x} \text{ for } x \geq 0$$

Then

$$EX = \int_0^\infty x\,e^{-x}\,dx = 1 \qquad \text{(integral tables)}$$

$$E(X^5) = \int_0^\infty x^5\,e^{-x}\,dx = 5! \quad \text{(integral tables)}$$

Example 2

If X and Y have joint density

$$f(x, y) = e^{-(x+y)} \text{ for } x \geq 0, y \geq 0$$

then

$$E(XY) = \int_{\text{plane}} xy\,f(x, y)\,dA = \int_{x=0}^\infty \int_{y=0}^\infty xy\,e^{-x}\,e^{-y}\,dy\,dx$$

$$\text{inner integral} = \int_{y=0}^\infty y\,e^{-y}\,dy = 1$$

$$\text{outer integral} = \int_{x=0}^\infty x\,e^{-x}\,dx = 1$$

Expected Value of max (X,Y)

Let X and Y have joint density

$$f(x, y) = \frac{5}{162}\,x^2 y \text{ for } x \geq 0, y \geq 0, x + 2y \leq 6$$

We'll set up the expected value of the maximum of X and Y. We have

$$E(\text{max}) = \int_{\text{plane}} \max(x, y)\,f(x, y)\,dA$$

The formula for $\max(x, y)$ is

$$\max(x, y) = \begin{cases} x & \text{if } x \geq y \\ y & \text{if } y \geq x \end{cases}$$

$$= \begin{cases} x & \text{in region I} \\ y & \text{in region II} \end{cases} \qquad \text{(Fig. 2)}$$

So

$$E(\max) = \int_I x\, f(x,y)\, dA + \int_{II} y\, f(x,y)\, dA$$

$$= \int_{y=0}^{y=2} \int_{x=y}^{x=6-2y} x\, \frac{5}{162}\, x^2 y\, dx\, dy + \int_{x=0}^{2} \int_{x=y}^{y=3-x/2} y\, \frac{5}{162}\, x^2 y\, dy\, dx$$

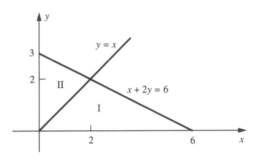

Figure 2

Warning

$E(\max)$ is a *sum* of two integrals. *Don't* do it like this:

$$\text{WRONG} \qquad E(\max) = \begin{cases} \displaystyle\int_I x\, f(x,y)\, dA & \text{if } x \geq y \\[3mm] \displaystyle\int_{II} y\, f(x,y)\, dA & \text{if } x \leq y \end{cases} \qquad \text{WRONG}$$

In general, when you find the expected value of a random variable, the answer is a *number*; the answer has no letters in it and doesn't come with cases or intervals.

This might be a good time to pull yourself together and review the max/min problems we've done already:

    probabilities involving max and min

        Section 5.1, Problems 1(c), 2(b), 3(c))

    distribution of the max and min

        text of Section 5.3

        Section 5.3, Problems 3(a), 7

        review for Chapters 4 & 5, Problems 18(a)(c)

    special case of iid random variables

        Section 6.2, Problems 1(d), 2(b), 4

Properties of Expectation (for Continuous, Discrete, and Mixed Random Variables)

(2)	$Ek = k$	(constant k)
(3)	$E(X + Y) = EX + EY$	
(4)	$E(kX) = k\,EX$	(constant k)
(5)	If X and Y are independent then $E(XY) = (EX)\,(EY)$	

For example, by (2) – (4),

$$E(2X - 3Y + 5) = 2\,EX - 3\,EY + 5$$

The rule in (5) goes only one way: if X and Y are *not* independent, then $E(XY)$ may or not equal $(EX)\,(EY)$ (Fig. 3).

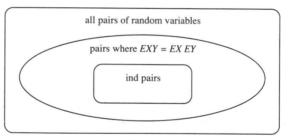

Figure 3

Proof of (2)

If X is a random variable that only takes on the value k, then X is discrete and

$$EX = k\,P(X = k) = k \cdot 1 = k$$

Proof of (3) in the Continuous Case

Let $f(x, y)$ be the joint density of X and Y. Then

$$E(X + Y) = \int_{\text{plane}} (x + y)\, f(x, y)\, dA$$

$$= \int_{\text{plane}} x\, f(x, y)\, dA + \int_{\text{plane}} y\, f(x, y)\, dA$$

$$= EX + EY$$

Proof of (4) in the Continuous Case

Let X have density $f(x)$. Then

$$E(kX) = \int_{-\infty}^{\infty} kx \, f(x) \, dx$$

$$= k \int_{-\infty}^{\infty} x \, f(x) \, dx$$

$$= k \, EX$$

Proof of (5) in the Continuous Case

Let X have density $f(x)$ and Y have density $f(y)$. If X and Y are independent then the joint density is

$$f(x, y) = f(x) f(y)$$

and

$$E(XY) = \int_{\text{plane}} xy \, f(x, y) \, dA = \int_{x=-\infty}^{\infty} \int_{y=-\infty}^{\infty} xy \, f(x) f(y) \, dy \, dx$$

$$\text{inner integral} = \int_{y=-\infty}^{\infty} y \, f(y) \, dy = EY$$

$$\text{outer integral} = \int_{x=-\infty}^{\infty} x \, f(x) \, EY \, dx$$

But EY is a constant that can be pulled out of the integral so

$$E(XY) = EY \int_{x=-\infty}^{\infty} x \, f(x) \, dx = (EY)(EX)$$

On the other hand, problem 11 gives one set of non-independent random variables X and Y for which $E(XY) = (EX)(EY)$ and another set for which $E(XY) \neq (EX)(EY)$, proving that if X and Y are not independent, then anything can happen.

Problems for Section 7-1

1. If X has density $f(x) = \frac{1}{4}xe^{-x/2}$ for $x \geq 0$, find EX.

2. Find EX if X has density $f(x) = 1 - |x|$ for $-1 \leq x \leq 1$.

3. If X has density $f(x) = \frac{1}{4}xe^{-x/2}$ for $x \geq 0$, find the expected value of X^3.

4. If X has a unit normal distribution, find $E|X|$.

5. If X has a binomial distribution with the parameters $n = 4$, p, find $E(\sin \frac{1}{2}\pi x)$.

6. The function Int (greatest integer function) is defined like this:

Int x is the largest integer which is $\leq x$.

For example, Int $4.3 = 4$, Int $\pi = 3$, Int $e = 2$, Int $6 = 6$, Int $.5 = 0$. If X has an exponential distribution with $\lambda = 1$, find $E(\text{Int } X)$.

7. Let X and Y each have an exponential distribution with parameter $\lambda = 1$.
(a) Find $E(X - Y)$.
(b) If X and Y are independent, find $E|X - Y|$.

8. Pick a point at random in a circle of radius a. Find the expected distance to the center. Try it twice.
(a) Use the law of the unconscious statistician.
(b) Let R be the distance to the center. Find $F_R(r)$ and $f_R(r)$ and use the definition in (1).

9. Pick a point (X, Y) at random in the upper half of the circle with center at the origin and radius R. Find $E(Y)$.

10. Let X be uniform on $[0,10]$. Let

$$Y = \begin{cases} X^2 & \text{if } X \leq 6 \\ 12 & \text{if } X \geq 6 \end{cases}$$

Find EY.

11. Use the following pairs of random variables to show that if X and Y are not independent, then $E(XY)$ may or may not be equal to $(EX)(EY)$.
(a) $X = \cos \Theta$, $Y = \sin \Theta$, where Θ is uniformly distributed in $[0,2\pi]$.
(b) X is uniform on $[0,2]$, $Y = X^2$.

12. Let X and Y have joint density

$$f(x, y) = \begin{cases} \dfrac{1}{2} & \text{in region } G \\[2mm] \dfrac{3}{2} & \text{in region } H \end{cases}$$

Set up the integral(s) for the expected value of the max of X and Y.

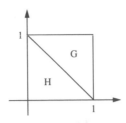

Figure P.12

13. In a game you begin with a score of 5.

    Pick X at random between 0 and 8 (X is not necessarily an integer).

    If $X \leq 5$, then you get X as your new score. Otherwise, you keep the old score (so that your original score can go down but not up).

    Find your expected score in the game.

14. True or False?
    (a) If X and Y are independent, then $E(XY) = (EX)(EY)$.
    (b) If X and Y are not independent, then $E(XY) \neq (EX)(EY)$.
    (c) If $E(XY) = (EX)(EY)$, then X and Y are independent.
    (d) If $E(XY) \neq (EX)(EY)$, then X and Y are not independent.

15. Suppose the lifetime T (in months) of a bulb is exponential with $\lambda = 1$. When the bulb burns out, you buy a new one. The price of a bulb now is \$2, but because of inflation, in t months the price will be \$2 $\times e^{.1t}$. Find the expected price of the replacement bulb.

SECTION 7-2   VARIANCE

Definition of Variance

Let X be a random variable (continuous, discrete, mixed) with mean μ_X. Remember (Section 3.4) that to find the *variance* of X, you look at the differences between the possible values of X and the mean of X, square them so that a positive over-the-mean doesn't cancel a negative under-the-mean and take a weighted average:

(1)
$$\text{Var } X = E(X - \mu_X)^2$$

Var X is intended to be a measure of the dispersion of X from its mean. If Var X is large, then it is likely that X will be far from its mean; if Var X is small, it is likely that X will be close to its mean.

The variance is frequently denoted by σ^2. Its positive square root, σ, is called the *standard deviation* of X; that is, the standard deviation is $\sqrt{\text{Var}X}$.

Mean and Variance of the Normal Distribution

Ever since we defined the normal distribution in Chapter 4, we've been thinking of the parameter μ as the mean and the parameter σ^2 as a measure of spread. Let's make it official.

(2)

> If X is normal with parameters μ, σ^2, then
> $EX = \mu$ and Var $X = \sigma^2$.

Tedious Proof

First, we'll find EX:

$$EX = \int_{x=-\infty}^{\infty} x\, f(x)\, dx = \int_{x=-\infty}^{\infty} x\, \frac{1}{\sigma\sqrt{2\pi}} e^{-(x-\mu)^2/2\sigma^2}\, dx$$

Let $v = x - \mu, dv = dx$. Then

$$EX = \frac{1}{\sigma\sqrt{2\pi}} \int_{v=-\infty}^{\infty} (v+\mu)e^{-v^2/2\sigma^2}\, dv$$

$$= \frac{1}{\sigma\sqrt{2\pi}} \int_{v=-\infty}^{\infty} v e^{-v^2/2\sigma^2}\, dv + \mu \cdot \frac{1}{\sigma\sqrt{2\pi}} \int_{v=-\infty}^{\infty} e^{-v^2/2\sigma^2}\, dv$$

$$= \frac{1}{\sigma\sqrt{2\pi}} \cdot -\sigma^2 e^{-v^2/2\sigma^2} \Big|_{v=-\infty}^{\infty} + \mu \cdot \text{integral of the normal density}$$

$$= 0 + \mu \cdot 1$$

$$= \mu$$

Now let's try Var X:

$$\text{Var } X = E(X-\mu)^2 = \int_{x=-\infty}^{\infty} (x-\mu)^2 f(x)\, dx \text{ (law of the unconscious stat)}$$

$$= \int_{x=-\infty}^{\infty} (x-\mu)^2 \frac{1}{\sigma\sqrt{2\pi}} e^{-(x-\mu)^2/2\sigma^2}\, dx$$

Let

$$t = \frac{x-\mu}{\sigma}, \quad dt = \frac{1}{\sigma}\, dx$$

Then

$$\text{Var } X = \frac{\sigma^2}{\sqrt{2\pi}} \int_{t=-\infty}^{\infty} t^2 \, e^{-t^2/2} \, dt$$

Now use integration by parts with

$$u = t, \quad dv = te^{-t^2/2}, \quad du = dt, \quad v = -e^{-t^2/2}$$

to get

$$\text{Var } X = \frac{\sigma^2}{\sqrt{2\pi}} \cdot -te^{-t^2/2} \Big|_{t=-\infty}^{\infty} + \frac{\sigma^2}{\sqrt{2\pi}} \int_{t=-\infty}^{\infty} e^{-t^2/2} \, dt$$

$$= 0 + \frac{\sigma^2}{\sqrt{2\pi}} \ \sqrt{2\pi} \ \text{(use integral tables)}$$

$$= \sigma^2$$

More Practical Way (Usually) to Find Var X

For most distributions (the normal distribution was an exception), it is not convenient to use (1) directly. Here's the better way.

(3)
$$\boxed{\text{Var } X = E(X^2) - (EX)^2}$$

Here's why (3) holds:

$$\begin{aligned}
\text{Var } X &= E(X - \mu_X)^2 && \text{(by (1))} \\
&= E(X^2 - 2\mu_X X + \mu_X^2) \\
&= E(X^2) - 2\mu_X E(X) + \mu_X^2 && \text{(by (2), (3), (4) in Section 7.1)} \\
&= E(X^2) - 2\mu_X^2 + \mu_X^2 \\
&= E(X^2) - \mu_X^2 \\
&= E(X^2) - (EX)^2
\end{aligned}$$

Example 1

Let X have an exponential distribution with parameter λ. We already know that

$$E(X) = \frac{1}{\lambda}$$

Now we'll find Var X. We know that X has density

$$f(x) = \lambda e^{-\lambda x} \text{ for } x \geq 0$$

To use (3) we need $E(X^2)$:

$$E(X^2) = \int_0^\infty x^2 \lambda e^{-\lambda x}\, dx = \lambda \frac{2!}{\lambda^3} \text{ (integral tables)} = \frac{2}{\lambda^2}$$

So

$$\text{Var } X = E(X^2) - (EX)^2 = \frac{2}{\lambda^2} - \left(\frac{1}{\lambda}\right)^2 = \frac{1}{\lambda^2}$$

Covariance

When we come to the rule for $\text{Var}(X + Y)$, we'll find a new idea turning up, called the covariance of X and Y. And in the next section, you'll see that positive (resp. negative) covariance indicates that if X increases, Y tends to increase (resp. decrease). In the meantime, here's the definition of the covariance of two random variables:

(4)
$$\boxed{\text{Cov } (X, Y) = E[(X - \mu_X)(Y - \mu_Y)]}$$

We can use properties of expectation to convert this to a more practical formula:

$$
\begin{aligned}
\text{Cov } (X, Y) &= E(X - \mu_X)(Y - \mu_Y) && \text{(by definition)} \\
&= E(XY - \mu_X Y - \mu_Y X + \mu_X \mu_Y) \\
&= E(XY) - \mu_X EY - \mu_Y EX + \mu_X \mu_Y \; (\mu_X \text{ and } \mu_Y \text{ are constants} \\
&\qquad\qquad\qquad\qquad\qquad\qquad\qquad\qquad \text{that can be pulled out)} \\
&= E(XY) - \mu_X \mu_Y - \mu_Y \mu_X + \mu_X \mu_Y \\
&= E(XY) - \mu_X \mu_Y \\
&= E(XY) - (EX)(EY)
\end{aligned}
$$

So

(5)
$$\boxed{\text{Cov } (X, Y) = E(XY) - (EX)(EY)}$$

From Section 7.1 we know that if X and Y are independent, then $E(XY) = (EX)(EY)$, so in that case, Cov $(X, Y) = 0$. But if X and Y are not independent, then $E(XY)$ may or may not equal $(EX)(EY)$, so in that case Cov (X, Y) may or may not be 0 (Fig. 1)

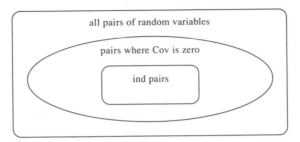

Figure 1

Variance of a Sum

$$\text{(6)} \quad \text{Var}(X_1 + \cdots + X_n) = \text{Var } X_1 + \cdots + \text{Var } X_n + 2\underbrace{\sum_{i<j} \text{Cov } (X_i, X_j)}_{\binom{n}{2} \text{ terms}}$$

The sum $\sum_{i<j} \text{Cov }(X_i, X_j)$ means the sum of all the "2-at-a-time" terms, terms such as $\text{Cov }(X_1, X_5)$, $\text{Cov}(X_4, X_6)$, and so on. For example,

$$\text{Var}(X + Y) = \text{Var } X + \text{Var } Y + 2\text{Cov }(X, Y)$$
$$\text{Var}(X_1 + X_2 + X_3) = \text{Var}(X_1 + X_2 + X_3)$$
$$+ 2[\text{Cov }(X_1, X_2) + \text{Cov}(X_2, X_3) + \text{Cov}(X_1, X_3)]$$

$$\text{(7)} \quad \boxed{\begin{array}{l} \text{If } X_1, \ldots, X_n \text{ are independent, then (6) becomes} \\[6pt] \text{Var}(X_1 + \cdots + X_n) = \text{Var } X_1 + \cdots + \text{Var } X_n \end{array}}$$

Proof of (6) and (7)

$$\text{Var }(X_1 + \cdots + X_n)$$
$$= E(X_1 + \cdots + X_n)^2 - [E(X_1 + \cdots + X_n)]^2 \qquad \text{(by (3))}$$

$$= E\left(X_1^2 + \cdots + X_n^2 + 2\sum_{i<j} X_i X_j\right) - (EX_1 + \cdots + EX_n)^2$$

$$= E(X_1^2) + \cdots + E(X_n^2) + 2\sum_{i<j} E(X_i X_j)$$

$$-(EX_1)^2 - \cdots - (EX_n)^2 - 2\sum_{i<j}(EX_i)(EX_j)$$

Now group the terms to see variances and covariances appear:

$$\text{Var}\,(X_1 + \cdots + X_n)$$

$$= [E(X_1^2) - (EX_1)^2] + \cdots + [E(X_n^2) - (EX_n)^2]$$

$$+2\sum_{i<j}[E(X_iX_j) - (EX_i)(EX_j)]$$

$$= \text{Var}\,X_1 + \cdots + \text{Var}\,X_n + 2\sum_{i<j}\text{Cov}\,(X_i, X_j)$$

which proves (6).

If X_i and X_j are independent, then Cov $(X_i, X_j) = 0$ (Fig. 1), so all the covariance terms in (6) drop out and you get (7).

Some More Properties of Variance

(8)   Var $k = 0$ (constant k)

(9)   Var $(aX + b) = a^2$ Var X (constant a, b)

(10)   Var $(-X) = $ Var X

(11)   If X and Y are independent then

Var $(X - Y) = $ Var $X + $ Var Y

(MINUS)   (PLUS)

The converse of (8) is also true (if Var $X = 0$ then X is a constant) so all in all:

Constant random variables and only constant random variables have zero variance.

For example, by (8) and (9),

$$\text{Var}\,(X + Y + Z + 7) = \text{Var}\,(X + Y + Z)$$
$$\text{Var}\,(3X) = 9\,\text{Var}\,X$$

Proof of (8) and Its Converse

By (1),

$$\text{Var } k = E(k - Ek)^2 = E(k - k)^2 = E(0) = 0$$

Conversely, if Var $X = 0$, then, in the discrete case,

$$\sum (X - \mu_X)^2 P(X = x) = 0$$

But a sum of non-negative terms can't be 0 unless all of them are 0. So X is always μ_X; that is, X is a constant. The proof in the continuous case is similar.

Proof of (9) and (10)

First, we'll work on Var (aX) using the fact that $E(kX) = k(EX)$:

$$\begin{aligned}
\text{Var } (aX) &= E(aX)^2 - [E(aX)]^2 \\
&= E(a^2 X^2) - [aE(X)]^2 \\
&= a^2 E(X^2) - a^2 [E(X)]^2 \\
&= a^2 [E(X^2) - (EX)^2] \\
&= a^2 \text{ Var } X
\end{aligned}$$

Now look at Var $(aX + b)$. The constant random variables b and aX are independent (knowing aX doesn't change the odds on b), so

$$\begin{aligned}
\text{Var } (aX + b) &= \text{Var } (aX) + \text{ Var } b \quad \text{(by (7))} \\
&= a^2 \text{ Var } X + 0 \\
&= a^2 \text{ Var } X
\end{aligned}$$

This proves (9).

The rule in (10) is the special case of (9) with $a = 1$ and $b = 0$.

Proof of (11)

If X and Y are independent, then X and $-Y$ are independent and

$$\begin{aligned}
\text{Var } (X - Y) = \text{Var } (X + -Y) &= \text{ Var } X + \text{ Var } (-Y) \quad \text{(by (7))} \\
&= \text{Var } X + \text{ Var } Y \quad \text{(by (10))}
\end{aligned}$$

Variance of the Binomial Distribution

Let X be the number of successes in n Bernoulli trials where $P(\text{success}) = p$. We know (Section 3.2) that $EX = np$. In Section 4.4 we predicted that

$$\boxed{\text{Var } X = npq}$$

(which is why the approximating normal distribution has $\sigma^2 = npq$). Now we'll check this out.

For $i = 1, 2, ..., n$ define indicators

$$X_i = \begin{cases} 1 & \text{if there is a success on the } i\text{th trial} \\ 0 & \text{otherwise} \end{cases}$$

Then

$$X = X_1 + \cdots + X_n$$

The indicators are independent, so

$$\text{Var } X = \text{Var } X_1 + \cdots + \text{Var } X_n$$

Now we need $\text{Var } X_i$. We have

$$\text{Var } X_i = E(X_i^2) - (EX_i)^2$$

Since X_i takes on only the values 0 and 1, we have $X_i^2 = X_i$, so

$$E(X_i^2) = E(X_i) = P(X_i = 1) = P(\text{Success on } i\text{th trial}) = p$$

So

$$\text{Var } X_i = E(X_i^2) - (EX_i)^2 = p - p^2 = p(1 - p) = pq$$

and

$$\text{Var } X = \text{sum of } n \, pq\text{'s} = npq$$

Example 2

Draw 20 times from a box containing 30 reds and 70 whites.

Let X be the number of reds in the sample.

You already know (Section 3.2) that $EX = .3$ (since 30% of the balls in the box are red, on the average 30% of any sample is red, whether the drawing is with or without replacement).

Now let's find the variance of X.

If the drawing is *with* replacement then the drawings are independent, X has a binomial distribution with $n = 20$, $p = 30/100$, and

$$\text{Var } X = npq = 4.2$$

Suppose the drawing is with*out* replacement. We can use indicators to find $\text{Var } X$.

Let

$$X_i = 1 \text{ if the } i\text{th draw is red}$$

Then

$$X = X_1 + \cdots + X_{20}$$

The indicators are *not* independent (getting a red on one draw changes the odds on another draw), so

$$\text{Var } X = \text{Var } X_1 + \cdots + \text{Var } X_{20} + 2 \sum_{i<j} \text{Cov}(X_i, X_j)$$

First, find Var X_i.

Since X_i takes on only the values 0 and 1, $X_i^2 = X_i$ and

$$E(X_i^2) = EX_i = P(\text{red on } i\text{th draw}) = P(\text{red on 1st}) = .3$$
$$\text{Var } X_i = E(X_i^2) - (EX_i)^2 = .3 - (.3)^2 = .21$$

Next find Cov (X_i, X_j):

$$\text{Cov } (X_i, X_j) = E(X_i X_j) - (EX_i)(EX_j) = E(X_i X_j) - (.3)(.3)$$

The product $X_1 X_j$ works like this:

$$X_i X_j = \begin{cases} 1 & \text{if } i\text{th and } j\text{th draws are red} \\ 0 & \text{otherwise} \end{cases}$$

So

$$
\begin{aligned}
E(X_i X_j) = P(X_i X_j = 1) &= P(\text{red on } i\text{th and red on } j\text{th}) \\
&= P(\text{red on 1st and red on 2nd}) \quad \text{(by symmetry)} \\
&= \frac{30}{100} \frac{29}{99}
\end{aligned}
$$

So

$$\text{Cov } (X_i, X_j) = \frac{30}{100} \frac{29}{99} - (.3)^2$$

Put it all together:

$$\text{Var } X = 20(.21) + 2 \binom{20}{2} \left[\frac{30}{100} \frac{29}{99} - (.3)^2 \right] = 3.39$$

(The variance is smaller when the drawing is without replacement.)

Summary of *EX*, Var *X*, Cov *(X,Y)*

$$E(X + Y) = EX + EY \text{ whether or not } X \text{ and } Y \text{ are ind}$$

$\mathrm{Var}(X + Y) = \mathrm{Var}\, X + \mathrm{Var}\, Y$ if X and Y are ind (but *not* only if)

$E(X - Y) = EX - EY$ whether or not X and Y are ind
$\mathrm{Var}(X - Y) = \mathrm{Var}\, X + \mathrm{Var}\, Y$ if X and Y are ind (but *not* only if)

$E(kX) = k\, EX$
$\mathrm{Var}\, kX = k^2\, \mathrm{Var}\, X$

$E\, k = k$
$\mathrm{Var}\, k = 0$

$E(kX + b) = k\, EX + b$
$\mathrm{Var}(kX + b) = k^2\, \mathrm{Var}\, X$

$E(XY) = (EX)(EY)$ if X and Y are independent (but *not* only if)

$$\mathrm{Var}\,(X_1 + \cdots + X_n) = \mathrm{Var}\, X_1 + \cdots + \mathrm{Var}\, X_n + 2\sum_{i<j} \mathrm{Cov}\,(X_i, X_j)$$

$\mathrm{Var}\, X = E(X - EX)^2$
$\mathrm{Var}\, X = E(X^2) - (EX)^2$

$\mathrm{Cov}\,(X, Y) = E[(X - EX)(Y - EY)]$
$\mathrm{Cov}\,(X, Y) = E(XY) - (EX)(EY)$

Some Famous Expectations and Variances

Distribution of X	Mean	Variance	Section Reference
uniform on $[a,b]$	$\dfrac{a+b}{2}$	$\dfrac{(b-a)^2}{12}$	7.1, 7.2, Problem 2
binomial with parameters n, p	np	npq	3.2, 7.2
geometric with parameter p	$\dfrac{1}{p}$	$\dfrac{q}{p^2}$	3.2, 7.2, Problem 19(a)
negative binomial with parameters r, p	$\dfrac{r}{p}$	$\dfrac{rq}{p^2}$	7.2, Problem 19(b)
Poisson with parameter λ	λ	λ	3.1, 7.2, Problem 6
exponential with parameter λ	$\dfrac{1}{\lambda}$	$\dfrac{1}{\lambda^2}$	7.1, 7.2, Example 1
normal with parameters μ, σ^2	μ	σ^2	7.2

Problems for Section 7-2

1. Let X be the face value when a die is tossed once. Find EX and Var X.

2. Let X be uniform on $[a,b]$. Find EX and Var X.

3. A couple decides to have children until they have a girl or until they have a total of 3 children. Find the expectation and variance of
   (a) the number of boys they have
   (b) the number of girls they have

4. If the density of X is

$$f(x) = 1 - |x| \qquad \text{for } -1 \le x \le 1$$

find EX and Var X.

5. Find EY and Var Y if the density of X is

$$f_X(x) = \frac{1}{x^2}, \quad x \ge 1$$

and

$$Y = \begin{cases} X^3 & \text{if } X \le 2 \\ 8 & \text{if } X \ge 2 \end{cases}$$

6. (a) Guess the variance of the Poisson distribution by using its connection with the binomial where the variance is npq.
   (b) Find the variance of the Poisson officially.

7. If $EX = 10$ and Var $X = 2$, find
   (a) Var $10X$
   (b) Var$(10X + 3)$
   (c) $E(-X)$
   (d) Var$(-X)$
   (e) $E(X^2)$
   (f) Var$\left(\dfrac{2X + 4}{3}\right)$

8. True or False?
   (a) If X, Y are not independent, then Var$(X + Y) \ne$ Var $X +$ Var Y.
   (b) If X and Y are independent, then Var$(X + Y) =$ Var $X +$ Var Y.
   (c) If Var$(X + Y) \ne$ Var $X +$ Var Y, then X and Y are not independent.
   (d) If Var$(X + Y) =$ Var $X +$ Var Y, then X and Y are independent.
   (e) If X and Y are independent, then $E(X + Y) = EX + EY$.
   (f) If X and Y are not independent, then $E(X + Y) \ne EX + EY$.

9. Express Var$(X - Y)$ in terms of Var X, Var Y, and Cov(X, Y). Try it three times as follows.

(a) Use the definition of variance in (1).
(b) Use (3).
(c) Use the rule for variance of a sum.

10. If $EX = \mu$ and Var $X = \sigma^2$ find $E(aX^2 + bX + c)$.

11. Given random variables $X_1, \ldots, X_n$ where

$$E(X_i) = 3 \text{ for all } i$$
$$E(X_i X_j) = 5 \text{ for } i \neq j$$
$$E(X_i^2) = 11 \text{ for all } i$$

Find $\text{Var}(X_1 + \cdots + X_n)$.

12. True or False?
(a) If X and Y are independent, then $\text{Cov}(X, Y) = 0$.
(b) If X and Y are not independent, then $\text{Cov}(X, Y) \neq 0$.
(c) If $\text{Cov}(X, Y) = 0$, then X and Y are independent.
(d) If $\text{Cov}(X, Y) \neq 0$, then X and Y are not independent.

13. A box contains 3 slips of papers numbered 1, 2, 3. Draw twice.

Let X be the first number drawn and let Y be the second number drawn.

Find $\text{Cov}(X, Y)$ if the drawing is (a) without and (b) with replacement.

14. A box contains 3 red balls and 2 black balls. Draw 2 without replacement.

Let X be the number of reds and let Y be the number of blacks.

Find Var X and $\text{Cov}(X, Y)$.

15. Let X and Y be independent, each with mean 2 and variance 1.

Let $U = 3X + 2Y$, $V = 2X - 3Y$.

Find Var U and $\text{Cov}(U, V)$.

16. Toss a die 10 times. Let X be the number of 2's. Find EX and Var X by inspection.

17. Let X be the number of matches when two decks of n cards each are turned over simultaneously. Find EX and Var X using these indicators

$$X_i = \begin{cases} 1 & \text{if } i\text{th cards match} \\ 0 & \text{otherwise} \end{cases}$$

18. Ten men $M_1, \ldots, M_{10}$ and 10 women $W_1, \ldots, W_{10}$ are arranged at random into 10 (not necessarily male/female) pairs.

Let X be the number of male/female pairs. Find EX and Var X using the following indicators:

$$X_i = \begin{cases} 1 & \text{if } W_i \text{ is paired with a man} \\ 0 & \text{otherwise} \end{cases}$$

19. (a) Let X have a geometric distribution with parameter p (X is the number of tosses to get the first H where $P(H) = p$). Find Var X.

(b) Let X have a negative binomial distribution with parameters r, p (X is the number of tosses to get r heads where $P(H) = p$). Find Var X.

SECTION 7-3   CORRELATION

Definition

If Var X (denoted σ_X^2) and Var Y (denoted σ_Y^2) are not 0 and not ∞, then the correlation ρ of X and Y is given by

$$\rho(X, Y) = \frac{\text{Cov}(X, Y)}{\sqrt{\text{Var } X} \sqrt{\text{Var } Y}} = \frac{\text{Cov}(X, Y)}{\sigma_X \, \sigma_Y}$$

We'll prove later that

(1)
$$-1 \le \rho \le 1$$

so the correlation is just the covariance "normalized" to be in the range $[-1, 1]$.

An Application of Correlation—The Least Squares Estimate

Let X and Y be random variables. We would like to use X to estimate Y (can an admissions officer predict your college grade point average Y from your SAT score X). For simplicity we consider only estimates for Y of the form $aX + b$. The problem is to find a and b so that we get the best estimate possible.

First, how do you decide whether an estimate is good or not.

To measure the error when you use $aX + b$ to estimate Y, look at the differences between the values of Y and the values of $aX + b$, square them so that a positive estimation error doesn't cancel a negative estimation error, and take a weighted average: the value of

$$E[Y - (aX + b)]^2$$

is called the *mean square error* of the estimate. Of all estimates of the form $aX + b$, the "best" is the one that *minimizes* the mean square error; it's called the *linear least squares estimate* of Y based on X.

Here's how to find the linear least squares estimate.

Let X and Y be random variables with means μ_X, μ_Y, variances σ_X^2, and σ_Y^2 and correlation ρ.

The linear least squares estimate for Y based on X is

(2) $aX + b$ where $a = \rho\dfrac{\sigma_Y}{\sigma_X}, \quad b = \mu_Y - a\mu_X$

The graph of $y = ax + b$ is called the *line of regression*.

The mean square error corresponding to this best estimate in (2), that is, the minimum value of $E[Y - (aX + b)]^2$, is

(3) $$\sigma_Y^2(1 - \rho^2)$$

---

(4)   If $\rho = \pm 1$, then Y actually equals the least squares estimate in (2).

In particular if $\rho = 1$, then Y is $aX + b$, where $a > 0$, so Y increases with X; if $\rho = -1$, then $Y = aX + b$, where $a < 0$, so Y decreases as X increases.

If ρ is near ± 1, then Y is very likely to be close to its linear least squares estimate.

Independent Random Variables versus Uncorrelated Random Variables

If $\rho = 0$, then the best mean square error in (3) is at its worst. In this case X and Y are said to be *uncorrelated*.

If X and Y are independent, then $E(XY) = (EX)(EY)$, $\text{Cov}(X, Y) = 0$, and $\rho(X, Y) = 0$.

So independent pairs are uncorrelated.

But X and Y can be uncorrelated even if X and Y are not independent (this goes back to the fact that $E(XY)$ can equal $(EX)(EY)$ even if X and Y are not independent) (Fig. 1).

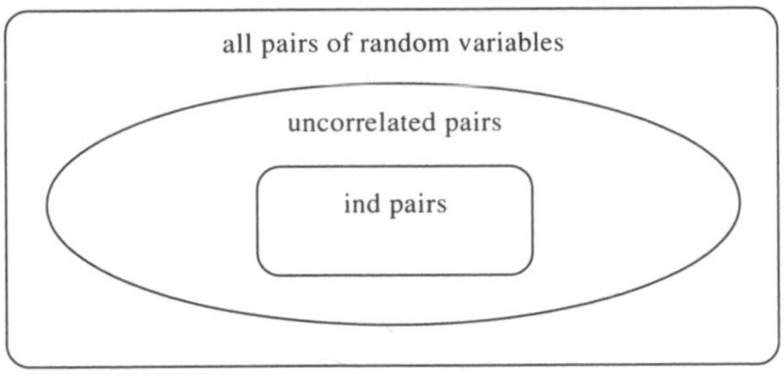

Figure 1

Proof of (2) and (3)

Part I.   First we'll consider the special case where

$$\mu_X = 0 \quad \text{and} \quad \mu_Y = 0$$

Then

$$E[Y - (aX + b)]^2 = E[Y^2 - 2Y(aX + b) + (aX + b)^2]$$
$$= EY^2 - 2a\,EXY + a^2\,EX^2 + b^2$$

(drop all terms involving EX and EY since they are 0)

In this special case where $EX = 0$, $EY = 0$, we have

$$EX^2 = \text{Var}\,X = \sigma_X^2, \quad EY^2 = \text{Var}\,Y = \sigma_y^2,$$
$$EXY = \text{Cov}\,(X, Y) = \rho\sigma_X\sigma_Y$$

so

(5) $$E[Y - (aX + b)]^2 = \sigma_Y^2 - 2a\rho\,\sigma_X\sigma_Y + a^2\sigma_X^2 + b^2$$

Now let's find the a and b that make the expectation in (5) minimum. By inspection, we need

(6) $$b = 0$$

That leaves

$$\sigma_Y^2 - 2a\rho\,\sigma_X\sigma_Y + a^2\sigma_X^2$$

which you can think of as a quadratic in the variable a, namely,

$$Aa^2 + Ba + C$$

where

$$A = \sigma_X^2$$
$$B = -2\rho\,\sigma_X\sigma_Y$$
$$C = \sigma_Y^2$$

To find the min, take the derivative with respect to a and set it equal to 0. The min occurs when

$$2Aa + B = 0$$

(7)
$$a = -\frac{B}{2A} = \frac{2\rho\,\sigma_X\sigma_Y}{2\sigma_X^2} = \rho\,\frac{\sigma_Y}{\sigma_X}$$

The value of a in (7) matches the value claimed in (2), and the value $b = 0$ in (6) matches the value $b = \mu_Y - a\mu_X$ claimed in (2) (remember that this part of the proof assumes $\mu_X = 0$ and $\mu_Y = 0$). This proves (2) in the special case.

With the a and b in (6) and (7), the mean square error itself, from (5), is

$$\sigma_Y^2 - 2\rho\,\frac{\sigma_Y}{\sigma_X}\,\rho\,\sigma_X\sigma_Y + \rho^2\,\frac{\sigma_Y^2}{\sigma_X^2}\,\sigma_X^2$$

(8)
$$= \sigma_Y^2 - 2\rho^2\,\sigma_Y^2 + \rho^2\,\sigma_Y^2$$
$$= \sigma_Y^2 - \rho^2\,\sigma_Y^2$$
$$= \sigma_Y^2\,(1 - \rho^2)$$

This proves (3) in the special case.

Part II.   Now consider the general case where X and Y have respective means μ_X and μ_Y.

First, we'll define some new random variables and find their means, variances, and correlation.

Let

$$X_{new} = X - \mu_X$$
$$Y_{new} = Y - \mu_Y$$

The new random variables have zero means but the same variances and correlation as the original X and Y; in other words,

(9)
$$\mu_{Xnew} = 0, \quad \mu_{Ynew} = 0$$

(10)
$$\sigma_{Xnew}^2 = \sigma_X^2, \quad \sigma_{Ynew}^2 = \sigma_Y^2$$
$$\sigma_{Xnew} = \sigma_X, \quad \sigma_{Ynew} = \sigma_Y$$

(11)
$$\rho(X_{new}, Y_{new}) = \rho(X, Y)$$

Here's why. We have

$$E_{(x\text{new})} = E(X - \mu_X) = EX - \mu_X = \mu_X - \mu_X = 0$$

and similarly for Y, so (9) holds.
Also

$$\sigma^2_{X\text{new}} = \text{Var}\,(X_{\text{new}}) = \text{Var}(X - \mu_X) = \text{Var}\,X$$
$$[\text{remember the rule Var}\,(aX + b) = a^2\,\text{Var}\,X]$$

and similarly for Y, so (10) holds.
And

$$\text{Cov}\,(X_{\text{new}}, Y_{\text{new}}) = E(X_{\text{new}}\,Y_{\text{new}}) - E(X_{\text{new}})E(Y_{\text{new}})$$
$$= E[(X - \mu_X)(Y - \mu_Y)] - 0 \cdot 0 \quad \text{(by (9))}$$
$$= \text{Cov}(X, Y)$$

This result together with (10) shows that (11) holds.
Now we're ready to minimize

(12) $$E[Y - (aX + b)]^2$$

Use algebra to rewrite (12) like this:

(13)
$$E[Y - (aX + b)]^2$$
$$= E\{Y - \mu_Y - [a(X - \mu_X) + b - \mu_Y + a\mu_X]\}^2$$
$$= E[Y_{\text{new}} - (aX_{\text{new}} + c)]^2 \quad \text{where } c = b - \mu_Y + a\mu_X.$$

The key is that $\mu_{X\text{new}} = 0$ and $\mu_{Y\text{new}} = 0$ so you can apply Part I of the proof with X_{new} playing the role of X, Y_{new} playing the role of Y and c playing the role of b.
The minimum value in (13) is achieved with

$$a = \rho_{\text{new}}\frac{\sigma_{Y\text{new}}}{\sigma_{X\text{new}}} \quad \text{(by (7) in Part I)}$$

(14) $$= \rho\frac{\sigma_Y}{\sigma_X} \quad \text{(by (10) and (11))}$$

and

$$c = 0 \qquad \text{(by (6) in Part I)}$$
$$b - \mu_Y + a\mu_X = 0$$
(15) $$b = \mu_Y - a\mu_X$$

The values of a and b in (14) and (15) prove (2) in the general case.

And

$$\text{min value of (12)} = \text{min value of (13)}$$
$$= \sigma_{Y\text{new}}^2 (1 - \rho_{\text{new}}^2) \quad \text{(from (8))}$$
$$= \sigma_Y^2 (1 - \rho^2) \qquad \text{(by (10) and (11))}$$

This proves (3) in the general case.

Proof of (4)

If $\rho = \pm 1$, then the mean square error in (3) is 0, so Y equals its estimate.

Proof of (1)

The mean square error in (3) can't be negative, so $1 - \rho^2$ can't be negative, so $-1 \le \rho \le 1$.

Example

Toss a die 3 times.

Let X be the number of 1's.
Let Y be the number of 2's.

We'll find the correlation $\rho(X,Y)$ and the line of regression.
Each of X and Y has a binomial distribution with $n = 3, p = 1/6$, so

$$EX = EY = np = \frac{1}{2}, \quad \text{Var } X = \text{Var } Y = npq = \frac{5}{12}$$

and

$$E(XY) = 1P(X = 1, Y = 1) + 2P(X = 1, Y = 2 \text{ or vice versa })$$

$$= \frac{3!}{1!\,1!\,1!} \frac{1}{6} \frac{1}{6} \frac{4}{6} + 2 \cdot 2 \cdot \binom{3}{1} \frac{1}{6} \left(\frac{1}{6}\right)^2$$

$$= \frac{1}{6}$$

$$\text{Cov }(X,Y) = E(XY) - (EX)(EY) = -\frac{1}{12}$$

$$\rho(X,Y) = \frac{-1/12}{\sqrt{5/12}\,\sqrt{5/12}} = -\frac{1}{5}$$

(ρ is negative, indicating that if X goes up, Y tends to go down.)

The line of regression is

$$y = -\frac{1}{5}x + \frac{3}{5}$$

The linear least squares estimate for Y based on X is $-\frac{1}{5}X + \frac{3}{5}$.
By (3), the mean square error corresponding to this estimate is $2/5$.

CHAPTER 8

Conditional Probability

SECTION 8-1 CONDITIONAL DENSITIES

The Conditional Densities $f_{X|Y}$ and $f_{Y|X}$

The old rule

$$(1) \qquad P(A|B) = \frac{P(A \text{ and } B)}{P(B)}$$

still applies in the continuous case and can be used to find

$$P(a \le X \le b | c \le Y \le d)$$

But it can't be used to find

$$P(a \le X \le b | Y = y)$$

because $P(Y = y) = 0$ and (1) is intended only for events B where $P(B) \ne 0$.

To find conditional probabilities where the condition is $X = x$ or $Y = y$, events of probability 0, we'll invent conditional densities.

If $f(x|y)$ is a function (of x and y) such that

$$P(a \le X \le b | Y = y) = \int_a^b f(x|y)\, dx$$

then $f(x|y)$ is called the *conditional density of X given Y*. It is also denoted by $f_{X|Y}$.

Similarly, $f(y|x)$, the *conditional density of Y given X*, works like this:

$$P(a \le Y \le b | X = x) = \int_a^b f(y|x)\, dy$$

The units on $f(y|x)$ are probability per unit length so that

(2) $P(y \leq Y \leq y + dy | X = x) = P(Y \approx y | X = x) = f(y|x)\, dy$

and similarly for $f(x|y)$.

How the Conditional Densities, Joint Density, and Marginal Densities Are Related

The rule in (1) is for *probabilities*. There's a continuous version for *densities*:

(3) $$f(y|x) = \frac{f(x,y)}{f(x)} \quad \text{and} \quad f(x|y) = \frac{f(x,y)}{f(y)}$$

where

> $f(x,y)$ is the joint density of X and Y,
> $f(x)$ is the marginal density of X (i.e., $f_X(x)$),
> $f(y)$ is the marginal density of Y.

Here's why (3) holds.

$$f(y|x)\, dy = P(Y \approx y | X = x) \qquad \text{(by (2))}$$

$$= \frac{P(Y \approx y \text{ and } X \approx x)}{P(X \approx x)} \quad \text{(by (1))}$$

$$= \frac{f(x,y)\, dx\, dy}{f(x)\, dx}$$

Now cancel to get the first rule in (3).

Finding the Conditional Densities and the Conditional Universes

Let X and Y have joint density

(4) $f(x,y) = 2xy \qquad \text{for } 0 \leq y \leq 2x \leq 2$

First let's get the various universes straight.

> The joint universe is $0 \leq y \leq 2x \leq 2$, the triangular region in Fig. 1.
> The X universe is $0 \leq x \leq 1$ (look at the extreme x's in Fig. 1).
> The Y universe is $0 \leq y \leq 2$ (look at the extreme y's in Fig. 1).
> The $X|Y$ universe is $\frac{1}{2}y \leq x \leq 1$ (the horizontal segment in Fig. 2).
> The $Y|X$ universe is $0 \leq y \leq 2x$ (the vertical segment in Fig. 3).

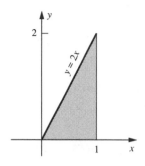

Figure 1   Joint universe

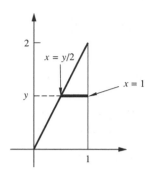

Figure 2   Conditional univ of $X|Y$

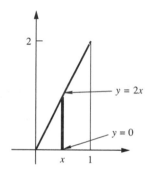

Figure 3   Conditional univ of $Y|X$

Here's how to use (3) to find the conditional densities $f(x|y)$ and $f(y|x)$.

To get $f(x|y)$, first you need the marginal $f(y)$. Use Fig. 2 to integrate out the other variable:

$$f(y) = \int_{x=-\infty}^{\infty} f(x, y)\, dx = \int_{x=y/2}^{x=1} 2xy\, dx = y - \frac{1}{4}y^3 \qquad \text{for } 0 \le y \le 2$$

Then by (3),

(5) $$f(x|y) = \frac{f(x, y)}{f(y)} = \frac{2x}{1 - \frac{1}{4}y^2} \qquad \text{for } 0 \le y \le 2, \frac{1}{2}y \le x \le 1$$

The conditional density $f(x|y)$ must be accompanied by *two* separate intervals which have *different* patterns.

First, the interval $0 \le y \le 2$ (the universe of Y) to explain what it is possible to condition on.

Second, the interval $\frac{1}{2}y \le x \le 1$, the universe of $X|Y$.

To find $f(y|x)$, first find $f(x)$. Use Fig. 3 to integrate out the other variable:

$$f(x) = \int_{y=-\infty}^{\infty} f(x, y)\, dy = \int_{y=0}^{2x} 2xy\, dy = 4x^3 \qquad \text{for } 0 \le x \le 1$$

Then

(6) $$f(y|x) = \frac{f(x|y)}{f(x)} = \frac{y}{2x^2} \qquad \text{for } 0 \le x \le 1, 0 \le y \le 2x$$

The conditional density $f(y|x)$ comes with *two* intervals:
First, the interval $0 \le x \le 1$ (the universe of X) to explain what it is possible to condition on.
Second, the interval $0 \le y \le 2x$, the universe of $Y|X$.

Finding Conditional Probabilities

To find $P(Y\text{-event}\,|X = x)$, integrate the density $f(y|x)$ over the favorable part of the (conditional) universe of $Y|X$:

$$P(Y\text{-event}|X = x) = \int_{\text{fav}} f(y|x)\, dy$$

Let's go back to X and Y with the joint density in (4) and the joint universe in Fig. 1, and find the conditional probability

$$P(Y \ge .4|X = x)$$

We found the conditional density $f(y|x)$ in (6).

The vertical segment in Fig. 3 is the universe of $Y|X$. The favorable part, where $y \ge .4$, depends on what x you condition on, so you need cases. Look at Fig. 4.

Case 1. $0 \le x \le .2$

Then

$$P(Y \ge .4|X = x) = 0$$

because in the conditional universe on the left in Fig. 4, there are *no* points where $y \ge .4$.

Case 2. $.2 \leq x \leq 1$

On the right in Fig. 4 is the conditional universe, showing the favorable part where $y \geq .4$. So

(7)
$$P(Y \geq .4|X = x) = \int_{y=.4}^{y=2x} \frac{y}{2x^2} \, dy = 1 - \frac{.04}{x^2}$$

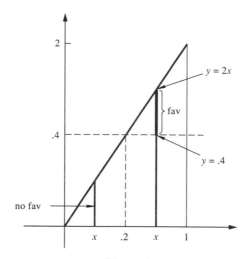

Figure 4

Warning

Be careful with the limits in (7).

The Y universe is [0,2], but the $Y|X$ conditional universe is [0,2x]. So to find $P(Y \geq .4|X = x)$, the upper limit of integration is $2x$, not 2.

Conditional Expectation

Remember that the *(plain)* expectation of the random variable Y is given by

(8)
$$EY = \int_{y=-\infty}^{\infty} y \, f(y) \, dy$$

Here's the definition of the *conditional* expectation of Y given X:

(9)
$$\boxed{E(Y|X = x) = \int_{\substack{\text{universe} \\ \text{of } Y|X}} y \, f(y|x) \, dy}$$

The definitions in (8) and (9) are similar. But for *conditional* expectation, use the *conditional* density $f(y|x)$ instead of the density $f(y)$ and integrate over the *conditional* universe of $Y|X$ instead of over the Y universe.

Let's go back to X and Y with the joint density in (4) and the joint universe in Fig. 1 and find the conditional expectations.

In (5) we found the conditional density

$$f(x|y) = \frac{2x}{1 - \frac{1}{4}y^2} \qquad \text{for } 0 \le y \le 2, \quad \tfrac{1}{2}y \le x \le 1$$

Figure 2 shows the conditional universe of $X|Y$ where $\frac{1}{2}y \le x \le 1$. Then

$$E(X|Y) = \int_{x=y/2}^{x=1} x\, f(x|y) = \int_{x=y/2}^{x=1} x\, \frac{2x}{1 - \frac{1}{4}y^2}\, dx$$

$$= \frac{y^2 + 2y + 4}{3(y + 2)} \qquad \text{for } 0 \le y \le 2$$

The formula for $E(X|Y = y)$ must come with the y-interval $0 \le y \le 2$ (the Y-universe) to describe the values of y it is possible to condition on.

In (6) we found the conditional density

$$f(y|x) = \frac{y}{2x^2} \text{ for } 0 \le x \le 1, \quad 0 \le y \le 2x$$

Figure 3 shows the conditional world of $Y|X$ where $0 \le y \le 2x$. Then

$$E(Y|X = x) = \int_{y=0}^{2x} y f(y|x)\,dy$$

$$= \int_{y=0}^{2x} y\, \frac{y}{2x^2}\,dy = \left.\frac{y^3}{6x^2}\right|_{y=0}^{2x} = \frac{4x}{3} \qquad \text{for } 0 \le x \le 1$$

Warning

1. In conditional problems you will be much better off if you *draw a picture* of the universe with the conditional universe inside.

2. The *conditional* expectation $E(X|Y = y)$ can have a y (but not an x) in it and must come with a y-interval (indicating what y's are possible). There may be cases; for example, you could have

$$E(X|Y = y) = \begin{cases} 5 & \text{if } 0 < y < 2 \\ 6y & \text{if } 2 \le y \le 8 \end{cases}$$

but the cases can't involve the letter x.

On the other hand, the *unconditional* expectation EX is a plain number with*out* an accompanying interval; it has *no* letters in it, and does *not* come with cases.

$$An\ answer\ like\ E(X) = \begin{cases} 5 & \text{if } \ldots \\ 6 & \text{if } \ldots \end{cases} \quad is\ wrong$$

3. The *conditional* probability $P(X \leq 3 | Y = y)$ can have a y (but not an x) in it and must come with a y-interval. There may be cases depending on y, that is, different answers for different y-intervals.

On the other hand, the *unconditional* probability $P(X \leq 3)$ has no letters in it and does not come with an interval or cases.

The Conditional Densities When the Joint Density Is Uniform

Here's an old result (see (9) in Section 5.2).

If X and Y have a joint uniform distribution on a *rectangle* of the form $a \leq x \leq b, c \leq y \leq d$, then X and Y are independent and each is uniform.

If X and Y have a joint uniform distribution but *not* on such a rectangle, then X and Y are not individually uniform (nor are they independent). But it *is* true that $X|Y$ and $Y|X$ are uniform. Here's an illustration of the general idea.

Suppose X and Y have a joint uniform distribution on the *non-rectangle* in Fig. 5, where $0 \leq x \leq \sqrt{2}, x^3 \leq y \leq 2x$.

Then $Y|X$ is uniform on the interval $[x^3, 2x]$ (Fig. 6).
And $X|Y$ is uniform on the interval $[y/2, \sqrt[3]{y}]$ (Fig. 7).

So by inspection,

$$f(y|x) = \frac{1}{2x - x^3} \quad \text{for } 0 \leq x \leq \sqrt{2},\ x^3 \leq y \leq 2x$$

$$f(x|y) = \frac{1}{\sqrt[3]{y} - \frac{1}{2}y} \quad \text{for } 0 \leq y \leq 2\sqrt{2},\ \tfrac{1}{2}y \leq x \leq \sqrt[3]{y}$$

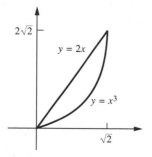

Figure 5   Joint universe

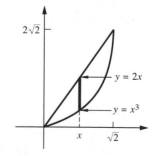

Figure 6   Universe for $Y|X$

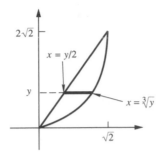

Figure 7   Universe for $X|Y$

This works because any one point in the region is as likely as any other point in the region. So, if $X = x$, then Y is just as likely to take on one value as another in the conditional universe in Fig. 6. So $Y|X = x$ is uniform on $[x^3, 2x]$.

Problems for Section 8-1

1. Let X and Y have joint density

$$f(x, y) = \frac{24}{5}(x + y) \qquad \text{for } 0 \le 2y \le x \le 1$$

Find

(a) $P(X \ge \frac{1}{2}|Y = y)$ **(c)** $E(X|Y = y)$
(b) $P(.1 \le X \le .2|Y = y)$

2. Let X and Y have joint density

$$f(x, y) = e^{-y} \qquad \text{for } x \ge 0,\ y \ge x$$

Find

(a) $f(y|x)$ and $f(x|y)$ **(c)** $E(Y|X = x)$
(b) $P(X \le 2|Y = y)$ **(d)** $E(X|Y = y)$

3. Let X and Y be jointly uniform in the indicated parallelogram. Find (by inspection)

(a) $f(y|x)$ and $f(x|y)$
(b) $P(1 \le X \le 2|Y = \frac{1}{2})$
(c) $P(Y \le \frac{1}{2}|X = x)$
(d) $E(Y|X = x)$
(e) $E(X|Y = y)$

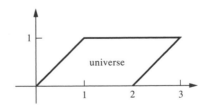

Figure P. 3

4. Let X and Y have joint density

$$f(x,y) = cxy^2 \qquad \text{for } 0 \le y \le 2x, \; 1 \le x \le 5$$

   (a) Find c but don't bother doing the integration.
   (b) Find $P(X \le 2 | Y = y)$.
   (c) Find $E(X | Y = y)$.

5. Pick a point (X, Y) at random inside the circle $x^2 + y^2 = 1$. Find $f(y|x)$ **(a)** by inspection and **(b)** the long way.

SECTION 8-2   2-STAGE EXPERIMENTS

Finding a Second Stage Probability

(1)

   Suppose at the first stage of an experiment the result is X where

$$f_X(x) = xe^{-x} \qquad \text{for } x \ge 0$$

   Then, if $X = x$, at the second stage Y is chosen at random between 0 and x. In other words

$$f(y|x) = \frac{1}{x} \qquad \text{for } x \ge 0 \text{ and } 0 \le y \le x$$

I'll find $P(Y \le 2)$ with two methods.

Using the joint density. When X is chosen at the first stage, its value x can be anything ≥ 0. Once $X = x$, the value y of Y must be between 0 and x. So the universe consists of those points (x, y) in the plane where $x \ge 0$ and $0 \le y \le x$ (Fig. 1). Then

(2) $\qquad f(x,y) = f(x)f(y|x) = e^{-x} \qquad \text{for } x \ge 0, 0 \le y \le x$

$$P(Y \le 2) = \int_{\text{fav}} f(x,y) \, dA \quad \text{(Fig. 2)}$$

$$= \int_{y=0}^{2} \int_{x=y}^{\infty} e^{-x} dx \, dy$$

$$= \int_{y=0}^{2} e^{-y} \, dy = 1 - e^{-2}$$

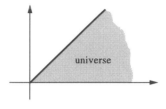

Figure 1

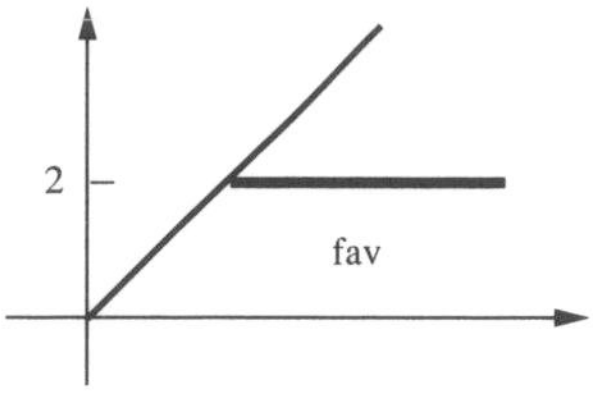

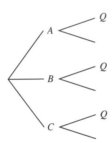

Figure 2

Using the theorem of total probability.   If the outcome of the first stage of an experiment is one of A, B, C, and Q is any event, then the *discrete* version of the theorem of total probability is

(3) $P(Q) = P(A)P(Q|A) + P(B)P(Q|B) + P(C)P(Q|C)$ (Fig. 3)

Figure 3

For the *continuous* case, suppose the result of the first stage is the random variable X, where X takes on a continuum of values. (We can't draw a legal tree because there are uncountably many branches at the first stage, but Fig. 4 shows a typical branch). And suppose the result of the second stage is the random variable Y. If Q is any event, then by (3),

$$P(Q) = \text{ sum of favorable branches } = \sum_x \underbrace{P(X \approx x)}_{f_X(x)\,dx}\,P(Q|X = x)$$

which becomes the following *theorem of total probability*:

(4) $$P(Q) = \int_{x=-\infty}^{\infty} P(Q|X = x)\, f_X(x)\, dx$$

In particular,

$$P(a \le Y \le b) = \int_{x=-\infty}^{\infty} P(a \le Y \le b|X = x)\, f_X(x)\, dx$$

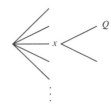

Figure 4

For the 2-stage experiment in (1), $Y|X$ is uniformly distributed in $[0, x]$ so

$$P(Y \le 2|X = x) = \frac{\text{fav length}}{\text{total length}} = \begin{cases} 1 & \text{if } 0 \le x \le 2 \\ 2/x & \text{if } x \ge 2 \end{cases} \quad \text{(Fig. 5)}$$

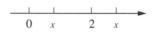

Figure 5

Then

$$P(Y \le 2) = \int_{x=-\infty}^{\infty} P(Y \le 2|X = x) f(x)\, dx$$

$$= \int_{x=0}^{2} 1 \cdot xe^{-x}\, dx + \int_{x=2}^{\infty} \frac{2}{x} \cdot xe^{-x}\, dx$$

$$= (-xe^{-x} - e^{-x})\Big|_{0}^{2} - 2e^{-x}\Big|_{2}^{\infty} = 1 - e^{-2}$$

Using the theorem of total probability takes *two* 1-dimensional integrals: a preliminary integral (or fav/total calculation) to find $P(Q|X = x)$ and then the integral in (4). It amounts to exactly the same computation as the method of double integrating the joint density with inner y limits and outer x limits.

Unless the conditional prob $P(Q|X = x)$ can be found easily by inspection (in this example, it wasn't too bad), it is usually more straightforward to integrate the joint density than to use the theorem of total probability.

Finding a Second Stage Expectation

I'll find $E(Y)$ for the experiment in (1) with two methods.

Using the joint density and the law of the unconscious statistician.   The joint density is in (2) (the universe is in Fig. 1). Then

$$E(Y) = \int_{\text{universe}} y\, f(x,y)\, dA = \int_{x=0}^{\infty} \int_{y=0}^{x} y e^{-x}\, dx$$

$$= \int_{0}^{\infty} \frac{1}{2} x^2 e^{-x}\, dx = \frac{1}{2} 2! = 1$$

Using the theorem of total expectation.   The discrete version (Section 3.3) said that if the random variable Y is the (numerical) result of an experiment and the outcomes can be divided, say, into red, blue, and green, then

$$EY = P(\text{red})E(Y|\text{red}) + P(\text{blue})E(Y|\text{blue}) + P(\text{green})E(Y|\text{green}) \quad (\text{Fig. 6})$$

Figure 6   **Figure 7**

For the continuous case, if X is the result of the first stage of an experiment and Y is the result of the second stage (Fig. 7), then by analogy

$$EY = \sum_{X} P(X \approx x)\, E(Y|X = x) = \sum_{X} f_X(x)\, dx\, E(Y|X = x)$$

so here's the continuous version of the *theorem of total expectation*:

(5)
$$EY = \int_{x=-\infty}^{\infty} E(Y|X = x)\, f_X(x)\, dx$$

For the experiment in (1), $Y|X$ is uniform on $[0, x]$ so $E(Y|X = x) = x/2$ and

$$EY = \int_{x=0}^{\infty} \frac{x}{2} x e^{-x}\, dx = \frac{1}{2} \int_{0}^{\infty} x^2 e^{-x}\, dx = \tfrac{1}{2} 2! = 1$$

If $E(Y|X = x)$ can be found by inspection (as in this example), then (5) is the easiest way to find EY. Otherwise, use the joint density and the unconscious statistician.

Finding an a Posteriori Probability

Let's continue with the experiment in (1) and find

$$P(1 \le X \le 4|Y = y)$$

by first finding $f(x|y)$. We have

$$f(x,y) = f(x)f(y|x) = e^{-x} \qquad \text{for } x \geq 0,\ 0 \leq y \leq x \quad \text{(Fig. 8)}$$

$$f(y) = \int_{x=y}^{\infty} e^{-x}\,dx = e^{-y} \qquad \text{for } y \geq 0 \ \text{(Fig. 9)}$$

(6) $$f(x|y) = \frac{f(x,y)}{f(y)} = e^{y}\,e^{-x} \qquad \text{for } y \geq 0,\ x \geq y$$

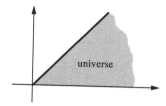

Figure 8

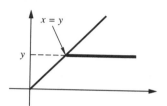

Figure 9

If $0 \leq y \leq 1$ (Fig. 10), then

$$P(1 \leq X \leq 4 | Y = y) = \int_{x=1}^{4} e^{y}\,e^{-x}\,dx = e^{y}\,(e^{-1} - e^{-4})$$

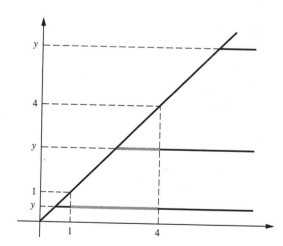

Figure 10

If $1 \leq y \leq 4$ (Fig. 10), then

$$P(1 \leq X \leq 4|Y = y) = \int_{x=y}^{4} e^y e^{-x} \, dx = e^y \left(e^{-y} - e^{-4} \right)$$

If $y \geq 4$ (Fig. 10), then

$$P(1 \leq X \leq 4|Y = y) = 0$$

Finding the a Posteriori Expectation

Continue with the experiment in (1). We have $f(x|y)$ in (6), and Fig. 9 shows a picture of the conditional world. So

$$E(X|Y = y) = \int_{x=y}^{\infty} xf(x|y) \, dx = \int_{x=y}^{\infty} x \, e^y e^{-x} \, dx$$

$$= e^y \left[-xe^{-x} - e^{-x} \right]_{x=y}^{\infty} = y + 1 \qquad \text{for } y \geq 0$$

Summary of What to Use as the Limits of Integration

Suppose X and Y have joint density $f(x, y)$ for the universe in Fig. 11.

If you look at *extreme* x's, you see $0 < x < \infty$ (the X universe).
If you look at *boundary* x's (Fig. 12), you see $y \leq x < \infty$ (the $X|Y$ universe.)

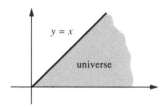

Figure 11

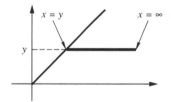

Figure 12

Here's when to use $\int_{x=0}^{\infty}$

1. To find EY with the theorem of total expectation, use

$$EY = \int_{x=0}^{\infty} E(Y|X = x)f(x) \, dx$$

2. To find $P(Y\text{-event})$ with the theorem of total probability, use

$$P(Y\text{-event}) = \int_{x=0}^{\infty} P(Y\text{-event}|X = x)f(x)\,dx$$

Here's when to use $\int_{x=y}^{\infty}$

1. To find $f(y)$ by integrating out the other variable, use

$$f(y) = \int_{x=y}^{\infty} f(x,y)\,dx \quad \text{for } y \geq 0$$

2. To find $E(X|Y = y)$, use

$$E(X|Y = y) = \int_{x=y}^{\infty} x\,f(x|y)\,dx \quad \text{for } y \geq 0$$

Problems For Section 8-2

1. Choose X at random between 0 and 8. If $X = x$, choose Y at random between 0 and x.
   (a) Find $f(x), f(y), f(x|y), f(y|x), f(x,y)$.
   (b) Find $P(3 \leq Y \leq 5)$ twice, with and without the theorem of total probability.
   (c) Find $P(X \leq 7|Y = y)$.
   (d) Find EY twice, with and without the theorem of total expectation.
   (e) Find $E(X|Y = y)$.

2. At the first stage of an experiment, the result is X with density $f(x) = e^{-x}$ for $x \geq 0$. Then, if $X = x$, at the second stage $Y|X$ has density x/y^2 for $y \geq x$.
   Find

   (a) $P(Y \leq 2)$ (c) EY
   (b) $P(X \geq 7|Y = y)$ (d) $E(X|Y = y)$

   For reference: $\int xe^{-x}dx = -xe^{-x} - e^{-x}$

3. Let X have density $f(x) = 1/x^2$ for $x \geq 1$. If $X = x$, pick Y at random between 0 and x. Find

   (a) $P(Y \leq 5)$ (c) EY
   (b) $P(2 \leq X \leq 7|Y = y)$ (d) $E(X|Y = y)$

4. Pick X at random between 0 and 5. If $X = x$, pick Y at random between x and $2x$. Find

   (a) $P(Y \leq 2)$

   (b) $P(1 \leq X \leq 2|Y = 3)$

   (c) $P(X \geq 4|Y = y)$

   (d) EY

   (e) $E(X|Y = y)$

5. Pick R at random between 0 and 10. If $R = r$, pick a point at random in a circle of radius r. Let Z be its distance to the center.

   (a) Find $P(Z \leq 5)$ and EZ using the theorems of total prob and total expectation.

   (b) Find $f(z|r)$ by first finding the distribution function $F(z|r)$.

   (c) Find $f(r, z)$ and use it to find $P(Z \leq 5)$ and EZ again.

   (d) Find $P(R \leq 5|Z = z)$.

6. The lifetime (in years) of a gadget has an exponential distribution with parameter $\lambda = 1/V^2$, where V is uniformly distributed between 0 and v_0.

   (a) Find the expected lifetime.

   (b) Find the probability that a gadget lasts at most 3 years but leave the integral(s) unevaluated.

7. Let X be uniform on [0,10]. If $X = x$, let Y be uniform on $[0, x]$. Find EY and Var Y.

SECTION 8-3   MIXED 2-STAGE EXPERIMENTS

In Section 2.3 we did 2-stage experiments where each stage was discrete. In the preceding section we did 2-stage experiments where each stage was continuous. Now we'll try it with one stage discrete and the other stage continuous.

Example 1 first stage continuous, second stage discrete

Pick X at random between 0 and 1.

   If $X = x$, toss a coin 5 times where $P(\text{heads}) = x$. Let N be the number of heads.

   I'll find all the densities and then use them to find expectations and probabilities.

   First,

$$f(x) = 1 \qquad \text{for } 0 \leq x \leq 1$$

The random variable N is discrete, so the conditional "density" $f(n|x)$ is just a plain probability function:

$$f(n|x) = P(N = n|X = x)$$
$$= P(n \text{ heads in 5 tosses where prob of H on any one toss is } x)$$
$$= \binom{5}{n} x^n (1 - x)^{5-n} \qquad \text{for } 0 \leq x \leq 1 \text{ and } n = 0, 1, 2, 3, 4, 5$$

The joint universe of X and N consists of those points (x, n) where $0 \leq x \leq 1$ and $n = 0, 1, 2, 3, 4, 5$ (the line segments in Fig. 1). The joint density is

$$f(x, n) = f(x)f(n|x) = \binom{5}{n} x^n (1-x)^{5-n} \qquad \text{for } 0 \leq x \leq 1 \text{ and } n = 0, 1, 2, 3, 4, 5$$

and the marginal density of N is

$$f_N(n) = \int_0^1 f(x, n)\, dx \qquad \text{(integrate out the other variable, Fig. 2)}$$

(1)
$$= \binom{5}{n} \int_0^1 x^n (1 - x)^{5-n}\, dx$$

$$= \binom{5}{n} \frac{n!\,(5 - n)!}{6!} \qquad \text{(see (*))}$$

$$= \frac{1}{6} \qquad \text{for } n = 0, 1, \ldots, 5$$

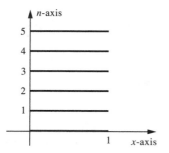

Figure 1   The joint universe

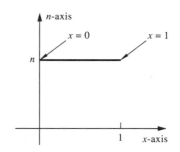

Figure 2

In (1) we used the formula

(*)
$$\int_0^1 x^n (1 - x)^m\, dx = \frac{n!\,m!}{(n + m + 1)!}$$

Finally,

$$f(x|n) = \frac{f(x, n)}{f(n)} = 6\binom{5}{n} x^n (1 - x)^{5-n} \qquad \text{for } n = 0, \ldots, 5; 0 \leq x \leq 1$$

Here's how to use all the densities.

Finding the second stage expectation $E(N)$ using the theorem of total expectation.
$N|X$ has a binomial distribution with $n = 5, p = x$ so $E(N|X = x) = 5x$.

Then

$$E(N) = \int_0^1 E(N|X = x)f(x)\,dx = \int_0^1 5x \cdot 1\,dx = \frac{5}{2}$$

Finding the second stage expectation $E(N)$ again, using $f(n)$.   We have

$$E(N) = \sum_n n\,P(N = n)$$

N is discrete, so $P(N = n)$ is just the density $f_N(n)$ from (1) and

$$E(N) = \sum_n n\,f(n) = 1 \cdot \frac{1}{6} + 2 \cdot \frac{1}{6} + \cdots + 5 \cdot \frac{1}{6} = \frac{5}{2}$$

Finding the second stage probability $P(N = n)$.

$$P(N = n) = f(n) = \frac{1}{6} \qquad \text{for } n = 0, 1, \ldots, 5$$

Finding an a posteriori probability.

$$P(X \le .5 | N = 0) = \int_0^{.5} f(x|0)\,dx \quad \textbf{(Fig. 3)}$$

$$= 6\binom{5}{0} \int_0^{.5} (1 - x)^5\,dx$$

$$= 1 - (.5)^6$$

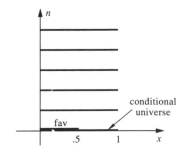

Figure 3

Finding the a posteriori expectation $E(X|N = n)$.

$$E(X|N = n) = \int_0^1 x\, f(x|n)\, dx = 6\binom{5}{n} \int_0^1 x^{n+1}\, (1-x)^{5-n}\, dx$$

$$= 6\binom{5}{n} \frac{(n+1)!\,(5-n)!}{7!} \quad (\text{use } (*))$$

$$= \frac{n+1}{7} \qquad \text{for } n = 0, 1, \ldots, 5$$

For example, if you get no heads, then the expected value of X is $1/7$.

Example 2 first stage discrete, second stage continuous

Toss a 3-sided die with faces 4, 5, 10 where

$$P(4) = .5$$
$$P(5) = .2$$
$$P(10) = .3$$

Let N be the face value of the die.

If $N = n$, let Y have an exponential distribution with parameter $\lambda = n$. (You can think of Y as the waiting time for the first arrival where particles arrive at the rate of 4 or 5 or 10 per second, depending on the die toss.)

The random variable N is discrete, so

$$f_N(n) = P(N = n) = \begin{cases} .5 & \text{if } n = 4 \\ .2 & \text{if } n = 5 \\ .3 & \text{if } n = 10 \end{cases}$$

And we are given that

$$f(y|n) = ne^{-ny} \qquad \text{for } n = 4, 5, 10;\ y \geq 0$$

Finding the second stage expectation EY with the theorem of total expectation. $Y|N$ has an exponential distribution with parameter n so by inspection, $E(Y|N = n) = 1/n$. Then

$$E(Y) = \sum_n E(Y|N = n)\, P(N = n) = \sum_n E(Y|N = n)\, f_N(n)$$

$$= \frac{1}{4}(.5) + \frac{1}{5}(.2) + \frac{1}{10}(.3) = .195$$

Finding the second stage expectation EY again, using $f(y)$. First, we need $f(n, y)$ and then $f(y)$.

The joint universe of N and Y consists of the three half lines in Fig. 4, where $n = 4, 5, 10$ and $y \geq 0$. The joint density is

$$f(n, y) = f_N(n) f(y|n) = f(n) \, n e^{-ny} \qquad \text{for } n = 4, 5, 10; \; y \geq 0$$

To get the marginal density $f_Y(y)$ from the joint density, *sum* out the other variable n (Fig. 5):

$$f(y) = \sum_n f(n, y) = f(4, y) + f(5, y) + f(10, y)$$

(2)
$$= .5(4e^{-4y}) + .2(5e^{-5y}) + .3(10e^{-10y})$$

$$= 2e^{-4y} + e^{-5y} + 3e^{-10y} \qquad \text{for } y \geq 0$$

Then

$$EY = \int_0^\infty y f(y) \, dy = \int_0^\infty y(2e^{-4y} + e^{-5y} + 3e^{-10y}) \, dy = .195$$

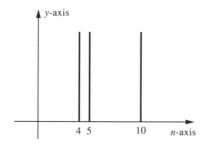

Figure 4

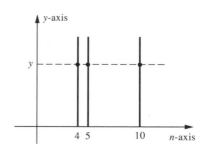

Figure 5

Finding a second stage probability using the theorem of total probability.

$$P(Y \leq 3) = P(N = 4) \, P(Y \leq 3|N = 4) + P(N = 5) \, P(Y \leq 3|N = 5)$$

$$+ P(N = 10) \, P(Y \leq 3|N = 10)$$

$$= .5 \int_0^3 4e^{-4y} \, dy + .2 \int_0^3 5e^{-5y} \, dy + .3 \int_0^3 10e^{-10y} \, dy$$

Finding a second stage probability again using $f(y)$.   We found $f(y)$ in (2). Then

$$P(Y \leq 3) = \int_0^3 f(y) \, dy = \int_0^3 (2e^{-4y} + e^{-5y} + 3e^{-10y}) \, dy$$

Finding an a posteriori probability.   I'll find $P(N = 4|Y = y)$.
The random variable N is discrete, so $P(N = n|Y = y)$ is just the density $f(n|y)$:

$$f(n|y) = \frac{f(n, y)}{f(y)} = \frac{f(n)\, ne^{-ny}}{2e^{-4y} + e^{-5y} + 3e^{-10y}} \qquad \text{for } y \geq 0; \qquad n = 4, 5, 10$$

So

$$P(N = 4|Y = y) = f(4|y) = \frac{(.5)4e^{-4y}}{2e^{-4y} + e^{-5y} + 3e^{-10y}} \qquad \text{for } y \geq 0$$

Finding the a posteriori expectation $E(N|Y = y)$.   The random variable N is discrete so $P(N = n|Y = y)$ is $f(n|y)$ and

$$E(N|Y = y) = \sum_n n\, P(N = n|Y = y)$$

$$= \sum_n n\, f(n|y) = 4f(4|y) + 5f(5|y) + 10f(10|y)$$

$$= \frac{4(.5)\, 4e^{-4y} + 5(.2)\, 5e^{-5y} + 10(.3)\, 10e^{-10y}}{2e^{-4y} + e^{-5y} + 3e^{-10y}} \qquad \text{for } y \geq 0$$

Problems for Section 8-3

1. Let X have density $f_X(x) = 6x(1 - x)$ for $0 \leq x \leq 1$. If $X = x$, toss a coin 10 times where $P(H) = x$. Let N be the number of heads.
   (a) Find $f_N(n), f(x, n), f(n|x)$, and $f(x|n)$.
   (b) Find $P(10H)$.
   (c) Find $E(N)$.
   (d) Find $E(X|N = n)$ in general and $E(X|N = 10)$ in particular.
   (e) Find $P(X \leq \frac{1}{2}|N = n)$ leaving the integral unevaluated.

2. Let X be uniform on [0,1]. If $X = x$, toss a coin with $P(\text{heads}) = x$. Let N be the number of tosses needed to get a head.
   (a) Find $P(N = n)$ in general and $P(N = 7)$ in particular.
   (b) Find $E(N)$.
   (c) Find $P(X \leq .5|N = n)$ but leave the integral unevaluated.
   (d) Find $E(X|N = n)$.

3. Toss a fair 3-sided die with faces 1, 2, 3. Let N be the resulting face value. If $N = n$, pick a number X at random between 0 and n.
   (a) Find $f_X(x), f_N(n), f(n, x), f(n|x)$, and $f(x|n)$.
   (b) Look at $f(n|x)$. To check that you understand it, decide what is supposed to add up or integrate to 1 as befits a legal (conditional) density.

  (c) Find $E(X)$.
  (d) Find $P(0 \leq X \leq 1)$.
  (e) Find $E(N|X = x)$.
  (f) Find $P(N = 2|X = x)$.

4. Toss a fair 4-sided die with faces 1, 2, 3, 4. Call the resulting value N.
    If $N = n$, pick a number Y so that Y has density n/y^2 for $y \geq n$.
    (a) Show that $f(y|n)$ is a legitimate density.
    (b) Find $P(Y \geq \pi)$.
    (c) Find EY.
    (d) Find $E(N|Y = y)$.
    (e) Find $P(N = 3|Y = y)$.

Review Problems for Chapters 6, 7, 8

1. Let X and Y have joint density

$$f(x, y) = \begin{cases} \dfrac{3}{2} & \text{for points in region I} \\[2mm] \dfrac{1}{2} & \text{for points in region II} \end{cases}$$

  (a) Set up the integrals for EX and Var X.
  (b) Find $E(X|Y = y)$ and then $E(X|Y = \frac{1}{2})$.
  (c) Find $P(X \leq \frac{3}{4}|Y = y)$.

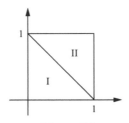

Figure P.1

2. The length of a telephone call is a random variable with density

$$f(x) = xe^{-x}, \; x \geq 0$$

Any call up to 3 minutes in length costs a flat fee of $2. After 3 minutes, the meter starts running steadily and there is an additional cost at the rate of 50 cents per minute (e.g., 25 cents for a half minute). Find the expected cost of a telephone call (set up the integral but don't evaluate).

3. Let X and Y have joint density

$$f(x,y) = ke^{-y} \qquad \text{for } 0 \le x \le y \le 1$$

where k is a fixed constant. Find $E(X|Y=y)$.

4. Pick X at random between 0 and 10. If $X = x$, let Y have density $2x^2/y^3$ for $y \ge x$. Find

   (a) $P(2 \le Y \le 3)$ **(c)** EY
   (b) $P(X \ge 2|Y=y)$ **(d)** $E(X|Y=y)$

5. Let X have an exponential distribution with $\lambda = 1$. If $X = x$, let Y have an exponential distribution with $\lambda = 1/x$. Find EY and Var Y.

6. Let $X_1, \ldots, X_n$ be iid, each uniform on $[0,3]$. Find the expected value of the maximum of $X_1, \ldots, X_n$.

7. Let X be uniform on $[0,1]$. If $X = x$, toss a coin 10 times where $P(\text{heads}) = x$. Let

$$N = \begin{cases} 1 & \text{if there is at least one H} \\ 0 & \text{otherwise} \end{cases}$$

Find

   (a) $P(N=0)$ and $P(N=1)$ **(c)** $E(N)$
   (b) $P(0 \le X \le \frac{1}{2}|N=0)$ **(d)** $E(X|N=0)$

8. Toss a fair coin. If it comes up heads, let $Y = 3$. If it comes up tails, pick Y at random between 0 and 5. Find EY and Var Y.

9. Toss two dice. Find the expectation and variance of the maximum of the two faces.

10. Let X have density

$$f(x) = \begin{cases} \dfrac{1}{3} & \text{if } -1 \le x \le 0 \\[2mm] \dfrac{2}{3} & \text{if } 0 \le x \le 1 \end{cases}$$

Let $Y = e^X$. Find $E(Y)$ and Var Y.

11. Toss a coin with $P(\text{heads}) = .2$.

   If the result is heads, pick X at random between 0 and 5.
   If the result is tails, pick X at random between 0 and 10.

Find
   (a) $P(2 \le X \le 6)$

(b) EX

(c) $P(\text{heads}|X = x)$ and $P(\text{tails}|X = x)$

12. Simplify
    (a) $\text{Cov}(X, X)$ **(c)** $\text{Var}(X - \mu_X)$

    (b) $E(X - \mu_X)$ **(d)** $\text{Cov}(X, 3)$

13. Express $\text{Cov}(X + Y, Z)$ in terms of $\text{Cov}(X, Y)$, $\text{Cov}(Y, Z)$, $\text{Cov}(X, Z)$. Try it twice.
    (a) Use the definition of covariance.
    (b) Use the "more practical" way to find covariance.

14. Pick a point at random in the indicated triangle. Find
    (a) $E(Y|X = x)$
    (b) $E(X|Y = y)$
    (c) EY
    (d) $P\left(X \le \frac{1}{2} \middle| Y = \frac{1}{10}\right)$
    (e) the expected value of $\min(X, Y)$ (just set up the integral)

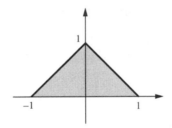

Figure P.14

15. Pick R at random between 0 and 10.

    If $R = r$, pick a point at random in a sphere of radius r.

    Let Dist be its distance to the center of the sphere. Use the letter ρ for values of Dist.
    (a) Find $f(\rho|r)$ by first finding the distribution function $F(\rho|r)$.
    (b) Find $f(r, \rho)$.
    (c) Find $P(\text{Dist} \le 5)$. You can use the theorem of total probability, but I think it's easier to use the joint density from part (b).
    (d) Find $E(\text{Dist})$. You can use the theorem of total expectation, but it's probably easier to use the joint density from part (b).
    (e) If the point's distance to the center is ρ, find the expected radius of the sphere.

    In particular, if the point's distance to the center is 7, what is the expected radius of the sphere?

CHAPTER 9

Limit Theorems

SECTION 9-1   THE CENTRAL LIMIT THEOREM

Sums of random variables turn up a lot. We already know something about sums of independent binomials, Poissons, exponentials, and normals (Section 6.2). Now we have a famous result about the sum of a large number of arbitrary iid random variables.

The Central Limit Theorem

(1)

> Let $X_1, \ldots, X_n$ be iid random variables with common mean μ and common variance σ^2.
>
> (Think of the X_i's as n independent observations of the same random variable or as n independent samples or as the results of performing the same experiment independently n times.)
>
> If n is large, then $X_1 + \cdots + X_n$ is approximately normal with mean $n\mu$ and variance $n\sigma^2$.

It follows from properties of expectation and variance that the sum has mean $n\mu$ and variance $n\sigma^2$. The fact that the sum is approximately *normal* is much harder to prove and can't be done in the context of a first course in probability.

We know (Section 4.4) that if a random variable is normal, then dividing it by n keeps it normal but divides the mean by n and the variance by n^2. So here's a corollary of (1):

> Let $X_1, \ldots, X_n$ be iid random variables with common mean μ and common variance σ^2. If n is large, then the sample average
>
> (2)
> $$\frac{X_1 + \cdots + X_n}{n}$$
>
> is approximately normal with mean μ and variance σ^2/n.

Special Case of the Normal Approximation to the Binomial

In Section 4.4 we made this claim:

(3)   If X has a binomial distribution with large n, then X is approximately normal.

Here's why (3) is just a special case of the central limit theorem.
    Suppose X is the number of heads in n tosses where $P(\text{head}) = p$. Then

$$X = X_1 + \cdots + X_n$$

where

$$X_i = \begin{cases} 1 & \text{if } i\text{th toss is a head} \\ 0 & \text{otherwise} \end{cases}$$

The X_i's are iid with common mean

$$EX_i = P(X_i = 1) = p$$

and common variance

$$\begin{aligned} \text{Var } X_i &= E(X_i^2) - (EX_i)^2 \\ &= EX_i - (EX_i)^2 \quad (\text{because } X_i^2 = X_i) \\ &= p - p^2 \\ &= p(1-p) \\ &= pq \end{aligned}$$

So by (1), X is approximately normal, with mean np and variance npq.

Example 1

A computer adds 1000 random numbers that have been rounded off to the nearest tenth. Find the probability that the total roundoff error is ≥ 1 (that the sum is at least 1 larger than it should be).

Let X_i be the ith roundoff error. The X_i's are iid, and each is uniform on $[-.05, .05]$. Their common mean is 0 and common variance is

$$\frac{(b-a)^2}{12} = \frac{1}{1200} \qquad \text{(see Section 7.2, problem 2)}$$

By (1), the sum $X_1 + \cdots + X_{1000}$ is approximately normal with mean 0 and variance $1000 \cdot 1/1200 = .83$. So

$$P(\text{total roundoff error } \geq 1) = P(X_1 + \cdots + X_{1000} \geq 1)$$

$$= P\left(X^* \geq \frac{1}{\sqrt{.83}}\right)$$

$$= 1 - F^*(1.1)$$

$$= .136 \qquad \text{(approximately)}$$

Generalized Central Limit Theorem

The sum of a large number of independent but *not necessarily identically distributed* random variables is approximately normal provided that no one random variable contributes appreciably to the sum; that is, no term dominates the others.

Problems for Section 9-1

1. The average time to service a customer at a checkout counter is 1.5 minutes with variance 4. Find the probability that 100 customers can be serviced in less than 2 hours.

2. When the meter says you pumped a gallon of gas, it's making an error that has mean 0 (so on the average you do get your gallon) and variance .01.

   Find the probability that when the meter reads 400 gallons, the actual amount pumped is within 4 gallons of 400.

3. Let $X_1, \ldots, X_n$ be iid, each exponentially distributed with parameter λ, so that each has mean $1/\lambda$ and variance $1/\lambda^2$.
   (a) Find the distribution of $X_1 + \cdots + X_n$ (no approximations).
   (b) If n is large, for what a_n and b_n will

   $$\frac{X_1 + \cdots + X_n - a_n}{b_n}$$

   have approximately the unit normal distribution?

4. One hundred people each toss a coin where $P(\text{heads}) = 1/3$. Each tosses until he or she gets 5 heads. Find the probability that it takes 1550 or fewer tosses total.

For reference: If X has a negative binomial distribution with parameters r and p (X is the number of tosses to get r heads where $P(\text{heads}) = p$), then $EX = r/p$ and $\text{Var } X = rq/p^2$.

5. Suppose the average height of an American woman is 65 inches with variance 2. How large a sample of women do you have to take to be 90% sure that the sample's average height is within 1 inch of the true average 65.

SECTION 9-2   THE WEAK LAW OF LARGE NUMBERS

Here's what we've been thinking all along about the connection between the expected value of X and the physical world:

(1)   Suppose the result of an experiment is the random variable X with mean μ. If you perform the experiment many times and average the results, it is very likely that the average will be near μ.

The weak law of large numbers tries to capture this idea. We'll state the law first, provide a running commentary to show that its physical interpretation is (1), and then prove it.

The Weak Law of Large Numbers	Physical Interpretation
Let $X_1, \ldots, X_n$ be iid with common mean μ and common variance σ^2.	The result of an experiment is a random variable with mean μ and variance σ^2. Perform the experiment independently n times (take a sample of size n).
Pick any number $\epsilon > 0$ and look at the event $$\left| \frac{X_1 + \cdots + X_n}{n} - \mu \right| < \epsilon$$	See if the sample average is ϵ- close to the true mean μ, a closeness we're rooting for.
If $n \to \infty$, then $$P\left(\left| \frac{X_1 + \cdots + X_n}{n} - \mu \right| < \epsilon \right) \to 1$$ We say that $X_1 + \cdots + X_n/n$ *converges in probability* to μ.	If you perform the experiment many times, you are likely to get that arbitrary closeness.

Here's a familiar special case of the weak law of large numbers.
Toss a coin n times where $P(\text{head}) = p$. Let

$$X_i = 1 \quad \text{if the } i\text{th toss is a head}$$

The X_i's are iid with common mean p, so for any $\epsilon > 0$,

$$P\left(\left|\frac{X_1 + \cdots + X_n}{n} - p\right| < \epsilon\right) \to 1 \qquad \text{as } n \to \infty$$

In this case $(X_1 + \cdots + X_n)/n$ is the relative frequency of heads in the sample, so the interpretation is that *if* P*(head)* = p *and you toss many times, the percentage of heads is very likely to be very close to* p.

The proof of the weak law is based on two inequalities that we have to prove first.

Markov's Inequality

Suppose X is a random variable and you're interested in knowing how likely it is that X will be very large.

If you know the distribution of X, you can find $P(X \geq k)$ easily:

$$P(X \geq k) = 1 - F_X(k)$$

But suppose you know only the mean of X and not its specific distribution. Then you can't find $P(X \geq k)$ precisely, but if X is non-negative, you can find a bound on it as follows:

If X is a non-negative random variable and $k > 0$, then

$$P(X \geq k) \leq \frac{EX}{k}$$

Markov's inequality gives a universal bound, good no matter what distribution X has.

For example, suppose X has mean 3 (and you know nothing else about the distribution of X). Then by Markov's inequality,

$$P(X \geq 10) \leq \frac{3}{10}$$

On the other hand, suppose X has mean 3 and you know further that it has an exponential distribution. Then $\lambda = 1/3$,

$$f_X(x) = \frac{1}{3}e^{-x/3} \quad \text{for } x \geq 0$$

and

$$P(X \geq 10) = \int_{x=10}^{\infty} \tfrac{1}{3}e^{-x/3} \, dx = e^{-10/3} \approx .036$$

a much better answer than the universal bound .3.

Proof of Markov's Inequality

$$EX = \int_{x=0}^{\infty} x \, f(x) \, dx$$

$$\geq \int_{x=k}^{\infty} x \, f(x) \, dx \quad \text{(Since } xf(x) \text{ is positive in } [0, \infty), \text{ cutting down to a smaller interval makes the integral smaller.)}$$

$$\geq \int_{x=k}^{\infty} k \, f(x) \, dx \quad \text{(Since } f(x) \text{ is non-negative, changing its ``amplitude'' from } x \text{ where } x \text{ is in } [k, \infty) \text{ to the lowest value, } k, \text{ makes the integral still smaller.)}$$

$$= k \int_{x=k}^{\infty} f(x) \, dx$$

$$= k \, P(X \geq k)$$

So

$$P(X \geq k) \leq \frac{EX}{k}$$

Chebychev's Inequality

Suppose X is a random variable and you're interested in knowing how likely it is that X will be far from its mean.

It's easy if you know the distribution of X. But suppose you know only the mean and variance of X and not its specific distribution. Then you can't find $P(|X - \mu_X| \geq k)$ precisely, but you can find a bound on it as follows:

Let X be any random variable. If $k > 0$, then

$$P(|X - \mu_X| \geq k) \leq \frac{\operatorname{Var} X}{k^2}$$

Proof of Chebychev's Inequality

$$P(|X - \mu_X| \geq k) = P\left((X - \mu_X)^2 \geq k^2\right) \quad \text{(algebra)}$$

$$\leq \frac{E(X - \mu_X)^2}{k^2} \qquad \text{(Markov's inequality)}$$

$$= \frac{\text{Var } X}{k^2} \qquad \text{(definition of Var } X\text{)}$$

Proof of the Weak Law of Large Numbers

By Chebychev's inequality, with ϵ acting as k and $(X_1 + \cdots + X_n)/n$ playing the role of X,

$$(2) \quad P\left(\left|\frac{X_1 + \cdots + X_n}{n} - E\left(\frac{X_1 + \cdots + X_n}{n}\right)\right| \geq \epsilon\right) \leq \frac{\text{Var}\left(\frac{X_1 + \cdots + X_n}{n}\right)}{\epsilon^2}$$

By properties of expectation and variance,

$$E\left(\frac{X_1 + \cdots + X_n}{n}\right) = \frac{1}{n}\left(EX_1 + \cdots + EX_n\right) = \frac{1}{n}n\mu = \mu$$

$$\text{Var}\left(\frac{X_1 + \cdots + X_n}{n}\right) = \frac{1}{n^2}\left(\text{Var } X_1 + \cdots + \text{Var } X_n\right)$$
$$\text{(since the } X_i\text{'s are ind)}$$

$$= \frac{1}{n^2}n\sigma^2$$

$$= \frac{\sigma^2}{n}$$

Substitute into (2) to get

$$P\left(\left|\frac{X_1 + \cdots + X_n}{n} - \mu\right| \geq \epsilon\right) \leq \frac{\sigma^2}{n\epsilon^2}$$

But

$$\frac{\sigma^2}{n\epsilon^2} \to 0 \qquad \text{as } n \to \infty$$

so

$$P\left(\left|\frac{X_1 + \cdots + X_n}{n} - \mu\right| \geq \epsilon\right) \to 0 \qquad \text{as } n \to \infty$$

and, switching to the opposite event, we have

$$P\left(\left|\frac{X_1 + \cdots + X_n}{n} - \mu\right| < \epsilon\right) \to 1 \qquad \text{as } n \to \infty, \text{ QED}$$

Using Chebychev and Markov to Get Bounds

Chebychev's inequality and Markov's inequality are mainly used to prove other theorems such as the weak law of large numbers. But they can be used on their own to find bounds on probabilities when we don't know the underlying distribution. Here's an example.

Suppose, on the average, the daily demand for an item is 28 with variance 16. How many items should you make so that supply meets the demand at least 90% of the time?

Let X be the daily demand. We want to find k so that

$$P(X \leq k) \geq .9$$

or, equivalently,

$$P(X \geq k) \leq .1$$

Method 1   By Markov's inequality

$$P(X \geq k) \leq \frac{EX}{k} = \frac{28}{k}$$

The solution to

$$\frac{28}{k} = .1$$

is $k = 280$, so

$$P(X \geq 280) \leq .1$$

If you make 280 items, you'll meet the demand at least 90% of the time.

Method 2

$$P(X - 28 \geq c) \leq P(|X - 28| \geq c)$$

$$\leq \frac{\text{Var } X}{c^2} \qquad \text{(Chebychev)}$$

$$= \frac{16}{c^2}$$

The solution to

$$\frac{16}{c^2} = .1$$

is $c = 4\sqrt{10}$. So

$$P(X - 28 \geq 4\sqrt{10}) \leq .1$$
$$P(X \geq 28 + 4\sqrt{10}) \leq .1$$

If you make $28 + 4\sqrt{10} \approx 41$ items, you'll meet the demand at least 90% of the time.

In this example, Chebychev's inequality gave a better answer than Markov's inequality. (And someone else's inequality may come along and give a still better answer.)

Review Problems for Chapters 4–9

1. Let X and Y have joint density $f(x, y)$ with universe $0 \leq y \leq 3, x \geq y$. Set up the integrals for the following.

   (a) $P(XY \leq 2)$ (e) $E(Y|X = x)$
   (b) the distribution of $X - Y$ (f) the distribution of the max
   (c) $P(X \leq 2|Y = y)$ (g) $E(\max)$
   (d) EX

2. Let X be uniform on $[0, 1]$. Find the distribution of $1/(X + 1)$.

3. The lifetime of a bulb (measured in days) has mean 10.2 and variance 9. When a bulb burns out, it is replaced by a similar bulb. Find the prob that more than 100 bulbs are needed in the next 3 years.

Solutions

Solutions Section 1-1

1. There are 6 favorable outcomes, (3,4), (4,3), (6,1), (1,6), (5,2), (2,5), so prob is 6/36.

2. 8/36

3. There are 6 outcomes where the two dice are equal. Of the remaining 30 outcomes, half have second > first. Answer is 15/36.

4. Fav outcomes are (6,1), ... , (6,6), (1,6), ... , (6,6), but don't count (6,6) twice. Answer is 11/36.

5. Fav outcomes are (5,5), (5,6), (6,5), (6,6). Answer is 4/36.

6. The fav outcomes lie in columns 5 and 6 and in rows 5 and 6 in (4). Prob is 20/36.

7. P(neither over 4) $= 1 - P$(at least one ≥ 5)

   $= 1-$ answer to problem 6 $= 16/36$.

8. 9/36

9. P(at least one odd) $= 1 - P$(both even) $= 1 - \dfrac{9}{36} = \dfrac{27}{36}$

Solutions Section 1-2

1. (a) $\dfrac{\binom{8}{4} \text{ pick 4 others}}{\binom{9}{5} \text{ total}}$ (b) $\dfrac{\binom{7}{5} \text{ pick 5 from the others}}{\binom{9}{5}}$

2. For the favs, pick 9 more cards from the 39 non-spades. Prob is $\binom{39}{9}/\binom{52}{13}$.

3. $P(\text{not both C}) = 1 - P(\text{both C}) = 1 - \dfrac{\binom{7}{2}\ \text{pick 2 from the 7 C's}}{\binom{18}{2}\ \text{pick 2 people}}$

4. **(a)** $\dfrac{\binom{39}{5}}{\binom{52}{5}}$ **(b)** $\dfrac{\binom{50}{3}}{\binom{52}{5}}$ **(c)** $\dfrac{\binom{48}{3}}{\binom{52}{5}}$

5. **(a)** The total number of ways in which the coats can be returned is 4! (think of each woman as a slot; the first can get any of 4 coats, the second any of 3 coats, etc.). Only one way is fav, so prob is 1/4!.

(b) Fill Mary's slot. Total number of ways is 4, fav is 1. Answer is 1/4.

6. **(a)** $7 \cdot 6 = 42$ **(b)** $\dfrac{7!}{2!\ 5!} = \dfrac{7 \cdot 6}{1 \cdot 2} = 21$

(c) $\dfrac{8!}{5!\ 3!} \Big/ \dfrac{4!}{3!\ 1!} = \dfrac{8!\ 3!}{5!\ 3!\ 4!} = \dfrac{8 \cdot 7 \cdot 6}{4 \cdot 3 \cdot 2 \cdot 1} = 14$

7. Assume John has chosen his name. Find the prob that Mary makes the same choice. When Mary picks her name, only one is favorable (John's choice).

To count the total number of names, consider names of length 1, of length 2, ..., of length 7 and add.

For names of length 1, there is one spot to fill. It must be done with a letter, so there are 26 possibilities.
For names of length 2 there are two spots to fill. The first must be a letter, and the second can be any of 36 symbols (26 letters, 10 digits). So there are 26 · 36 possibilities.
For names of length 3, there are 3 spots to fill. The first must be a letter, and the next two spots can be any of 36 symbols. So there are $26 \cdot 36^2$ names of length 3.
And so on.

All in all,

$P(\text{same name})$

$= \dfrac{1}{26 + 26 \cdot 36 + 26 \cdot 36^2 + 26 \cdot 36^3 + 26 \cdot 36^4 + 26 \cdot 36^5 + 26 \cdot 36^6}$

Why is there a *sum* in the denominator? Because the *total* number of names is the number of names of length 1 *plus* the number of names of length 2 *plus* the number of names of length 3 *plus*, and so on.

8. **(a)** $7 \cdot 7 \cdot 7$ **(b)** $7 \cdot 6 \cdot 5$ **(c)** $\dbinom{7}{3}$

(d) With replacement, unordered, that is, a committee where someone can serve more than once (a histogram).

For example, $A_2A_2A_5$ is one such sample and is the *same* as $A_2A_5A_2$.

9. There are 4 fav outcomes (spade flush, heart flush, diamond, clubs).
Answer is $4/\binom{52}{5}$.

10. (a) Total number of outcomes is 7^3 (each person is a slot). There are
7 favorable outcomes (all go to church 1, all go to church 2, etc.).
Answer is $7/7^3 = 1/49$.

(b) $1 -$ answer to (a) $= 48/49$.

(c) $(7 \cdot 6 \cdot 5)/(7 \cdot 7 \cdot 7)$.

(d) $1 -$ answer to (c) $= 1 - (7 \cdot 6 \cdot 5)/7^3$.

(e) The total number of draws is $\binom{54}{6}$ and the favorable number is 2.
Answer is $2/\binom{54}{6}$, which is approximately $1/13{,}000{,}000$.

Solutions Section 1-3

1. (a) $\dfrac{\binom{13}{3}\binom{13}{2} \begin{smallmatrix}\text{pick 3 diamonds}\\\text{pick 2 hearts}\end{smallmatrix}}{\binom{52}{5}}$

(b) $\dfrac{12 \cdot \binom{39}{3} \begin{smallmatrix}\text{pick non-ace spade}\\\text{pick 3 non-spades}\end{smallmatrix}}{\binom{52}{5}}$

(c) $\dfrac{\binom{26}{4} \cdot 26}{\binom{52}{5}}$

(d) $\dfrac{\binom{4}{2}\binom{48}{3} \begin{smallmatrix}\text{pick 2 aces}\\\text{pick 3 non-aces}\end{smallmatrix}}{\binom{52}{5}}$

(e) $\dfrac{\binom{50}{4} \begin{smallmatrix}\text{pick 4 more cards to go with the}\\\text{spade ace but don't pick spade king}\end{smallmatrix}}{\binom{52}{5}}$

2. (a) $\dfrac{\binom{3}{2}\binom{14}{2} \begin{smallmatrix}\text{pick 2 A's}\\\text{pick 2 non-A's non-}R_3\text{'s}\end{smallmatrix}}{\binom{18}{5}}$

(b) $\dfrac{8 \cdot \binom{10}{4}}{\binom{18}{5}}$

3. Total is $\binom{8}{3}$. For the fav, pick 3 couples. Then pick one spouse from
each couple.

$$\frac{\binom{4}{3} \cdot 2^3}{\binom{8}{3}}$$

4. For the total, pick a committee of 12 symbols from the 36. For the fav,
pick a subcommittee of 3 evens and a subcommittee of 9 others.

$$\frac{\binom{5}{3}\binom{31}{9}}{\binom{36}{12}}$$

5. $\dfrac{\binom{25}{3}\binom{30}{2}\binom{90}{6}}{\binom{145}{11}}$

6. There are $\binom{7}{4}$ ways of picking 4 seats. Only 4 ways have the seats together (namely, the blocks S_1–S_4, S_2–S_5, S_3–S_6, S_4–S_7). Answer is $4/\binom{7}{4}$.

7. $\dfrac{13\binom{4}{3}\ \begin{smallmatrix}\text{pick the face value}\\\text{pick 3 cards from that face}\end{smallmatrix}}{\binom{52}{3}}$

8. (a) $\dfrac{\binom{13}{2}\binom{4}{2}\binom{4}{2}\ \begin{smallmatrix}\text{pick 2 faces for the pairs}\\\text{pick 2 cards from each face}\end{smallmatrix}}{\binom{52}{4}}$

(b) For the fav, pick a face for the 3 of a kind and then a face for the pair. Then pick 3 cards from the first face and 2 from the second face. Answer is

$$\frac{13 \cdot 12\binom{4}{3}\binom{4}{2}}{\binom{52}{5}}$$

Why did part (a) have $\binom{13}{2}$ in the numerator while part (b) has $13 \cdot 12$?
In (b) there are two slots: the face-for-the-3-of-a-kind and the face-for-the-pair. In (a) we need 2 faces but they are not slots because there isn't a first pair and a second pair—we want a committee of 2 faces.

9. (a) $\dfrac{4\binom{13}{5}\ \begin{smallmatrix}\text{pick a suit}\\\text{pick 5 cards from that suit}\end{smallmatrix}}{\binom{52}{5}}$

(b) $\dfrac{48\ \text{pick one remaining card}}{\binom{52}{5}}$

(c) $\dfrac{13 \cdot 48\ \begin{smallmatrix}\text{pick a face value for the four}\\\text{pick a fifth card}\end{smallmatrix}}{\binom{52}{5}}$

(d) For the fav, pick a face value for the pair and then pick 2 cards in that face. To make sure you don't get any more of that face value and don't get any other matching faces (i.e., to avoid a full house, 2 pairs, 3 or 4 of a kind), pick 3 other faces and pick a card from each of those faces. Answer is

$$\frac{13 \cdot \binom{4}{2}\binom{12}{3} \cdot 4^3}{\binom{52}{5}}$$

10. Lining up objects is the same as drawing without replacement. By symmetry,

$$P(\text{girl in the } i\text{th spot}) = P(\text{girl in the first spot}) = \frac{g}{b+g}$$

11. (a) This method counts the outcomes J_S, J_H, J_C, A_S, 3_H and J_S, J_H, J_C, 3_H, A_S as different when they are the same hand. In general this method counts every outcome twice.
Here's the correct version.

Pick a face value and 3 cards from that value.
Pick 2 more faces from the remaining 12 (a *committee* of 2 faces).
Pick a card in each of those 2 faces.

Answer is $13 \cdot \binom{4}{3}\binom{12}{2} \cdot 4^2$.
So the prob of 3 of a kind is

$$\frac{13 \cdot \binom{4}{3}\binom{12}{2} \cdot 4^2}{\binom{52}{5}}$$

(b) It counts the following outcomes as different when they are really the same.

outcome 1	outcome 2
Pick spot 1 for an A.	Pick spot 2 for an A.
Pick spot 2 to get an A.	Pick spot 1 for an A.
Pick spot 7 to get an A.	Pick spot 7 for an A.
Fill other spots with Z's.	Fill other spots with Z's.

Here's the correct version.

Pick a committee of 3 spots for the A's.
Fill the other places from the remaining 25 letters.

Answer is $\binom{7}{3} \cdot 25^4$.
So the prob of 3 A's in a 7-letter word is

$$\frac{\binom{7}{3} \cdot 25^4}{26^7}$$

(c) It counts the hand $J_H\, 3_S$ as different from the hand $3_S\, J_H$. In fact it counts every outcome exactly twice. It uses "first card" and "second card" as slots, but there is no such thing as a first or a second card in a hand.

Here's the correct version. Pick a committee of 2 faces. Pick a card from each face. Answer is $\binom{13}{2} \cdot 4^2$.

So the prob of a pair is

$$\frac{\binom{13}{2}4^2}{\binom{52}{2}}$$

Solutions Section 1-4

1. **(a)** $P(2A)+P(2K)-P(2A \text{ and } 2K) = \dfrac{\binom{4}{2}\binom{48}{3} + \binom{4}{2}\binom{48}{3} - \binom{4}{2}\binom{4}{2}\cdot 44}{\binom{52}{5}}$

(b) (mutually exclusive events) $P(3A) + P(3K) = \dfrac{\binom{4}{3}\binom{48}{2} + \binom{4}{3}\binom{48}{2}}{\binom{52}{5}}$

(c) *Method 1.*

$$P(A_S) + P(K_S) - P(A_S \text{ and } K_S) = \frac{\binom{51}{4} + \binom{51}{4} - \binom{50}{3}}{\binom{52}{5}}$$

Method 2.

$$1 - P(\text{not spade ace and not spade king}) = 1 - \frac{\binom{50}{5}}{\binom{52}{5}}$$

2. **(a)** $P(3W) + P(2R) + P(5G) - P(3W \text{ and } 2R)$

$$\left(\begin{array}{c}\text{Can leave out the rest of the terms since} \\ \text{an event such as "3W and 5G" is impossible.}\end{array}\right)$$

$$= \frac{\binom{10}{3}\binom{50}{2} + \binom{20}{2}\binom{40}{3} + \binom{30}{5} - \binom{10}{3}\binom{20}{2}}{\binom{60}{5}}$$

(b) $P(5W \text{ or } 5R \text{ or } 5G)$

$$= P(5W) + P(5R) + P(5G) = \frac{\binom{10}{5} + \binom{20}{5} + \binom{30}{5}}{\binom{60}{5}}$$

3. **(a)** Counting these terms is like counting twosomes from a population of size 8. There are $\binom{8}{2}$ such terms.

   (b) Count committees of size 3 from a pop of 8. There are $\binom{8}{3}$ such terms.

4. **(a)** $P(\text{no women}) + P(\text{no H}) - P(\text{no W and no H})$

$$= \frac{\binom{17}{12} + \binom{37}{12} - \binom{15}{12}}{\binom{42}{12}}$$

   (b) $P(\text{only non-H men}) = \dfrac{\binom{15}{12}}{\binom{42}{12}}$

5. *Method 1.*   XOR = OR − BOTH, so

$$P(\text{J XOR Q}) = P(\text{J OR Q}) - P(\text{both})$$
$$= P(\text{J}) + P(\text{Q}) - P(\text{J and Q}) - P(\text{J and Q}) \text{ again}$$
$$= \frac{\binom{51}{4} + \binom{51}{4} - 2\binom{50}{3}}{\binom{52}{5}}$$

Method 2.

$$P(\text{J and not Q}) + P(\text{Q and not J}) = \frac{\binom{50}{4} + \binom{50}{4}}{\binom{52}{5}}$$

6. **(a)** $1 - P(\text{no S}) = 1 - \dfrac{\binom{39}{5}}{\binom{52}{5}}$

   (b) *Method 1.*

$$P(\text{3S or 4S or 5S}) = P(\text{3S}) + P(\text{4S}) + P(\text{5S})$$
$$= \frac{\binom{13}{3}\binom{39}{2} + \binom{13}{4} \cdot 39 + \binom{13}{5}}{\binom{52}{5}}$$

Method 2.

$$1 - [P(\text{no S}) + P(\text{1S}) + P(\text{2S})] = 1 - \frac{\binom{39}{5} + 13\binom{39}{4} + \binom{13}{2}\binom{39}{3}}{\binom{52}{5}}$$

(c) *Method 1.*

$$1 - P(3A) - P(4A) = 1 - \frac{\binom{4}{3}\binom{48}{2} + 48}{\binom{52}{5}}$$

Method 2.

$$P(\text{no A}) + P(1A) + P(2A) = \frac{\binom{48}{5} + 4\binom{48}{4} + \binom{4}{2}\binom{48}{3}}{\binom{52}{5}}$$

(d) *Method 1.*

$$P(4 \text{ pics}) - P(4 \text{ pics with no aces}) - P(4 \text{ pics with 1 ace})$$

$$= \frac{\binom{16}{4}36 - \binom{12}{4}36 - 4\binom{12}{3}36}{\binom{52}{5}}$$

Method 2.

$$P(4 \text{ pics with 2 aces}) + P(4 \text{ pics with 3 aces})$$

$$+ P(4 \text{ pics with 4 aces}) = \frac{\binom{4}{2}\binom{12}{2}36 + \binom{4}{3} \cdot 12 \cdot 36 + 36}{\binom{52}{5}}$$

7. (a) $P(\text{spades or hearts or diamonds or clubs})$

$$= P(S) + P(H) + P(D) + P(C) - \underbrace{[P(S \text{ and } H) + \cdots]}$$

$$\binom{4}{2}\text{terms in here}$$

$$(\text{no room for 3 or more royal flushes})$$

$$= \frac{4\binom{47}{8} - \binom{4}{2}\binom{42}{3}}{\binom{52}{13}}$$

(b) $P(4 \text{ 2's or 4 3's or } \dots \text{ or 4 aces})$

$$= P(4 \text{ 2's}) + \cdots + P(4 \text{ aces})$$

$$-\underbrace{[P(4 \text{ 2's and 4 3's}) + \text{ other 2-at-a-time-terms}]}$$

$$\binom{13}{2} \text{ terms, each is } \binom{44}{5} \Big/ \binom{52}{13}$$

$$+\underbrace{[P(4 \text{ 2's and 4 3's and 4 J's}) + \cdots + \text{ other 3-at-a-time-terms}]}$$

$$\binom{13}{3} \text{ terms, each is } 40 \Big/ \binom{52}{13}$$

$$= \frac{13\binom{48}{9} - \binom{13}{2}\binom{44}{5} + \binom{13}{3}40}{\binom{52}{13}}$$

8. *Method 1.*

$$P(\text{happy Smith family}) = 1 - P(\text{no Smiths win}) = 1 - \frac{\binom{96}{3}}{\binom{100}{3}}$$

Method 2.

$$P(\text{happy Smiths}) = P(1 \text{ Smith wins or 2 Smiths win or}$$
$$3 \text{ Smiths win or 4 Smiths win})$$
$$= P(1S) + P(2S) + P(3S) + P(4S)$$

To compute $N(1 \text{ Smith wins})$, pick the Smith in 4 ways, the other 2 winners in $\binom{96}{2}$ ways.

To compute $N(2 \text{ Smiths win})$, pick the 2 Smiths in $\binom{4}{2}$ ways, the other winner in 96 ways.

To compute $N(3 \text{ Smiths win})$, pick the 3 Smiths in $\binom{4}{3}$ ways. So

$$P(\text{happy Smith family}) = \frac{4\binom{96}{2} + 96\binom{4}{2} + \binom{4}{3}}{\binom{100}{3}}$$

Method 3.

$$P(\text{happy Smith family}) = P(\text{John wins or Mary wins or Bill wins}$$
$$\text{or Henry wins})$$

$$= P(\text{John}) + P(\text{Mary}) + P(\text{Bill}) + P(\text{Henry})$$

$$\left(4 \text{ terms, each is } \binom{99}{2} / \binom{100}{3}\right)$$

$$- [P(\text{JohnMary}) + P(\text{JohnBill}) + \cdots]$$

$$\left(\binom{4}{2} \text{ terms, each is } 98/\binom{100}{3}\right)$$

$$+ [P(\text{JMB}) + P(\text{JMH}) + \cdots]\left(\binom{4}{3} \text{ terms, each is } 1/\binom{100}{3}\right)$$

$$- P(\text{JMBH}) \quad (\text{this term is 0 since there are only 3 winners})$$

$$= \frac{4\binom{99}{2} - \binom{4}{2}98 + \binom{4}{3}}{\binom{100}{3}}$$

9. (a) $\dfrac{7\binom{11}{3}}{\binom{18}{4}}$

(b) $1 - P(\text{no women}) = 1 - \binom{11}{4} \bigg/ \binom{18}{4}$

(c) $P(\text{no women}) + P(1 \text{ woman}) = \dfrac{\binom{11}{4} + 7\binom{11}{3}}{\binom{18}{4}}$

(d) $1 - P(\text{no M or no W or no C})$

$$= 1 - \begin{bmatrix} P(\text{no M}) + P(\text{no W}) + P(\text{no C}) \\ -[P(\text{no M \& no W}) + \cdots] \end{bmatrix}$$

$$= 1 - \dfrac{\binom{12}{4} + \binom{11}{4} + \binom{13}{4} - \left[\binom{5}{4} + \binom{6}{4} + \binom{7}{4}\right]}{\binom{18}{4}}$$

(e) $P(\text{no women}) - P(\text{no women and no men}) = \dfrac{\binom{11}{4} - \binom{5}{4}}{\binom{18}{4}}$

10. (a) $1 - P(\text{no H or no M or no P})$

$$= 1 - \begin{bmatrix} P(\text{no H}) + P(\text{no M}) + P(\text{no P}) \\ -[P(\text{no H, no M}) + P(\text{no H, no P}) + P(\text{no M, no P})] \\ + P(\text{no H, no M, no P}) \end{bmatrix}$$

$$= 1 - \dfrac{3\binom{98}{15} - 3\binom{96}{15} + \binom{94}{15}}{\binom{100}{15}}$$

(b) $1 - P(\text{no H and no M and no P}) = 1 - \binom{94}{15} \big/ \binom{100}{15}$

(c) *Method 1.*

$$P(1\text{H}) - P(1\text{H and no M}) = \dfrac{2\binom{98}{14} - 2\binom{96}{14}}{\binom{100}{15}}$$

Method 2.

$$P(1\text{H and 1M}) + P(1\text{H and 2M}) = \dfrac{2 \cdot 2\binom{96}{13} + 2\binom{96}{12}}{\binom{100}{15}}$$

11. (a) Think of the husbands as slots. The total number of ways they can be filled is 7! For the favorable, fill H_3 in one way, the others in 6! ways. Answer is $6!/7! = 1/7$.

(You can get this immediately by considering the prob that the H_3 slot gets the 1 favorable out of 7 total possibilities.)

(b) For the favorable: The H_2, H_5, H_7 slots are determined. Fill the others in 4! ways. Answer is $4!/7! = 1/(7 \cdot 6 \cdot 5)$.

(c) $P(H_1$ or H_2 or $\ldots$ or H_7 is matched with his wife)

$$= P(H_1) + \cdots + P(H_7) - [P(H_1 \,\&\, H_2) + \cdots]$$

$$+ \left(3\text{-at-a-time-terms—there are } \binom{7}{3} \text{ of them}\right.$$

$$\text{and by part (b) each is } 4!/7!)$$

$$- (4\text{-at-a-time-terms}) + (5\text{-at-a-time-terms}) - (6\text{-at-a-time})$$

$$+ P(\text{all match})$$

$$= 7\frac{6!}{7!} - \binom{7}{2}\frac{5!}{7!} + \binom{7}{3}\frac{4!}{7!} - \binom{7}{4}\frac{3!}{7!} + \binom{7}{5}\frac{2!}{7!} - \binom{7}{6}\frac{1!}{7!}$$

$$+ \binom{7}{7}\frac{0!}{7!} \qquad (\text{the last term is } 1/7!, \text{ the prob they all match})$$

$$= 1 - \frac{1}{2!} + \frac{1}{3!} - \frac{1}{4!} + \frac{1}{5!} - \frac{1}{6!} + \frac{1}{7!}$$

(d) $1 -$ answer to (c)

Solutions Section 1-5

1. (a) and   **(b)** fav length/total length $= \frac{1}{2}/2 = 1/4$

(c) $.2/2 = 1/10$

(d) To find the sol to $3x^2 > x$, first solve $3x^2 = x$ to get $x = 0$, 1/3. Then look in between at intervals $(-1,0)$, $(0,\frac{1}{3})$, $(\frac{1}{3},1)$ to see where $3x^2 > x$. The inequality is satisfied in $(-1,0)$ and $(\frac{1}{3}, 1)$. So fav/total $= (5/3)/2 = 5/6$.

2. $\dfrac{\text{fav area}}{\text{total area}} = \dfrac{4\pi}{81\pi} = \dfrac{4}{81}$

Figure P.2

3. The roots of $ax^2 + bx + c = 0$ are real iff $b^2 - 4ac \geq 0$, so in this case we need

$$16Q^2 - 4 \cdot 4(Q + 2) \geq 0$$
$$Q^2 - Q - 2 \geq 0$$
$$Q \geq 2 \text{ or } Q \leq -1$$

So the fav Q's in $[0,5]$ are $2 \leq Q \leq 5$ and the answer is fav/total $= 3/5$.

4. $\sin \theta > \frac{1}{3}$ where $-\frac{1}{2}\pi \leq \theta \leq \frac{1}{2}\pi$ iff $\theta > \sin^{-1} \frac{1}{3}$

$$\text{prob} = \frac{\text{fav length}}{\text{total}} = \frac{\frac{1}{2}\pi - \sin^{-1} \frac{1}{3}}{\pi}$$

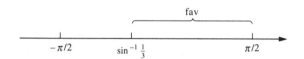

Figure P.4

5. **(a)** The diagram shows several chords and a conveniently placed equilateral triangle for comparison. The favorable θ's, that correspond to long enough chords, are between $60°$ and $120°$, so

$$\text{prob} = \frac{\text{fav length}}{\text{total}} = \frac{60}{180} = \frac{1}{3}$$

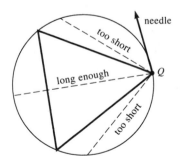

Figure P.5a

(b) Look at the diagram to see that favorable d's lie between 0 and $\frac{1}{2}R$, so

$$\text{prob} = \frac{\text{fav length}}{\text{total}} = \frac{\frac{1}{2}R}{R} = \frac{1}{2}$$

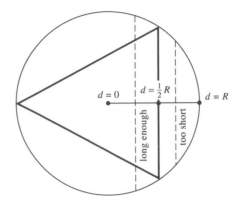

Figure P.5b

6. Let x be the first number; let y be the second number. Then the pair (x, y) is uniformly distributed in the rectangle $0 \leq x \leq 1, 1 \leq y \leq 3$.

   (a) $P(x + y \leq 3) = \frac{\text{fav area}}{\text{total}} = 1 - \frac{\text{unfav}}{\text{total}} = 1 - \frac{1/2}{2} = \frac{3}{4}$

   (b) $\text{total} = 2, \quad \text{fav} = \int_{x=1/3}^{1} (3 - \frac{1}{x}) \, dx = 2 + \ln \frac{1}{3} = 2 - \ln 3$

   $P(xy > 1) = 1 - \frac{1}{2} \ln 3$

7. **(a)** Let x be John's arrival time and let y be Mary's arrival time. Then (x, y) is uniformly distributed in the indicated square and

   $P(\text{first to arrive must wait at least 10 minutes for the other})$
   $= P(|x - y| > 10)$ (i.e., the arrival times differ by more
   than 10)

   $= \frac{\text{fav area}}{\text{total}} = \frac{2500}{3600} = \frac{25}{36}$

   (b) $P(y - x \geq 20) = \frac{\text{fav}}{\text{total}} = \frac{\frac{1}{2} \cdot 40 \cdot 40}{3600} = \frac{2}{9}$

   (c) The event consists of all points on a segment. The fav *area* is 0. $P(y = x) = 0$.

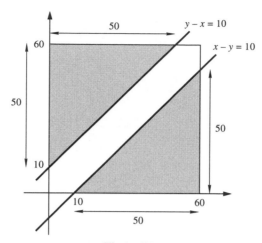

Figure P.7a

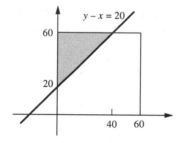

Figure P.7b

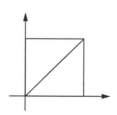

Figure P.7c

Solutions Section 2-1

1. **(a)** 3/51   **(b)** 4/51
   (c) P(2nd is king or lower|1st is K) = 47/51
   (d) P(second is non-ace|first is ace) = 48/51

2. **(a)** $1 - P$(rain|Jan 7) = .7
   (b) Can't do with the information given.

3. The new universe of 8-sums contains the five equally likely points (4,4), (5,3), (3,5), (6,2), (2,6). So P(first die is 6) = 1/5.

4. P(at least one K|at least one A)

$$= \frac{P(\text{at least one K and at least one A})}{P(\text{at least one A})}$$

$$\text{denominator} = 1 - P(\text{no aces}) = 1 - \frac{\binom{48}{5}}{\binom{52}{5}}$$

$$\text{numerator} = 1 - P(\text{no K or no A})$$

$$= 1 - [P(\text{no K}) + P(\text{no A}) - P(\text{no K and no A})]$$

$$= 1 - \left[2\frac{\binom{48}{5}}{\binom{52}{5}} - \frac{\binom{44}{5}}{\binom{52}{5}} \right]$$

5. By symmetry, the prob that E gets 3 of the spades is the same as the prob that W gets 3 spades, so just find the prob that E gets 3 spades and double it. In the new universe (after N and S get their hands with the 9 spades) we want the prob that E gets 3 spades from the 26 remaining cards (which include 4 spades).

$$P(\text{3–1 split}) = 2P(E \text{ gets 3 spades})$$

$$= \frac{\binom{4}{3}\binom{22}{10} \begin{array}{l} \text{pick 3 spades} \\ \text{pick 10 others} \end{array}}{\binom{26}{13}}$$

6. *Method 1.*

$$P(ABD|ABC) = \frac{P(\text{in } ABD \text{ and in } ABC)}{P(\text{in } ABC)}$$

$$\text{numerator} = \frac{\text{fav area}}{\text{total}} = \frac{\text{area } ABE}{1} = \frac{1}{4}$$

$$\text{denom} = \frac{\text{fav area}}{\text{total}} = \frac{1/2}{1} = \frac{1}{2}$$

$$\text{Answer} = \frac{1/4}{1/2} = \frac{1}{2}$$

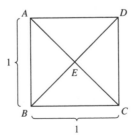

Figure P.6

Method 2.   In the new universe ABC,

$$P(\text{ in } ABC) = \frac{\text{fav area}}{\text{total}} = \frac{\text{area } ABE}{\text{area } ABC} = \frac{1/4}{1/2} = \frac{1}{2}$$

7. **(a)** $P(\text{B R B W}) = P(\text{B})P(\text{R}|\text{B})P(\text{B}|\text{BR})P(\text{W}|\text{BRB}) = \dfrac{9}{24}\dfrac{5}{23}\dfrac{8}{22}\dfrac{10}{21}$

   (b) Same as (a) by symmetry.

   (c) $\dfrac{9}{24}\dfrac{5}{23}\dfrac{8}{24}\dfrac{10}{23}$

   (d) $\dfrac{\binom{9}{2} \cdot 5 \cdot 10}{\binom{24}{4}}$

   (e) By symmetry, $P(\text{W on 4th}) = P(\text{W on 1st}) = 10/24$.

   (f) By symmetry,

   $$P(\text{W on 3rd, W on 4th}) = P(\text{W on 1st, W on 2nd}) = \frac{10}{24}\frac{9}{23}$$

8. $P(\text{saved}) = P(\text{at least one message gets through})$.

   Method 1.

   $$1 - P(\text{no smoke and no bottle and no pigeon})$$
   $$= 1 - P(\text{no s})P(\text{no b})P(\text{no p}) = 1 - (.9)(.8)(.7)$$

   Method 2.

   $P(\text{smoke or bottle or pigeon})$
   $$= P(\text{s}) + P(\text{b}) + P(\text{p})$$
   $$-[P(\text{s and b}) + P(\text{s and p}) + P(\text{b and p})] + P(\text{sbp})$$
   $$= .1 + .2 + .3 - [(.1)(.2) + (.1)(.3) + (.2)(.3)] + (.1)(.2)(.3)$$

9. $P(\text{II open}|\text{at least one open}) = \dfrac{P(\text{II open and at least one open})}{P(\text{at least one open})}$

   numerator $= P(\text{II open}) = .8$

   denominator method 1

   $= 1 - P(\text{both closed})$

   $= 1 - P(\text{I closed})P(\text{II closed}) = 1 - (.7)(.2)$

   denominator method 2

   $= P(\text{I open or II open})$

   $= P(\text{I open}) + P(\text{II open}) - P(\text{I and II open})$

   $= .3 + .8 - (.3)(.8)$

10. **(a)** The interpretation is that we want $P(2B|\text{at least one B})$.

Method 1.

$$\frac{P(2B \text{ and at least 1B})}{P(\text{at least one B})} = \frac{P(2B)}{1 - P(WW)} = \frac{P(B)P(B)}{1 - P(W)P(W)}$$

$$= \frac{1/4}{1 - 1/4} = \frac{1}{3}$$

Method 2. In the new universe of at least one B, there are 3 equally likely points, BB, BW, WB. One of them is fav, so the prob is 1/3.

(b) This is interpreted as $P(\text{2nd is B}|\text{1st is B})$, which is $P(\text{2nd is B})$, since the balls are painted independently. Answer is 1/2.

11. $P(\text{need more than 3 missiles}) = P(\text{first 3 missiles miss})$
$$= P(\text{miss})P(\text{miss})P(\text{miss}) = (.2)^3$$

12. **(a)** $P(\text{W on 1st and 4th}) = P(\text{W on 1st and 2nd})$ by symmetry

$$= \frac{10}{15}\frac{9}{14}$$

(b) $\dfrac{10}{15}\dfrac{10}{15}$

13. Use the notation HTH for the event "A tosses H, B tosses T, C tosses H."

Method 1. The prob that A is odd man out is the prob of

HTT or THH on round 1
    OR
(HHH or TTT on round 1) and (HTT or THH on round 2)
    OR
(HHH or TTT on round 1) and (HHH or TTT on round 2)
and (HTT or THH on round 3)
    OR

$\vdots$

On any one round, events like HHH, HHT, and so on, are mutually exclusive, and the three coin tosses are independent. And the rounds themselves are independent. So

$$P(\text{A is odd man out}) = p_a q_b q_c + q_a p_b p_c \ (\text{where } q_a = 1 - p_a, \text{ etc.})$$
$$+ (p_a p_b p_c + q_a q_b q_c)(p_a q_b q_c + q_a p_b p_c)$$
$$+ (p_a p_b p_c + q_a q_b q_c)^2 (p_a q_b q_c + q_a p_b p_c)$$
$$+ \cdots$$

The series is geometric with $a = p_a q_b q_c + q_a p_b p_c$, $r = p_a p_b p_c + q_a q_b q_c$. So

$$P(\text{A is odd man out}) = \frac{a}{1 - r} = \frac{p_a q_b q_c + q_a p_b p_c}{1 - (p_a p_b p_c + q_a q_b q_c)}$$

Method 2.　　Let p be the prob that A is odd man out. Then

$$p = P(\text{HTT or TTH}) + P(\text{HHH or TTT, and then A is odd}$$
$$\text{man out in the rest of the game})$$
$$= p_a q_b q_c + q_a p_b p_c + (p_a p_b p_c + q_a q_b q_c)p$$

Solve for p to get the same answer as in method 1.

14. Imagine drawing from a deck containing just hearts, spades, clubs.

$$P(\text{heart before black}) = P(\text{heart in 1 draw from the new deck})$$
$$= \frac{13}{39} = \frac{1}{3}$$

15. *Method 1.*　　Consider a new universe where only 5 and 7 are possible. This cuts down to outcomes (1,4), (4,1), (3,2), (2,3), (3,4), (4,3), (6,1), (1,6), (5,2), (2,5), all equally likely.

$$P(\text{5 before 7 in repeated tosses}) = P(\text{5 in restricted universe})$$
$$= \frac{\text{fav}}{\text{total}} = \frac{4}{10}$$

Method 2.

$$P(5) = \frac{4}{36}, P(7) = \frac{6}{36}, P(\text{5 before 7}) = \frac{4/36}{4/36 + 6/36} = \frac{4}{10}$$

16. (a)　Consider a new universe with just C_1, C_2.
　　　　Then $P(C_1 \text{ leaves first}) = 1/2$.

(b)　and (c)　Consider a new universe with just C_1, C_2, C_3.

$$P(C_1 \text{ first}) = \frac{1}{3}$$

$$P(C_1 \text{ then } C_2 \text{ then } C_3) = \frac{\text{fav permutations}}{\text{total perms}} = \frac{1}{3!} = \frac{1}{6}$$

17. $\dfrac{P(1)}{P(1) + P(2)} = \dfrac{2}{5}$

Solutions Section 2-2

1. (a) $\dfrac{40!}{30! \ 5! \ 3! \ 2!} \ (.6)^{30}(.3)^5(.07)^3(.03)^2$

(b) P(30 good, 3 fair, 6 others)= $\dfrac{40!}{30! \ 3! \ 6!} \ (.6)^{30}(.3)^4(.1)^6$

(c) $(.97)^{40}$

2. (a) $\left(\frac{1}{2}\right)^{16}$ **(b)** $\binom{16}{7}\left(\frac{1}{2}\right)^{16}$

(c) $P(15\text{H}) + P(16\text{H}) = \binom{16}{15}\left(\frac{1}{2}\right)^{16} + \left(\frac{1}{2}\right)^{16}$

3. $P(3 \text{ G \& 2R or 4G \& 1R or 5G}) = \binom{5}{3}(.3)^3(.7)^2 + \binom{5}{4}(.3)^4(.7) + (.3)^5$

4. (a) $(.3)^5$ **(b)** $\dfrac{5!}{1! \ 2! \ 2!} \ (.3)(.6)^2(.1)^2$

(c) The number Against has a binomial distribution where $P(\text{A}) = .6$.

$$P(\text{majority Against}) = P(3\text{A}) + P(4\text{A}) + P(5\text{A})$$

$$= \binom{5}{3}(.6)^3(.4)^2 + \binom{5}{4}(.6)^4(.4) + (.6)^5$$

5. The births are Bernoulli trials where we assume $P(\text{G}) = P(\text{B}) = 1/2$.

(a) $P(3\text{G}) = \binom{6}{3}\left(\frac{1}{2}\right)^6$ **(b)** $\left(\frac{1}{2}\right)^6$ **(c)** $\left(\frac{1}{2}\right)^6$

6. (a) This amounts to 6H in 9 tosses. Prob is $\binom{9}{6}(.6)^6(.4)^3$.

(b) $P(\text{at least 9H}|\text{at least 8H})$

$$= \frac{P(\text{at least 9H and at least 8H})}{P(\text{at least 8H})}$$

$$= \frac{P(\text{at least } 9)}{P(\text{at least } 8)} = \frac{P(9) + P(10)}{P(8) + P(9) + P(10)}$$

$$= \frac{\binom{10}{9}(.6)^9(.4) + (.6)^{10}}{\binom{10}{8}(.6)^8(.4)^2 + \binom{10}{9}(.6)^9(.4) + (.6)^{10}}$$

7. (a) The tosses are ind trials, each with 5 equally likely outcomes (the boxes).

$$P(2 \text{ of each}) = \frac{10!}{(2!)^5} \left(\frac{1}{5}\right)^{10}$$

(b) Each toss has two outcomes: B_2 with prob 1/5, and elsewhere with prob 4/5.

$$P(10 \text{ elsewheres}) = \left(\frac{4}{5}\right)^{10}$$

(c) Each toss has two outcomes, B_3 and non-B_3.

$$P(6 \ B_3\text{'s}) = \binom{10}{6} \left(\frac{1}{5}\right)^6 \left(\frac{4}{5}\right)^4$$

(d) $P(\text{each box gets at least one ball})$

$= 1 - P(B_1 = 0 \text{ or } B_2 = 0 \text{ or } \ldots \text{ or } B_5 = 0)$

$$= 1 - \left[\begin{array}{l} P(B_1 = 0) + \cdots + P(B_5 = 0) \left(5 \text{ terms, each is } \left(\frac{4}{5}\right)^{10}\right) \\[2mm] -P(B_1 = 0 \ \& \ B_2 = 0) \text{ and other 2-at-a-time terms} \\[2mm] \qquad\qquad \left(\binom{5}{2} \text{ terms, each is } \left(\frac{3}{5}\right)^{10}\right) \\[2mm] + \text{3-at-a-time terms} \\ - \text{4-at-a-time terms} \\ + P(\text{all 5 empty}) \qquad (\text{impossible}) \end{array} \right]$$

$$= 1 - \left[5\left(\frac{4}{5}\right)^{10} - \binom{5}{2}\left(\frac{3}{5}\right)^{10} + \binom{5}{3}\left(\frac{2}{5}\right)^{10} - \binom{5}{4}\left(\frac{1}{5}\right)^{10} \right]$$

8. $P(\text{machine fails}) = 1 - P(5 \text{ or } 6 \text{ or } 7 \text{ successful components})$

$$= 1 - \binom{7}{5}(.8)^5(.2)^2 - \binom{7}{6}(.8)^6(.2) - (.8)^7$$

9. (a) Each spot in the string is a trial with outcomes 4 versus non-4.

$$P(\text{two 4's}) = \binom{7}{2}(.1)^2(.9)^5$$

(b) Each spot in the string is a trial where the two outcomes are "> 5" with prob .4 and "≤ 5" with prob .6.

$$P(\text{one is} > 5) = \binom{7}{1}(.4)(.6)^6$$

10. $P(\text{pair}) = P(\text{two 6's, one each of } 3, 4, 5)$
$$+ P(\text{two 4's, one each of } 2, 3, 6) + \cdots$$

There are $6\binom{5}{3}$ terms in the sum (pick a face for the pair and then pick 3 more faces for the singletons). Each prob is

$$\frac{5!}{2! \; 1! \; 1! \; 1!} \left(\frac{1}{6}\right)^5$$

by the multinomial formula. Answer is

$$6\binom{5}{3} \frac{5!}{2!} \left(\frac{1}{6}\right)^5$$

11. (a) Draw 10 with replacement from a box of 7G, 14 others.

$$P(4G) = \binom{10}{4} \left(\frac{7}{21}\right)^4 \left(\frac{14}{21}\right)^6$$

(b) Draw 10 times *without* replacement.

$$P(4G) = \frac{\binom{7}{4}\binom{14}{6}}{\binom{21}{10}}$$

12. (a) $(.6)^2(.4)^6$

(b) $\binom{8}{2}(.6)^2(.4)^6$

13. The cars are 100 Bernoulli trials with $P(\text{doesn't stop}) = .05$.

$P(\text{at least 3 no-stops})$

$$= 1 - P(0 \text{ no-stops}) - P(1 \text{ no-stop}) - P(2 \text{ no-stops})$$

$$= 1 - (.95)^{100} - \binom{100}{1}(.05)(.95)^{99} - \binom{100}{2}(.05)^2(.95)^{98}$$

14. (a) Pick a first box to get 4 balls and then another box to get 2 balls. Then find the prob of getting 4 balls into the first box and 2 into the second.

Answer is

$$10 \cdot 9 \quad \frac{6!}{4! \, 2!} \left(\frac{1}{10}\right)^6$$

This is really using the OR rule for mutually exclusive events:

$P(4\text{–}2 \text{ split})$

$$= P(4 \text{ into } B_1, 2 \text{ into } B_6) + P(4 \text{ into } B_3, 2 \text{ into } B_1) + \cdots$$

There are $10 \cdot 9$ terms, and each term is $\dfrac{6!}{4! \, 2!}(1/10)^6$

(b) Pick a pair of boxes. Then find the prob of getting 3 balls in each.

Answer is

$$\binom{10}{2} \frac{6!}{3! \, 3!} \left(\frac{1}{10}\right)^6$$

Why use $10 \cdot 9$ in part (a) and $\binom{10}{2}$ in part (b)? Because in (a) there are 2 slots, the box to get the 4 balls and the box to get 2 balls. But in (b), the 2 boxes aren't distinguished from one another since each gets 3 balls; they are just a committee of 2 boxes.

(c) *Method 1.*

$P(\text{all different}) = P(\text{any first})P(\text{second different}|\text{first}) \ldots$

$$= 1 \cdot \frac{9}{10} \frac{8}{10} \frac{7}{10} \frac{6}{10} \frac{5}{10}$$

Method 2. Pick 6 boxes. Then use the multinomial to find the prob that each of the 6 gets 1 ball. Answer is

$$\binom{10}{6} P(1 \text{ each, say, of B}_1, \ldots, \text{B}_6 \text{ in 6 trials}) = \binom{10}{6} \frac{6!}{(1!)^6} \left(\frac{1}{10}\right)^6$$

15. A handful means without replacement. So the drawings are not Bernoulli trials and the binomial distribution doesn't apply.

$$P(2\text{L, 2R}) = \frac{\binom{10}{2}\binom{12}{2}}{\binom{22}{4}}$$

16. **(a)** $P(\text{first 19 are righties}) = (.85)^{19}$

   (b) $P(\text{first 19 are righties and 20th is a lefty}) = (.85)^{19}\,(.15)$

   (c) $P(2\text{L in 19 tries and then L on 20th})$
   $$= \text{P(2L in 19 tries) P(L on 20th)} = \binom{19}{2}(.15)^2(.85)^{17}(.15)$$

   (d) A long way (infinitely many terms) is

   $$P(20 \text{ tries to get 3L}) + P(21 \text{ tries to get 3L}) + \cdots$$

   $$= \binom{19}{2}(.15)^2(.85)^{17}.15 + \binom{20}{2}(.15)^2(.85)^{18}.15$$

   $$+ \binom{21}{2}(.15)^2(.85)^{19}.15 + \cdots$$

   Another fairly long way is

   $$1 - P(3 \text{ tries to get 3L})$$
   $$-P(4 \text{ tries to get 3L}) - \cdots - P(19 \text{ tries to get 3L})$$
   $$= 1 - (.15)^3 - \binom{3}{2}(.15)^2(.85).15 - \cdots - \binom{18}{2}(.15)^2(.85)^{16}.15$$

   The fastest way is to find

   $$P(\text{less than 3L in 19 tries})$$
   $$= P(\text{no L or 1L or 2L in 19 tries})$$
   $$= (.85)^{19} + \binom{19}{1}(.15)(.85)^{18} + \binom{19}{2}(.15)^2(.85)^{17}$$

(e) $P(R^4L \text{ or } R^9L \text{ or } R^{14}L \text{ or } \dots)$

$$= (.85)^4(.15) + (.85)^9(.15) + (.85)^{14}(.15) + \cdots$$

Geometric series with $a = (.85)^4(.15)$ and $r = (.85)^5$. Answer is

$$\frac{a}{1-r} = \frac{(.85)^4(.15)}{1-(.85)^4}$$

17. $P(T^9H \text{ or } H^9T) = q^9p + p^9q.$

Solutions Section 2-4

1. $P(CB) = P(M)P(CB|M) + P(W)P(CB|W) = \#1 + \#3$
$$= (.53)(.02) + (.47)(.001)$$

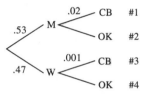

Figure P.1

2. $P(2\text{nd is } A_S)$
$$= P(1\text{st is spade}) \, P(2\text{nd is } A_S | 1\text{st is spade})$$
$$+ P(1\text{st is non-S})P(2\text{nd is } A_S | 1\text{st is non-S})$$
$$= \#1 + \#3 = \frac{13}{52}\frac{1}{52} + \frac{39}{52}\frac{1}{51}$$

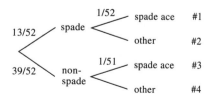

Figure P.2

3. (a) $P(\text{right}) = \#1 + \#3 = (.75)(.8) + (.25)(.2) = .65$

(expect 65% on the exam)

(b) $P(\text{guess}|\text{right}) = \dfrac{\#3}{\#1 + \#3} = \dfrac{(.25)(.2)}{65} \approx .07$

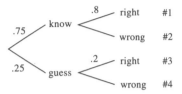

Figure P.3

4. $P(\text{B}|\text{W}) = \dfrac{P(\text{B and W})}{P(\text{W})} = \dfrac{\#3}{\#1 + \#3 + \#5} = \dfrac{\frac{1}{3}\frac{7}{15}}{\frac{1}{3}\frac{1}{3} + \frac{1}{3}\frac{7}{15} + \frac{1}{3}\frac{4}{9}}$

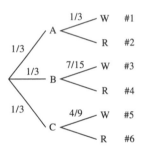

Figure P.4

5. $P(\text{at least 4H}) = \#1 + \#3$

$= P(\text{H on 1st})P(\text{3H or 4H or 5H later}|\text{H on 1st})$

$+ P(\text{T on 1st})P(\text{4H or 5H or 6H later}|\text{T on 1st})$

$$= \frac{2}{3}\left[\binom{5}{3}\left(\frac{2}{3}\right)^3\left(\frac{1}{3}\right)^2 + \binom{5}{4}\left(\frac{2}{3}\right)^4\frac{1}{3} + \left(\frac{2}{3}\right)^5\right]$$

$$+ \frac{1}{3}\left[\binom{6}{4}\left(\frac{2}{3}\right)^4\left(\frac{1}{3}\right)^2 + \binom{6}{5}\left(\frac{2}{3}\right)^5\frac{1}{3} + \left(\frac{2}{3}\right)^6\right]$$

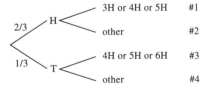

Figure P.5

6. $P(\text{tenth}|2B) = \dfrac{P(\text{tenth and 2B})}{P(2 \text{ bad})}$

$$= \dfrac{\#1}{\#1 + \#3} = \dfrac{(.1)\frac{10}{12}\frac{9}{11}}{(.1)\frac{10}{12}\frac{9}{11} + (.9)\frac{2}{12}\frac{1}{11}}$$

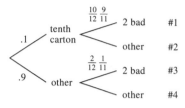

Figure P.6

7. $P(\text{at least one W on 2nd draw}) = \#1 + \#3 + \#5 + \#7$

$$= \dfrac{3}{5}\dfrac{1}{2}\left(1 - \dfrac{2}{4}\dfrac{1}{3}\right) + \dfrac{3}{5}\dfrac{1}{2}\dfrac{2}{4} + \dfrac{2}{5}(.8)1 + \dfrac{2}{5}(.2)\dfrac{3}{4}$$

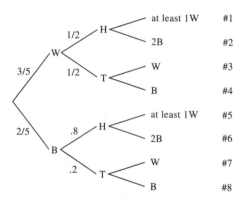

Figure P.7

8. *Method 1.*   Let A2 denote accident next year and let A1 denote accident this year.

$$P(A2|A1) = \dfrac{P(A2 \text{ and } A1)}{P(A1)} = \dfrac{\#3 + \#5}{\#3 + \#4 + \#5 + \#6} = \dfrac{\#3 + \#5}{\#1 + \#2}$$

$$= \dfrac{(.3)(.4)^2 + (.7)(.2)^2}{(.3)(.4) + (.7)(.2)} \approx .29$$

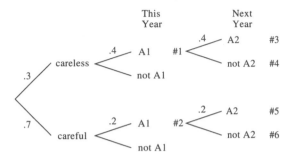

Figure P.8   Method 1

Method 2.   First find the prob of careless given A1.

$$P(\text{careless}|A1) = \frac{P(\text{careless and A1})}{P(A1)} = \frac{\#1}{\#1 + \#3} = \frac{12}{26}$$

Then make a new tree conditioning on A1. In this conditional world,

$$P(A2) = \#5 + \#6 = \frac{12}{26}(.4) + \frac{14}{26}(.2)$$

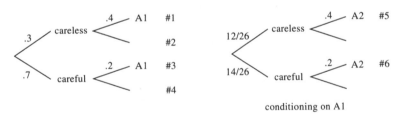

conditioning on A1

Figure P.8   Method 2

9. (a) $P(A)$ **(b)** $P(C|A)$ **(c)** $P(A \text{ and } C)$

10. $P(\text{first 2 tosses are less than third})$

$= P(T3 = 1)\ P(\text{other tosses smaller}|T3 = 1)$

$\quad + P(T3 = 2)\ P(\text{other tosses smaller}|T3 = 2)$

$\quad + P(T3 = 3)\ P(\text{other tosses smaller}|T3 = 3)$

$\quad + P(T3 = 4)\ P(\text{other tosses smaller}|T3 = 4)$

$\quad + P(T3 = 5)\ P(\text{other tosses smaller}|T3 = 5)$

$\quad + P(T3 = 6)\ P(\text{other tosses smaller}|T3 = 6)$

= sum of fav branches

$$= \frac{1}{6} \cdot 0 + \frac{1}{6} \left(\frac{1}{6}\right)^2 + \frac{1}{6} \left(\frac{2}{6}\right)^2 + \frac{1}{6} \left(\frac{3}{6}\right)^2 + \frac{1}{6} \left(\frac{4}{6}\right)^2 + \frac{1}{6} \left(\frac{5}{6}\right)^2$$

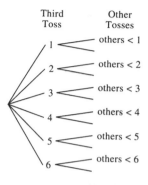

Figure P.10

Solutions Section 2-5

1. **(a)** Use the Poisson with $\lambda = 2$.

$$1 - P(0) - P(1) - P(2) = 1 - e^{-2} - 2e^{-2} - \frac{4e^{-2}}{2!}$$

(b) On the average, the community has 4 XYZ's, so $\lambda = 4$.

$$1 - P(0) - P(1) - \cdots - P(5) = 1 - e^{-4} \left(1 + 4 + \frac{4^2}{2!} + \frac{4^3}{3!} + \frac{4^4}{4!} + \frac{4^5}{5!}\right)$$

2. For the 5 flights, the average number of no-shows is 50, so use $\lambda = 50$.

(a) $P(\text{none}) = e^{-50}$

(b) $P(4) = \dfrac{50^4 e^{-50}}{4!}$

(c) $P(\text{none}) + P(1) + P(2) + P(3) + P(4)$

$$= e^{-50} \left(1 + 50 + \frac{50^2}{2!} + \frac{50^3}{3!} + \frac{50^4}{4!}\right)$$

3. **(a)** Use the binomial with $n = 100$, $p = .05 = P(\text{fail to stop})$.

$$P(\text{at least 2 fail}) = 1 - P(\text{none}) - P(1)$$
$$= 1 - (.95)^{100} - 100(.05)(.95)^{99}$$

(b) $\lambda = 3$, $P(\text{at least 2}) = 1 - P(\text{none}) - P(1) = 1 - e^{-3} - 3e^{-3}$

4. (a) $\lambda = 2$, $P(3 \text{ tickets}) = \dfrac{e^{-2} \, 2^3}{3!}$

    (b) When we chose to use the Poisson in part (a), we assumed that tickets are given out independently, so it is irrelevant that you got 2 in January. The average number of tickets in $\frac{11}{12}$ of a year is $\frac{11}{12} \cdot 2$, so $\lambda = 11/6$.

$$P(\text{no tickets in 11 months}) = e^{-11/6}$$

5. The average number of calls in 15 minutes is $\frac{1}{4} \cdot 2$, so $\lambda = 1/2$.

    (a) $P(\text{no calls in 15 minutes}) = e^{-1/2}$.

    (b) $P(\text{no more than 1 call in 15 minutes})$
$$= P(\text{none}) + P(1) = e^{-1/2} + \tfrac{1}{2} e^{-1/2}$$

6. On the average there are $\lambda_1 + \lambda_2 + \lambda_3$ disasters in a year, so $\lambda = \lambda_1 + \lambda_2 + \lambda_3$.

$$P(\text{at least 1 disaster}) = 1 - P(\text{none}) = 1 - e^{-(\lambda_1 + \lambda_2 + \lambda_3)}$$

7. *Step 1.*

The number of calls in a day is a Poisson random variable with $\lambda = 3$.

$P(\text{no calls in a day}) = e^{-3}$

Step 2.

There are 1825 days in 5 years (ignore leap days) and $P(\text{no calls in a day}) = e^{-3}$, so we have 1825 Bernoulli trials with $P(\text{success}) = e^{-3}, P(\text{failure}) = 1 - e^{-3}$.

$$P(\text{at least 1 success}) = 1 - P(\text{no successes}) = 1 - (1 - e^{-3})^{1825}$$

8. $\lambda = np = (1000)(.01) = 10$, Poisson approximation is $10e^{-10}$.

9. Bernoulli trials are independent repetitions of the same experiment where the experiment has two outcomes, success and failure (coin tosses).

Solutions Review Problems for Chapters 1, 2

1. (a) *With Replacement (multinomial)*

$$P(3\text{W}, 4\text{R}, 3 \text{ others}) = \frac{10!}{3! \, 4! \, 3!} \left(\frac{20}{140}\right)^3 \left(\frac{40}{140}\right)^4 \left(\frac{80}{140}\right)^3$$

W/O Replacement

$$P(\text{3W, 4R, 3 others}) = \frac{\text{fav}}{\text{total}} = \frac{\binom{20}{3}\binom{40}{4}\binom{80}{3}}{\binom{140}{10}}$$

(b) *With*

$$P(\text{WWW RRRR OOO}) = \left(\frac{20}{140}\right)^3 \left(\frac{40}{140}\right)^4 \left(\frac{80}{140}\right)^3$$

W/O

$$\frac{20}{140}\frac{19}{139}\frac{18}{138}\frac{40}{137}\frac{39}{136}\frac{38}{135}\frac{37}{134}\frac{80}{133}\frac{79}{132}\frac{78}{131}$$

(c) Same as part (b) by symmetry

2. Nine Bernoulli trials where on any one trial $P(2) = .1$.

$P(\text{at least four 2's})$

$$= 1 - P(\text{no 2's}) - P(\text{one 2}) - P(\text{two 2's}) - P(\text{three 2's})$$

$$= 1 - (.9)^9 - \binom{9}{1}(.1)(.9)^8 - \binom{9}{2}(.1)^2(.9)^7 - \binom{9}{3}(.1)^3(.9)^6$$

3. (a) By symmetry,

$P(\text{10th is king, 11th is non-K})$
$$= P(\text{K on 1st, non-K on 2nd})$$
$$= \frac{4}{52}\frac{48}{51}$$

(b) *Without Replacement*

$$P(\bar{\text{K}}^9\text{K}) = \frac{48}{52}\frac{47}{51}\frac{46}{50}\cdots\frac{40}{44}\frac{4}{43}$$

With Replacement

$$\left(\frac{48}{52}\right)^9\frac{1}{13}$$

(c) $P(2K \text{ in } 9 \text{ draws, then K on 10th})$

$$= P(2K \text{ in } 9 \text{ draws})P(K \text{ on 10th}|2K \text{ in first } 9 \text{ draws}) = \frac{\binom{4}{2}\binom{48}{7}}{\binom{52}{9}}\frac{2}{43}$$

(d) $P(\text{no K or 1K or 2K in first } 9 \text{ draws})$

$$= P(\text{no K}) + P(1K) + P(2K)$$

$$= \frac{\binom{48}{9} + 4\binom{48}{8} + \binom{4}{2}\binom{48}{7}}{\binom{52}{9}}$$

4. $P(\text{at least one card } < 6|\text{at least one card } > 9)$

$$= \frac{P(\text{at least one } < 6 \text{ and at least one } > 9)}{P(\text{at least one } > 9)}$$

denominator $= 1 - P(\text{all } \leq 9) = 1 - \dfrac{\binom{32}{13}}{\binom{52}{13}}$

numerator $= 1 - P(\text{all cards } \geq 6 \text{ or all } \leq 9)$

$$= 1 - [P(\text{all } \geq 6) + P(\text{all } \leq 9) - P(6 \leq \text{ all } \leq 9)]$$

$$= 1 - \frac{\binom{36}{13} + \binom{32}{13} - \binom{16}{13}}{\binom{52}{13}}$$

5. (a) The symbols in the string are Bernoulli trials where each trial results in vowel or non-vowel.

$$P(3 \text{ vowels}) = \binom{12}{3}\left(\frac{5}{36}\right)^3\left(\frac{31}{36}\right)^9$$

(b) $P(3 \text{ vowels}) = \dfrac{\text{fav}}{\text{total}} = \dfrac{\binom{5}{3}\binom{31}{9}}{\binom{36}{12}}$

6. This is like drawing 8 balls without replacement from a box containing 2L, 3I, 1N, 1S.

Method 1.

$P(\text{1st and last are L's})$

$= P(\text{1st and 2nd are L's}) \text{ (by symmetry)}$

$= P(\text{1st is L}) + P(\text{2nd is L}) - P(\text{1st and 2nd are L's})$

$= 2P(\text{1st is L}) - P(\text{1st and 2nd are L's}) \text{ (more symmetry)}$

$$= 2 \cdot \frac{2}{8} - \frac{2}{8}\frac{1}{7}$$

Method 2.

$$P(\text{1st and last are L's}) = P(\text{1st and 2nd are L's}) \text{ (by symmetry)}$$
$$= 1 - P(\text{1st and 2nd are non-L's}) = 1 - \frac{6}{8}\frac{5}{7}$$

7. **(a)** Here's one way to do it. The total number of ways to pick 2 of the 8 seats for J and M is $\binom{8}{2}$. There are 7 fav ways (seats 1 and 2, seats 2 and 3, ..., seats 7 and 8). Answer is $7/\binom{8}{2}$.

   (b) A circular table is tricky. For example, the circle $ABCDEFGH$ is the same as the circle $BCDEFGHA$. Here's one method. Put John down anywhere. When Mary sits down there are 7 seats available and 2 are fav. So prob = 2/7.

8. $P(\text{word contains } z) = 1 - P(\text{no } z\text{'s}) = 1 - \left(\frac{25}{26}\right)^3$.

9. There are 365 Bernoulli trials, and on any one trial $P(\text{towed}) = .1$.

   $$P(\text{at most one tow}) = P(\text{none}) + P(\text{one})$$
   $$= (.9)^{365} + \binom{365}{1}(.9)^{364}(.1)$$

10. This is drawing balls from a box without replacement. By symmetry,
   $$P(\text{last two are M}) = P(\text{first two are M}) = \frac{m}{m+w} \cdot \frac{m-1}{m+w-1}$$

11. **(a)** There are j married men and k single men, so the man can be picked in $j+k$ ways. Similarly, the woman can be picked in $j+n$ ways. Total number of ways of picking the pair is $(j+k)(j+n)$. There are $j \cdot j$ fav ways, so
   $$P(\text{both married}) = \frac{j^2}{(j+k)(j+n)}$$

   (b) $P(\text{man married, woman single}) + P(\text{man single, woman married})$
   $$= \frac{jn+kj}{(j+k)(j+n)}$$

   (c) There are j favs (the man and the woman have to be one of the j married couples).
   $$\frac{j}{(j+k)(j+n)}$$

12. **(a)** $P(\text{non-3 in one toss}) = 5/6$, so $P(\text{non-3 in 10 tosses}) = (5/6)^{10}$

(b) $(5/6)^{100000}$

(c) $\lim_{n\to\infty}(5/6)^n = (5/6)^\infty = 0$

13. $P(\text{match}) = P(2 \text{ black or } 2 \text{ blue or } 2 \text{ white})$
$$= P(2 \text{ black}) + P(2 \text{ blue}) + P(2 \text{ white})$$

Method 1.   Treat the pair of socks as a committee.

$$P(\text{match}) = \frac{\binom{5}{2} + \binom{6}{2} + \binom{7}{2}}{\binom{18}{2}}$$

Method 2.   We'll get the same answer if we let order count (as long as we do it consistently in the numerator and denominator).

$$P(\text{match}) = \frac{5}{18}\frac{4}{17} + \frac{6}{18}\frac{5}{17} + \frac{7}{18}\frac{6}{17}$$

14. *Method 1.*

$$P(\text{H on 8th}|6\text{H},4\text{T}) = \frac{P(6\text{H, 4T and H on 8th})}{P(6\text{H and 4T})}$$

$$\text{denom} = \binom{10}{6}\left(\frac{1}{2}\right)^{10} \quad \text{(binomial distribution)}$$

$$\text{numerator} = P(\text{H on 8th})P(5\text{H},4\text{T in 9 throws}) \quad \text{(by independence)}$$

$$= \frac{1}{2}\binom{9}{5}\left(\frac{1}{2}\right)^9$$

Method 2.   Think of an urn containing 6H and 4T. Draw w/o replacement.

$$P(\text{8th is H}) = P(\text{1st is H}) = \frac{6}{10}$$

15. (a) (multinomial)

$$P(1\text{A, 1B, 4 others in 6 trials}) = \frac{6!}{1!\,1!\,4!}\frac{1}{26}\frac{1}{26}\left(\frac{24}{26}\right)^4$$

(b) $1 - P(\text{no A or no B})$

$$= 1 - [P(\text{no A}) + P(\text{no B}) - P(\text{no A and no B})]$$

$$= 1 - \left[2\left(\frac{25}{26}\right)^6 - \left(\frac{24}{26}\right)^6\right]$$

(c) $1 - P(\text{no A and no B}) = 1 - (24/26)^6$

(d) $P(2\text{A's}) - P(2\text{A and no B}) - P(2\text{A and 1B})$

$$= \binom{6}{2}\left(\frac{1}{26}\right)^2\left(\frac{25}{26}\right)^4 - \frac{6!}{2!\,0!\,4!}\left(\frac{1}{26}\right)^2\left(\frac{24}{26}\right)^4$$

$$- \frac{6!}{2!\,1!\,3!}\left(\frac{1}{26}\right)^3\left(\frac{24}{26}\right)^3$$

(e) *Method 1.*

$$P(\text{any first})P(\text{different|first})\cdots = 1\cdot\frac{25}{26}\,\frac{24}{26}\,\frac{23}{26}\,\frac{22}{26}\,\frac{21}{26}$$

Method 2.   Pick 6 flavors. Then find the prob of getting one each of those 6.

$$\binom{26}{6}P(\text{one each of, say, A,B,C,D,E,F}) = \binom{26}{6}\frac{6!}{(1!)^6}\left(\frac{1}{26}\right)^6$$

(f) *Method 1.*

$$P(\text{any first})P(\text{same|first})\cdots = 1\left(\frac{1}{26}\right)^5$$

Method 2.

$$P(\text{all A or all B or } \ldots \text{ or all Z})$$

$$= P(\text{all A}) + \cdots + P(\text{all Z}) = 26\left(\frac{1}{26}\right)^6$$

16. (a) The numerator double counts. For instance, it counts the following outcomes as different when they are the same:

outcome 1 Pick spades to be the missing suit.
Pick all the hearts as your 13 non-spades.

outcome 2 Pick diamonds to be the missing suit.
Pick all the hearts as your 13 non-diamonds.

(b) $P(\text{at least one suit missing})$

$$= P(\text{no H or no S or no D or no C})$$

$$= P(\text{no H}) + P(\text{no S}) + P(\text{no D}) + P(\text{no C})$$

$- P(\text{no H and no S})$ and other 2-at-a-time terms

$$\left(\binom{4}{2} \text{ terms, each is } \binom{26}{13} \middle/ \binom{52}{13} \right)$$

$+ P(\text{no H, no S, no D})$ and other 3-at-a-time terms

$- P(\text{no H, no S, no C, no D})$ (impossible event)

$$= \frac{4\binom{39}{13} - \binom{4}{2}\binom{26}{13} + \binom{4}{3} \cdot 1}{\binom{52}{13}}$$

17. (a) Consider a new box containing only B_3 and B_5.

$P(B_3 \text{ before } B_5 \text{ from old box})$

$\qquad = P(B_3 \text{ in one draw from new box})$

$\qquad = \frac{1}{2}$

(b) Consider a new box containing only B_3 and the 5 whites.

$P(B_3 \text{ before white from old box})$

$\qquad = P(B_3 \text{ in one draw from new box}) = \frac{1}{6}$

18. $P(Z|\text{wrong}) = \dfrac{P(Z \text{ and wrong})}{P(\text{wrong})} = \dfrac{\#3}{\#1 + \#2 + \#3}$

$$= \frac{(.1)(.04)}{(.6)(.02) + (.3)(.03) + (.1)(.04)}$$

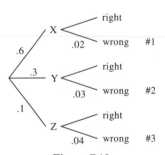

Figure P.18

19. $P(\text{at most 5H}|\text{at least 3H}) = \dfrac{P(\text{at most 5 and at least 3})}{P(\text{at least 3})}$

$$= \frac{P(3 \text{ or } 4 \text{ or } 5)}{1 - P(0 \text{ or } 1 \text{ or } 2)}$$

$$= \frac{\binom{10}{3}\left(\frac{1}{2}\right)^{10} + \binom{10}{4}\left(\frac{1}{2}\right)^{10} + \binom{10}{5}\left(\frac{1}{2}\right)^{10}}{1 - \left(\frac{1}{2}\right)^{10} - \binom{10}{1} - \left(\frac{1}{2}\right)^{10} - \binom{10}{2}\left(\frac{1}{2}\right)^{10}}$$

20. Let X and Y be the number of throws needed, respectively, by the two players.

$$P(X = Y) = P(X = 1, Y = 1) + P(X = 2, Y = 2) + \cdots$$
$$= P(X = 1)P(Y = 1) + P(X = 2)P(Y = 2) + \cdots$$
$$\text{(independence)}$$
$$= [P(X = 1)]^2 + [P(X = 2)]^2 + [P(X = 3)]^2 + \cdots$$

Here's how to calculate these probs.

The prob of a lucky throw on any one toss is $8/36 = 2/9$. So

$$P(X = 100) = P(\text{need 100 throws to get lucky})$$
$$= P(\text{99 unluckies followed by a lucky}) = \left(\frac{7}{9}\right)^{99} \frac{2}{9}$$

and

$$P(X = Y) = \left(\frac{2}{9}\right)^2 + \left[\frac{7}{9}\frac{2}{9}\right]^2 + \left[\left(\frac{7}{9}\right)^2 \frac{2}{9}\right]^2 + \left[\left(\frac{7}{9}\right)^3 \frac{2}{9}\right]^2 + \cdots$$

This is a geometric series with $a = (2/9)^2, r = (7/9)^2$. Answer is

$$\frac{4/81}{1 - 49/81} = \frac{1}{8}$$

21. The 10 foul shots are Bernoulli trials with $P(\text{success}) = .85$.

$$P(\text{at least 9 successes}) = P(9) + P(10) = \binom{10}{1}(.85)^9(.15) + (.85)^{10}$$

22. (a) $P(\text{each person gets at least one prize})$
$$= 1 - P(\text{none to P1 or none to P2 or } \ldots \text{ or none to P5})$$

$$= 1 - \begin{bmatrix} P(\text{none to P1}) + P(\text{none to P2}) + \cdots \\ - [P(\text{none to P1,P2}) + \cdots] \\ + [P(\text{none to P1,P2,P3}) + \cdots] \\ - [P(\text{none to P1,P2,P3,P4}) + \cdots] \\ + P(\text{none to P1,P2,P3,P4,P5}) \end{bmatrix}$$

$$= 1 - \left[5\left(\frac{4}{5}\right)^{10} - \binom{5}{2}\left(\frac{3}{5}\right)^{10} + \binom{5}{3}\left(\frac{2}{5}\right)^{10} - \binom{5}{4}\left(\frac{1}{5}\right)^{10} + 0 \right]$$

(b) $P(\text{no repeats}) = \dfrac{\text{fav}}{\text{total}} = \dfrac{10 \cdot 9 \cdot 8 \cdot 7 \cdot 6}{10^5}$

(c) $P(5M) = \left(\dfrac{6}{10}\right)^5$

23. *Method 1.*

$$P(11 \text{ penny heads}|17 \text{ total heads}) = \frac{P(11 \text{ penny H and 17 total H})}{P(17 \text{ total H})}$$

$$= \frac{P(11 \text{ penny H and 6 nickel H})}{P(17\text{H})}$$

$$\text{denominator} = \binom{40}{17}(.7)^{17}(.3)^{23}$$

$$\text{numerator} = \binom{20}{11}(.7)^{11}(.3)^9\binom{20}{6}(.7)^6(.3)^{14}$$

Method 2.   Think of a box with 17H and 23T. Draw 20 without replacement.

$$P(11\text{H in 20 draws}) = \frac{\binom{17}{11}\binom{23}{9}}{\binom{40}{20}}$$

24. (a) There are 400 Bernoulli trials with $P(\text{hits your block}) = 1/50 = .02$.

$$P(\text{at least 3 hits}) = 1 - P(\text{none}) - P(1) - P(2)$$

$$= 1 - (.98)^{400} - \binom{400}{1}(.98)^{399}(.02)$$

$$- \binom{400}{2}(.98)^{398}(.02)^2$$

(b) Use $\lambda = np = 8$. Prob is approximately $1 - e^{-8} - 8e^{-8} - \dfrac{64e^{-8}}{2!}$.

25. Let's call it a success on a round if there is no odd man out. On any round,

$$P(\text{success}) = P(\text{HHH or TTT}) = p^3 + q^3$$

$$P(\text{game lasts at least 6 rounds}) = P(S^5) = (p^3 + q^3)^5$$

26. (a) $P(A) + P(B) + P(C) = .8.$

(b) *Method 1.*

$$P(A) + P(B) + P(C) - [P(AB) + P(AC) + P(BC)] + P(ABC)$$
$$= .5 + .2 + .1 - [(.5)(.2) + (.5)(.1) + (.2)(.1)] + (.5)(.2)(.1) = .64$$

Method 2.

$$1 - P(\bar{A} \text{ and } \bar{B} \text{ and } \bar{C}) = 1 - (.5)(.8)(.9)$$

27. (a) Assuming 4 weeks to a month, on the average, there are 1/16 failures per week. Use the Poisson with $\lambda = 1/16$.

$P(\text{at least one failure during exam week})$
$$= 1 - P(\text{no failures})$$
$$= 1 - e^{-1/16}$$

(b) Use $\lambda = 1/4$. $P(\text{no failures in the next month}) = e^{-1/4}$.

28. The people are ind trials where each trial has 12 equally likely outcomes.

$$P(3 \text{ at one stop, 2 at another}) = P(3S_6\text{'s}, 2S_5\text{'s} + P(3S_1\text{'s}, 2S_7\text{'s}+) \cdots$$

There are $12 \cdot 11$ terms in the sum (pick a stop for the trio, pick a stop for the pair). Each prob is

$$\frac{5!}{2! \, 3!} \left(\frac{1}{12}\right)^5 \text{(multinomial formula)}.$$

Answer is

$$12 \cdot 11 \frac{5!}{2! \, 3!} \left(\frac{1}{12}\right)^5$$

29. (a) 999   **(b)** 1   **(c)** 1000

(d) $\dfrac{(n+m-1)!}{(n-1)! \, m!} \dfrac{n! \, m!}{(n+m)!} = \dfrac{n}{n+m}$

30. $P(\text{ends in 6 games})$
$$= P(A \text{ wins in 6 games})$$
$$+ P(B \text{ wins in 6 games})$$

$$= 2P(\text{A wins in 6 games})$$
$$= 2P(\text{3A's and 2B's in first 5 games})P(\text{A wins 6th})$$
$$= 2\binom{5}{3}\left(\frac{1}{2}\right)^5\frac{1}{2}$$

31. The side of the square is $R\sqrt{2}$. Its area is $2R^2$.

$$P(\text{shot lands in square}) = \frac{\text{fav area}}{\text{total area}} = \frac{2R^2}{\pi R^2} = \frac{2}{\pi}$$

$$P(\text{shot lands in I}) + P(\text{II}) + P(\text{III}) + P(\text{IV}) = 1 - \frac{2}{\pi} = \frac{\pi - 2}{\pi}$$

$$P(\text{I}) = P(\text{II}) = P(\text{III}) = P(\text{IV}) = \frac{1}{4}\frac{\pi - 2}{\pi}$$

The 5 shots are independent trials.

(a) $P(\text{all same zone})$

$$= P(\text{5 I or 5 II or 5 III or 5 IV or 5 V})$$
$$= P(5\text{I}) + P(5\text{II}) + P(5\text{III}) + P(5\text{IV}) + P(5\text{V})$$
$$= 4\left(\frac{1}{4}\frac{\pi - 2}{\pi}\right)^5 + \left(\frac{2}{\pi}\right)^5$$

(b) $P(\text{all different zones}) = P(\text{1 each of I,II,III,IV,V})$

$$= \frac{5!}{(1!)^5}\left(\frac{1}{4}\frac{\pi - 2}{\pi}\right)^4\frac{2}{\pi}$$

32. Let M stand for Mother Has XYZ .

Let C_1 stand for First Child Has XYZ, and so on.

Method 1.

$$P(\bar{C}_4|\bar{C}_1\bar{C}_2\bar{C}_3) = \frac{\#1 + \#3}{\#1 + \#2 + \#3 + \#4} = \frac{\#1 + \#3}{\#5 + \#6}$$

$$= \frac{1/32 + 1/2}{1/16 + 1/2} = \frac{17}{18}$$

Method 2.   First, find the prob of $\bar{\text{M}}$ given $\bar{C}_1\bar{C}_2\bar{C}_3$.

$$P(\bar{\text{M}}|\bar{C}_1\bar{C}_2\bar{C}_3) = \frac{\#8}{\#7 + \#8} = \frac{1/2}{1/16 + 1/2} = \frac{8}{9}$$

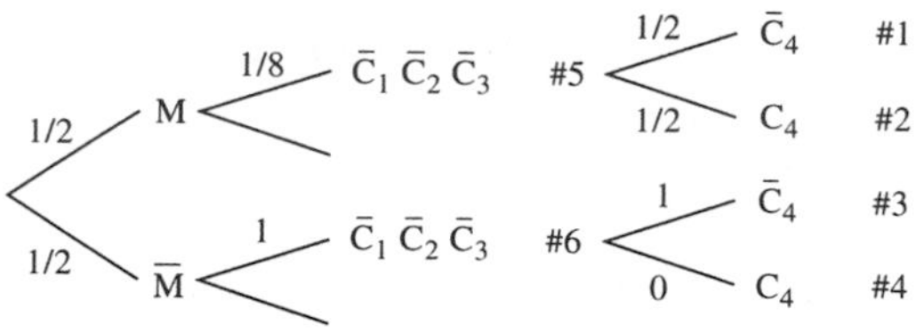

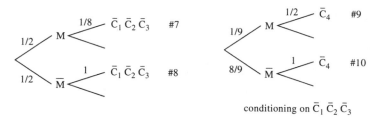

Figure P.32a Method 1

Now make a new tree conditioning on $\bar{C}_1\bar{C}_2\bar{C}_3$. In this conditional world,

$$P(\bar{C}_4) = \#9 + \#10 = \frac{1}{18} + \frac{8}{9} = \frac{17}{18}$$

Figure P.32b Method 2

33. $P(\text{at least 2 with the same birthday})$

$= 1 - P(\text{all different birthdays})$

$= 1 - \dfrac{365 \cdot 364 \cdot 363 \cdot \cdots \cdot (365 - n + 1)}{365^n}$

It turns out that if n is as small as 23, this prob is $\geq .5$. And if $n = 50$, the prob is .970. So it is more likely than you might think for people to share a birthday.

34. $P(\text{6 from one die in your 3 chances}) = 1 - P(\text{no 6's in 3 tosses})$

$= 1 - (5/6)^3 = 91/216.$

Now each of the 5 dice that can be tossed as many as 3 times each is a Bernoulli trial, where

$$P(\text{success}) = P(\text{6 from your 3 chances}) = \frac{91}{216}$$

$$P(\text{2 successes in 5 trials}) = \binom{5}{2}\left(\frac{91}{216}\right)^2\left(\frac{125}{216}\right)^3$$

35. By Bayes' theorem,

$$P(xxxxx \text{ sent}|2x, 3y \text{ received}) = \frac{\#1}{\#1 + \#2} = \frac{.6p_1}{.6p_1 + .4p_2}$$

Now we need p_1 and p_2. Each of the five symbols sent is a Bernoulli trial; the outcome is either error or no-error where $P(\text{error}) = .1$. So

$$p_1 = P(3 \text{ errors}) = \binom{5}{3}(.1)^3(.9)^2, \quad p_2 = P(2 \text{ errors}) = \binom{5}{2}(.1)^2(.9)^3$$

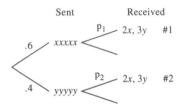

Figure P.35

36. Let x be her score; let y be his score.

(a) $P(y \geq 2x) = \dfrac{\text{fav area}}{\text{total}} = \dfrac{3 - \text{unfav}}{3} = \dfrac{3 - 1}{3} = \dfrac{2}{3}$

(b) $P\left(\max \leq \dfrac{1}{2}\right) = \dfrac{\text{fav}}{\text{total}} = \dfrac{1/4}{3} = \dfrac{1}{12}$

(c) $P\left(\min \geq \dfrac{1}{2}\right) = \dfrac{\text{fav}}{\text{total}} = \dfrac{5/4}{3} = \dfrac{3}{12}$

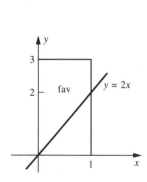

Figure P.36a

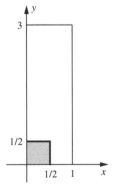

Figure P.36b

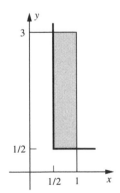

Figure P.36c

37. (a) $\dbinom{19}{4}\bigg/\dbinom{20}{5}$

(b) $P(\text{J on 1st draw}) + P(\text{J on 2nd draw}) + \cdots + P(\text{J on 5th draw})$

$$= 5P(\text{J on 1st draw}) \quad \text{(by symmetry)} = \frac{5}{20}$$

(c) $1 - \dfrac{\binom{19}{5}}{\binom{20}{5}}$

38. Each person is a Bernoulli trial where $P(\text{July 4}) = 1/365$.

$P(\text{at least 3 July 4's}) = 1 - P(\text{none}) - P(1) - P(2)$

$$= 1 - \left(\frac{364}{365}\right)^{30} - \binom{30}{1}\frac{1}{365}\left(\frac{364}{365}\right)^{29}$$

$$- \binom{30}{2}\left(\frac{1}{365}\right)^2\left(\frac{364}{365}\right)^{28}$$

39. (a) Each freshman is a Bernoulli trial where $P(\text{success}) = P(A) = .2$.

$$P(4A) = \binom{10}{4}(.2)^4(.8)^6$$

(b) Draw 6 times without replacement (no one can get two offices) from a population of 10F, 20S, 30J, 20G.

$$P(4F) = \frac{\binom{10}{4}\binom{70}{2}}{\binom{80}{6}}$$

40. We want the probability of a collision in a time slot. (If the prob of a collision is .7, then 70% of the time slots are wasted.) Consider a time slot. The n hosts are Bernoulli trials.

$$P(\text{success}) = P(\text{tries to use slot}) = p$$

$P(\text{collision}) = P(\text{at least 2S in } n \text{ trials}) = 1 - P(\text{none}) - P(\text{one})$

$$= 1 - q^n - \binom{n}{1}pq^{n-1}$$

41. Let his arrival time be x and her arrival time be y. Then (x, y) is uniformly distributed in a rectangle.

(a) $P(\text{meet}) = P(|y - x| \le 10) = \dfrac{\text{fav area}}{\text{total}} = \dfrac{1150}{5400}$

(b) $P(\text{J arrives first and then Mary arrives no more than}$

$10 \text{ minutes later}) = P(x \le y \le x + 10) = \dfrac{\text{fav area}}{\text{total}} = \dfrac{600}{5400}$

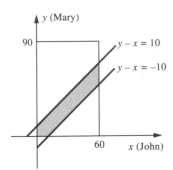

Figure P.41a   **Figure P.41b**

42. (a) Total is $(100)^{20}$ (each of the 20 spots can be filled in 200 ways). For the fav, we need the number of ways of lining up 20 out of the 100 numbers in increasing order. First, pick 20 (different) numbers out of the 100 to be in the lineup. Then there is only one way to line them up (namely, in increasing order). So the number of fav is $\binom{100}{20}$. Answer is

$$\dfrac{\binom{100}{20}}{(100)^{20}}$$

(b) Same fav as part (a). The total is $100 \cdot 99 \cdot 98 \cdot \; \cdots \; \cdot 81$.

Solutions Section 3-1

1. Let X be the number of trials it takes to locate the D's.

EX

$= 2P(X = 2) + 3P(X = 3) + 4P(X = 4)$

$= 2P(\text{DD}) + 3P(\text{DGD or GDD or GGG})$

$\qquad\qquad + 4P(\text{DGG or GDG or GGD})$

$= 2\left(\dfrac{2}{5}\dfrac{1}{4}\right) + 3\left(\dfrac{2}{5}\dfrac{3}{4}\dfrac{1}{3} + \dfrac{3}{5}\dfrac{2}{4}\dfrac{1}{3} + \dfrac{3}{5}\dfrac{2}{4}\dfrac{1}{3}\right) + 4\left(\dfrac{2}{5}\dfrac{3}{4}\dfrac{2}{3} + \dfrac{3}{5}\dfrac{2}{4}\dfrac{2}{3} + \dfrac{3}{5}\dfrac{2}{4}\dfrac{2}{3}\right)$

The three terms in the first parentheses are all equal (and the three terms in the second parentheses are all equal) (you could have seen this in advance using symmetry, see Section 1.3 (1) and (2)), so

$$EX = 2\left(\frac{2}{5}\frac{1}{4}\right) + 3\cdot 3\left(\frac{2}{5}\frac{3}{4}\frac{1}{3}\right) + 4\cdot 3\left(\frac{2}{5}\frac{3}{4}\frac{2}{3}\right)$$

2. $E(\text{winnings})$

$$= 10P(\text{within 1 inch}) + 5P(1 \text{ to } 3 \text{ inches}) + 2P(3 \text{ to } 5 \text{ inches})$$
$$- 4P(\text{more than 5 inches})$$

where

$$P(\text{within 1 inch}) = \frac{\text{fav area}}{\text{total}} = \frac{\pi}{64\pi} = \frac{1}{64}$$

$$P(1 \text{ to } 3 \text{ inches}) = \frac{9\pi - \pi}{64\pi} = \frac{1}{8}$$

$$P(3 \text{ to } 5 \text{ inches}) = \frac{25\pi - 9\pi}{64\pi} = \frac{1}{4}$$

$$P(\text{more than 5 inches away}) = \frac{64\pi - 25\pi}{64\pi} = \frac{39}{64}$$

So

$$E(\text{winnings}) = 10\cdot\frac{1}{64} + 5\cdot\frac{1}{8} + 2\cdot\frac{1}{4} - 4\cdot\frac{39}{64} = -\frac{74}{64}$$

(the game is stacked against you)

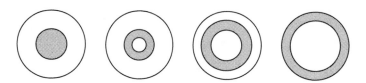

Figure P.2

3. (a) expected number of children

$$= 1P(1 \text{ child}) + 2P(2 \text{ C}) + 3P(3 \text{ C})$$

$$= 1P(\text{G}) + 2P(\text{BG}) + 3P(\text{BBG or BBB})$$

$$= 1\cdot\frac{1}{2} + 2\cdot\frac{1}{2}\frac{1}{2} + 3\cdot 2\cdot\frac{1}{8} = \frac{7}{4}$$

(b) expected number of girls $= 1P(\text{G or BG or BBG}) + 0 \cdot P(\text{BBB})$

$$= 1\left(\frac{1}{2} + \frac{1}{4} + \frac{1}{8}\right) = \frac{7}{8}$$

(c) expected number of boys $= 1P(\text{BG}) + 2P(\text{BBG}) + 3P(\text{BBB})$

$$= \frac{1}{4} + 2 \cdot \frac{1}{8} + 3 \cdot \frac{1}{8} = \frac{7}{8}$$

4. expected winnings $= 35P(\text{hit your number}) - 1 \cdot P(\text{another number})$

$$= 35 \cdot \frac{1}{38} - \frac{37}{38} = -\frac{2}{38}$$

5. $\displaystyle\sum_{i=1}^{\infty} P(X \geq i)$

$$= P(X \geq 1) + P(X \geq 2) + P(X \geq 3) + \cdots$$
$$= P(X = 1 \text{ or } 2 \text{ or } 3 \text{ or } \ldots) + P(X = 2 \text{ or } 3 \text{ or } 4 \text{ or } \ldots)$$
$$+ P(X = 3 \text{ or } 4 \text{ or } 5 \text{ or } \ldots) + \cdots$$
$$= P(X = 1) + 2P(X = 2) + 3P(X = 3) + 4P(X = 4) + \cdots$$
$$= EX, \text{ QED}$$

Solutions Section 3-2

1. (a) $EX = \%$ whites in box $\times n = \dfrac{w}{w + b} \cdot n$

(b) $X = X_1 + \cdots + X_w$

$$EX = EX_1 + \cdots + EX_w = \sum_{i=1}^{w} P(W_i \text{ is drawn})$$

$$P(W_i \text{ is drawn}) = \frac{\binom{w-1+b}{n-1} \text{ draw } n - 1 \text{ others}}{\binom{w+b}{n}} = \frac{n}{w + b}$$

$$EX = \text{ sum of } w\tfrac{n}{w+b}\text{'s} = \tfrac{nw}{w+b}$$

2. Let X be the number of fixed points. For $i = 1, \ldots, n$, let

$$X_i = \begin{cases} 1 & \text{if } a_i \text{ is in the } i\text{th spot} \\ 0 & \text{otherwise} \end{cases}$$

Then $X = X_1 + \cdots + X_n$,

$EX = EX_1 + \cdots + EX_n = \sum_{i=1}^{n} P(a_i \text{ is in the } i\text{th spot}),$

$$P(a_i \text{ is in the } i\text{th spot}) = P(a_1 \text{ is in the 1st spot}) \quad \text{(symmetry)}$$

$$= \frac{1}{n} \quad (n \text{ total ways to place } a_1, \text{ one is fav})$$

$$EX = \sum_{i=1}^{n} \frac{1}{n} = n \cdot \frac{1}{n} = 1$$

3. Let X be the number of wives next to their husbands. For $i = 1, \dots, 10$, let

$$X_i = \begin{cases} 1 & \text{if } W_i \text{ is next to her husband} \\ 0 & \text{otherwise} \end{cases}$$

Then $X = X_1 + \cdots + X_{10}$.

(a) To find $P(W_i \text{ next to her husband})$, put W_i anywhere on the circle. Then there are 19 other seats, 2 are fav. So $P(W_i \text{ next to her husband}) = 2/19$, and $EX = 10 \cdot 2/19 = 20/19$.

(b) If you seat W_i first and then H_i, the prob that H_i is next to W_i depends on whether W_i is at an end or a non-end. So by the theorem of total prob,

$$P(W_i \text{ next to her husband})$$

$$= P(W_i \text{ at end})P(H_i \text{ next to } W_i|W_i \text{at end})$$

$$+ P(W_i \text{ not at end})P(H_i \text{ next to } W_i|W_i \text{ not at end})$$

$$= \frac{2}{20}\frac{1}{19} + \frac{18}{20}\frac{2}{19} = \frac{1}{10}$$

For another method, there are $\binom{20}{2}$ ways to pick 2 seats for W_i and H_i, and 19 of those ways are fav, namely, seats 1 & 2, 2 & 3, $\dots$, 19 & 20. So prob is $19/\binom{20}{2} = 1/10$.

So $EX = 10 \cdot \frac{1}{10} = 1$.

4. Let X be the number of runs. For $i = 1, \dots, n$, let

$$X_i = \begin{cases} 1 & \text{if a run starts at } i\text{th toss} \\ 0 & \text{otherwise} \end{cases}$$

Then $X = X_1 + \cdots + X_n$, $EX = \sum_{i=1}^{n} P(\text{run starts at } i\text{th toss})$, where

$P(\text{run starts at 1st toss}) = P(H) = p$.
$P(\text{run starts at 2nd toss}) = P(TH) = pq$.
$P(\text{run starts at 3rd toss}) = P(T \text{ on 2nd, H on 3rd}) = pq$.
$P(\text{run starts at } n\text{th toss}) = pq$.

So $EX = p + (n-1)pq$.

5. Let X be the number of records. For $i = 1, \ldots, n$, let

$$X_i = \begin{cases} 1 & \text{if the } i\text{th number sets a record} \\ 0 & \text{otherwise} \end{cases}$$

Then $X = X_1 + \cdots + X_n$.

We know that $P(\text{1st sets a record}) = 1$. Here's how to find the other probs.

Consider the event 5th sets a record. We have 5 different numbers (this argument wouldn't work if any 2 of the numbers could be identical), and the largest of the 5 is just as likely to be in any one of the 5 spots. So prob that the largest is *last* is 1/5. In general,

$$P(n\text{th sets a record}) = \frac{1}{n} \text{ and } EX = 1 + \frac{1}{2} + \frac{1}{3} + \cdots + \frac{1}{n}$$

As $n \to \infty$, $EX \to \infty$ because the (harmonic) series $\sum_{i=1}^{\infty} 1/n$ diverges to ∞.

6. (a) Let X be the number of different letters in n draws.
Let $X_A = 1$ if A is drawn at least once.
Similarly, define X_B, X_C, and X_D.

Then $X = X_A + X_B + X_C + X_D$ and $E(X) = E(X_A) + E(X_B) + E(X_C) + E(X_D)$.

$$E(X_A) = P(X_A = 1) = P(\text{at least one } A \text{ in } n \text{ draws})$$
$$= 1 - P(\text{no } A\text{'s}) = 1 - \left(\tfrac{3}{4}\right)^n$$

and similarly for the other indicators. So

$$E(X) = 4\left[1 - \left(\tfrac{3}{4}\right)^n\right]$$

(b) Let X be the number of boxes that must be bought to get all 4 pictures.

Let X_1 = number of boxes needed to get the first picture $= 1$.
Let X_2 = number of boxes after a first picture to get a second picture to turn up.
Let X_3 = number of boxes after the second picture to get a third to turn up.
Let X_4 = number of boxes after the third picture to get the fourth to turn up.

For example, if the outcome is A A D C C A D C B then

$X = 9$ (it took 9 tries to get all 4 pictures)
$X_1 = 1$
$X_2 = 2$ (after getting A it took 2 more tries before a new one, D, turned up)
$X_3 = 1$ (after getting A and D it took only one try before a new one, C, turned up)
$X_4 = 5$ (after getting A, D, C it took 5 tries to get a new one)

Then $X = X_1 + X_2 + X_3 + X_4$, $EX = EX_1 + EX_2 + EX_3 + EX_4$.
We have $EX_1 = 1$. Here's how to get EX_2.
If the first picture is, say, B, then X_2 is the number of trials needed to get a success where $P(\text{success}) = P(\text{non-}B) = 3/4$.
So X_2 has a geometric distribution with $p = 3/4$.

$EX_2 = 1/p = 4/3$

Here's how to get EX_3. If the first two different pictures are, say, B and C, then X_3 is the number of trials needed to get a success where $P(S) = P(A \text{ or } D) = 2/4$.
So X_3 has a geometric distribution with $p = 2/4$ and $EX_3 = 4/2$.
Similarly, $EX_4 = 4/1$.
Finally, $EX = 1 + 4/3 + 4/2 + 4/1 = 25/3$.

7. The rounds are Bernoulli trials: At each trial the result is either odd man out or not where

$$P(\text{odd man out}) = 1 - P(\text{HHH or TTT}) = 1 - (p^3 + q^3)$$

The number of rounds in a game is the number of rounds until the first odd man out. So the number of rounds has a geometric distribution with parameter $1 - (p^3 + q^3)$ and

$$EX = \frac{1}{1 - (p^3 + q^3)}$$

Solutions Section 3-3

1. In the new universe where the sum is 4, there are 3 equally likely outcomes: (1,3), (3,1), (2,2).

$$E(\text{first die}|\text{sum} = 4) = 1P(1,3) + 3P(3,1) + 2P(2,2)$$

$$= 1 \cdot \frac{1}{3} + 3 \cdot \frac{1}{3} + 2 \cdot \frac{1}{3} = 2$$

2. Let X be the time it takes to solve the problem. The diagram shows the possible outcomes (successive methods tried).

$$EX = \frac{1}{6}(11 + 9 + 11) + \frac{1}{3} \cdot 4 = \frac{45}{6}$$

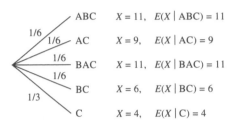

ABC $X = 11,$ $E(X \mid ABC) = 11$

AC $X = 9,$ $E(X \mid AC) = 9$

BAC $X = 11,$ $E(X \mid BAC) = 11$

BC $X = 6,$ $E(X \mid BC) = 6$

C $X = 4,$ $E(X \mid C) = 4$

Figure P.2

3. In the new universe we toss a die 4 times, but only 5 outcomes are possible (can't get any more 1's). So we're really tossing a 5-sided die with $P(2) = 1/5$. The number of 2's has a binomial distribution with $n = 4, p = 1/5$, so the expected number of 2's is $np = 4/5$.

4. Let X be the number of 5's. The number of 5's in k tosses has a binomial distribution with $p = 1/6$, so $E(X|k \text{ tosses}) = kp = k/6$. By the theorem of total expectation,

$$EX = P(\text{quad } 1)E(X|\text{quad } 1) + \cdots + P(\text{quad } 4)E(X|\text{quad } 4)$$

$$= \frac{1}{4}\frac{1}{6} + \frac{1}{4}\frac{2}{6} + \frac{1}{4}\frac{3}{6} + \frac{1}{4}\frac{4}{6} = \frac{10}{24}$$

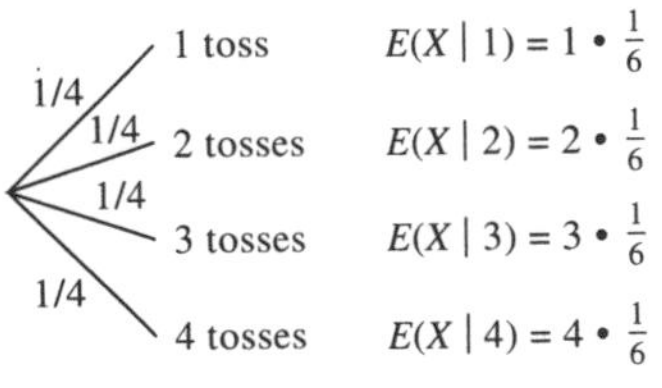

1 toss $E(X \mid 1) = 1 \cdot \frac{1}{6}$

2 tosses $E(X \mid 2) = 2 \cdot \frac{1}{6}$

3 tosses $E(X \mid 3) = 3 \cdot \frac{1}{6}$

4 tosses $E(X \mid 4) = 4 \cdot \frac{1}{6}$

Figure P.4

5. By symmetry, this is the same as finding the expected number of W in the last 3 draws given 1W in the first 5 draws. In turn, this is the same as the expected number of W in 3 draws from an urn with 9W, 16B. Answer is $3 \cdot 9/25$.

6. *Method 1.* Let

$$X_i = \begin{cases} 1 & \text{if } i\text{th is H} \\ 0 & \text{otherwise} \end{cases}$$

Then $X = X_1 + \cdots + X_n$.

$$E(X|X \geq 2) = E(X_1|X \geq 2) + \cdots + E(X_n|X \geq 2)$$

$$= \sum_{k=1}^{n} P(H \text{ on } i\text{th}|\text{at least 2H})$$

$$P(\text{H on } i\text{th}|\text{at least 2H}) = \frac{P(\text{H on } i\text{th and at least 2H})}{P(\text{at least 2H})}$$

$$= \frac{P(\text{H on } i\text{th})P(\text{at least 1 H in } n-1 \text{ tosses})}{1 - P(\text{no H}) - P(\text{1H})}$$

$$= \frac{p(1 - q^{n-1})}{1 - q^n - \binom{n}{1}pq^{n-1}}$$

So

$$E(X|X \geq 2) = n \cdot \frac{p(1 - q^{n-1})}{1 - q^n - npq^{n-1}}$$

Method 2.

$E(X|X \geq 2)$

$$= 2P(2\text{H}|\text{at least 2H}) + 3P(3\text{H}|\text{at least 2H})$$

$$+ \cdots + nP(n\text{H}|\text{at least 2H})$$

$$= 2\frac{P(2\text{H and at least 2H})}{P(\text{at least 2H})} + \cdots + n\frac{P(n\text{H and at least 2H})}{P(\text{at least 2H})}$$

$$= \frac{2P(2\text{H}) + 3P(3\text{H}) + \cdots + nP(n\text{H})}{1 - q^n - npq^{n-1}}$$

The clever way to do the numerator is to see that

$$\text{numerator} + 1P(1\text{H}) = \text{binomial mean} = np$$

so

$$\text{numerator} = np - \binom{n}{1}pq^{n-1}$$

and

$$E(X|X \geq 2) = \frac{np - npq^{n-1}}{1 - q^n - npq^{n-1}}$$

7. Let X be the number of tosses it takes to complete a run of 5H. We want EX, abbreviated E. The tree diagram shows all possible rounds at which the first T can occur.

If the first toss is a T, then the new situation (looking for 5 consecutive H) is the same as it was in the beginning, so after the first wasted toss, it will take on the average E more tries. So $E(X|T) = 1 + E$.

If you initially get HT, then the new situation (looking for 5 consecutive H) is the same as it was in the beginning, so after the first 2 wasted tosses we expected it will take another E tosses. So $E(X|HT) = 2 + E$.

$\vdots$

And if you initially get 5 consecutive H, then $X = 5$ and $E(X|H^5) = 5$.

By the theorem of total expectation,

$E = q(1+E) + pq(2+E) + p^2q(3+E) + p^3q(4+E) + p^4q(5+E) + p^5E$

Solve for E to get

$$E = \frac{q + 2pq + 3p^2q + 4p^3q + 5p^4q + 5p^5}{1 - q - pq - p^2q - p^3q - p^4q}$$

(The denominator contains a finite geometric series and the numerator contains a differentiated finite geometric series, so the answer simplifies.)

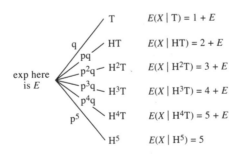

Figure P.7

Solutions Review Problems for Chapter 3

1. Let X be the number of empty boxes. For $i = 1, \ldots, 50$, let

$$X_i = \begin{cases} 1 & \text{if the } i\text{th box is empty} \\ 0 & \text{otherwise} \end{cases}$$

Then $X = X_1 + \ldots + X_{50}$, $EX_i = P(i\text{th box is empty}) = (49/50)^{100}$.

$$EX = EX_1 + \cdots + EX_{50} = 50 \left(\frac{49}{50}\right)^{100}$$

2. Think of a box containing 6 heads and 4 tails. Draw 5. The expected number of heads is $\frac{6}{10} \times 5 = 3$.

3. $E(\text{winnings}) = 1P(\text{one 5}) + 2P(\text{two 5's}) + 3P(\text{three 5's}) - 1P(\text{no 5's})$

$$= \binom{3}{1}\frac{1}{6}\left(\frac{5}{6}\right)^2 + 2\binom{3}{2}\left(\frac{1}{6}\right)^2\frac{5}{6} + 3\left(\frac{1}{6}\right)^3 - \left(\frac{5}{6}\right)^3 = -\frac{17}{216}$$

4. Let X be the number of objects between x_1 and x_2. For $i = 3, \ldots, n$, let

$$X_i = \begin{cases} 1 & \text{if } x_i \text{ is between } x_1 \text{ and } x_2 \\ 0 & \text{otherwise} \end{cases}$$

Then $X = X_3 + \cdots + X_n$, $EX = EX_3 + \cdots + EX_n$.

To find $P(X_i = 1) = P(x_i$ is between x_1 and $x_2)$, consider lining up just x_1, x_2, x_i. There are 3! total ways and 2 are fav. So $P(x_i$ is between x_1 and $x_2) = 2/3! = 1/3$.

So $E(X_i) = 1/3$ and $EX = $ sum of $n - 2$ indicator exps $= \frac{1}{3}(n - 2)$.

5. Let X be the number of heads.

Method 1.   (without indicators)

$$E(X|\text{at least 9H}) = 9P(9\text{H}|\text{at least 9H}) + 10P(10\text{H}|\text{at least 9})$$

$$P(9\text{H}|\text{at least 9}) = \frac{P(9\text{H and at least 9})}{P(\text{at least 9})} = \frac{P(9\text{H})}{P(9\text{H}) + P(10\text{H})}$$

$$= \frac{\binom{10}{9}\left(\frac{1}{2}\right)^{10}}{\binom{10}{9}\left(\frac{1}{2}\right)^{10} + \left(\frac{1}{2}\right)^{10}} = \frac{10}{11}$$

$$P(10\text{H}|\text{at least } 9) = 1 - \frac{10}{11} = \frac{1}{11}$$

$$E(X|\text{at least } 9\text{H}) = 9 \cdot \frac{10}{11} + 10 \cdot \frac{1}{11} = \frac{100}{11} \approx 9.1$$

Method 2.   Let $X_i = 1$ if ith toss is H. Then $X = X_1 + \cdots + X_{10}$. If A is the event "at least 9H" then

$$E(X|A) = E(X_1|A) + \cdots + E(X_{10}|A) = \sum_{i=1}^{10} P(\text{H on } i\text{th}|A)$$

$$P(\text{H on } i\text{th}|A) = \frac{P(\text{H on } i\text{th and at least 8H in the other 9 tosses})}{P(\text{at least 9H})}$$

$$= \frac{P(\text{H on } i\text{th})P(\text{at least 8H in 9 tosses})}{P(9\text{H}) + P(10\text{H})}$$

$$= \frac{\frac{1}{2}\left[\binom{9}{8}\left(\frac{1}{2}\right)^9 + \left(\frac{1}{2}\right)^9\right]}{\binom{10}{1}\left(\frac{1}{2}\right)^{10} + \left(\frac{1}{2}\right)^{10}} = \frac{10}{11}$$

So

$$E(X|A) = \text{sum of } 10 \ \frac{10}{11}\text{'s} = \frac{100}{11}$$

6. Let Y be the number of heads. By the theorem of total expectation,

$$EY = E(Y|X = 1)P(X = 1) + E(Y|X = 2)P(X = 2)$$
$$+ E(Y|X = 3)P(X = 3) + \cdots$$

In general, $E(Y|X = n) = np$ and $P(X = n) = e^{-\lambda}\lambda^n/n!$, so

$$EY = pe^{-\lambda}\left(1\frac{\lambda}{1!} + 2\frac{\lambda^2}{2!} + 3\frac{\lambda^3}{3!} + 4\frac{\lambda^4}{4!} + \cdots\right)$$

$$= \lambda pe^{-\lambda}\left(1 + \lambda + \frac{\lambda^2}{2!} + \frac{\lambda^3}{3!} + \frac{\lambda^4}{4!} + \cdots\right) = \lambda pe^{-\lambda} \cdot e^{\lambda} = \lambda p$$

7. **(a)** *Method 1.*

   expected number of trials

   $= 1P(\text{1st key unlocks}) + 2P(\text{2nd key unlocks})$

   $\qquad\qquad + \cdots + nP(\text{nth key unlocks})$

   $= 1 \cdot \dfrac{1}{n} + 2 \cdot \dfrac{1}{n} + \cdots + n \cdot \dfrac{1}{n}$

   $= \dfrac{1 + 2 + 3 + \cdots + n}{n} = \dfrac{\frac{1}{2}n(n+1)}{n} = \dfrac{n+1}{2}$

   Method 2.   Let X be the number of trials needed. Let K be the good key and $K_1, \ldots, K_{n-1}$ be the bad keys. For $i = 1, \ldots, n-1$, let

   $$X_i = 1 \text{ if } K_i \text{ is chosen before } K$$

   For example, if you try the keys in the order $K_3 K_2 K_5 K$, then

   $$X_3 = 1, X_2 = 1, X_5 = 1, \text{ other } X_i\text{'s are } 0$$

   Then $X = 1 + X_1 + \cdots + X_{n-1}$, $EX = 1 + EX_1 + \cdots + EX_{n-1}$.

   $$EX_i = P(K_i \text{ is chosen before } K) = \frac{1}{2}$$

   So

   $$EX = 1 + (n-1)\frac{1}{2} = \frac{n+1}{2}$$

   (b) Now we have Bernoulli trials with $P(\text{success}) = 1/n$. If X is the number of trials to get the first success, then X is geometric with parameter $p = 1/n$, and we know that $EX = 1/p = n$.

8. The people are Bernoulli trials. On any one trial, $P(\text{shutout}) = q^5$. The expected number of shutouts in 10 trials is $np = 10q^5$.

9. Think of the balls tossed as Bernoulli trials where

   $$P(\text{success}) = P(\text{ball goes into box 3}) = \frac{1}{50}$$

   (a) Let X be the number of balls that go into box 3 (the successes). Then X has a binomial distribution with $n = 100$, $p = 1/50$, and $EX = np = 2$.

   (b) Let X be the number of trials (balls tossed) to get a success (ball into box 3). Then X has a geometric distribution with $p = 1/50$ and $EX = 1/p = 50$.

10. Let X be the number of days when at least 2 monitors are needed. For $i = 1, \ldots, 365$, let

$$X_i = 1 \text{ if at least 2 monitors are needed on day } i$$

Then $X = X_1 + \cdots + X_{365}$ and $EX = EX_1 + \cdots + EX_{365}$.
The births are Bernoulli trials where $P(\text{needs monitor}) = .1$. So

$$EX_i = P(\text{at least 2 needed on } i\text{th day})$$
$$= 1 - P(\text{none needed}) - P(\text{one needed})$$

$$= 1 - (.9)^{20} - \binom{20}{1}(.1)(.9)^{19}$$

$$EX = 365 \left[1 - (.9)^{20} - \binom{20}{1}(.1)(.9)^{19} \right]$$

11. Let X be the number of tests needed. Let X_i be the number of tests needed in the ith group.

Then $X = X_1 + \cdots + X_5$ and $EX = EX_1 + \cdots + EX_5$.

Note that the X_i's are not indicators. They take on the values 1 and 21, not 0 and 1.

$$EX_i = 1P(\text{no well in } i\text{th group is polluted})$$
$$+ 21P(\text{at least one in the group is polluted})$$

The 20 wells in the group are Bernoulli trials with $P(\text{polluted}) = .1$.

$$EX_i = (.9)^{20} + 21[1 - (.9)^{20}] = 21 - 20(.9)^{20}$$

$$EX = \sum_{i=1}^{5} EX_i = 5[21 - 20(.9)^{20}] = 105 - 100(.9)^{20} \approx 93$$

12. Let X be the number of suits in a poker hand. Let $X_H = 1$ if the hand contains at least one heart and similarly define indicators X_S, X_C, X_D.
Then $X = X_H + X_S + X_C + X_D$, $EX = EX_H + EX_S + EX_C + EX_D$ where

$$EX_H = P(\text{at least 1H}) = 1 - P(\text{no H}) = 1 - \frac{\binom{39}{5}}{\binom{52}{5}}$$

And, similarly, for EX_S, EX_C, EX_D. So

$$EX = 4\left[1 - \frac{\binom{39}{5}}{\binom{52}{5}}\right] \approx 3.12$$

13. (a) Let X be the amount won. We'll find EX using the theorem of total expectation, conditioning on the first roll. Let $E(X)$ be denoted E. Conditional expectations are recorded on the tree diagram. Note that if the first toss is a 5 and the second toss is odd, then the game begins all over again. So at this point your expected winnings are the original E.

$$E = \frac{1}{2}(1) + \frac{1}{3}(-2) + \frac{1}{12}(-1) + \frac{1}{12}E$$

Solve for E to get the answer $E = -3/11$.

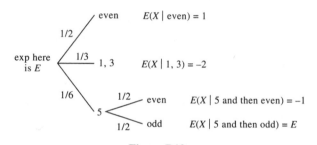

Figure P.13a

(b) Let Y be the number of tosses until you actually win or lose money and the game stops.

We'll find EY using the theorem of total expectation, conditioning on the first roll (see the diagram). Let EY be denoted E. If the first roll is a non-5, then the game is over after that initial roll, so $E(Y|\bar{5}) = 1$.

Suppose the first roll is 5 and your second toss is even. Then the game is over with these 2 rolls, so $E(Y|5$ and then even$) = 2$.

Suppose the first roll is 5 and your second toss is odd. After these 2 wasted rolls, the game begins again. So

$$E(Y|5 \text{ and then odd}) = 2 + \text{ expected length of a new game}$$
$$= 2 + \text{ the original } E$$

Then

$$E = \frac{5}{6} \times 1 + \frac{1}{12} \times 2 + \frac{1}{12}(2 + E)$$

Solve for E to get the answer $E = 14/11$.

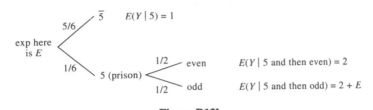

Figure P.13b

14. Let X be the number of pairs left. For $i = 1, \ldots, n$, let

$X_i = 1$ if the ith pair is left

Then $X = X_1 + \cdots + X_n$ and $EX = EX_1 + \cdots + EX_n$.

Now we need $EX_i = P(i\text{th pair is left})$.

We are drawing m balls from $2n$. The total number of ways is $\binom{2n}{m}$. For the favorable, we want to leave the two A_i's undrawn, so draw the m from the remaining $2n - 2$. There are $\binom{2n-2}{m}$ favs. So

$$EX_i = \frac{\binom{2n-2}{m}}{\binom{2n}{m}} = \frac{(2n - m)(2n - m - 1)}{2n(2n - 1)}$$

$$EX = \text{ sum of } n \text{ indicator expectations } = n\,\frac{(2n - m)(2n - m - 1)}{2n(2n - 1)}$$

15. By the theorem of total expectation

$$E(\text{heads}) = \frac{3}{10} \cdot 2 + \frac{7}{10}\frac{4}{3}$$

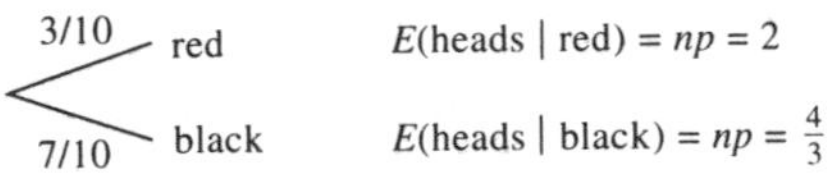

3/10 — red $E(\text{heads} \mid \text{red}) = np = 2$

7/10 — black $E(\text{heads} \mid \text{black}) = np = \frac{4}{3}$

Figure P.15

16. **(a)** The births are Bernoulli trials where $P(\text{success}) = P(\text{girl}) = q$. The expected number of trials to get the first success has a geometric distribution with parameter q. So the expected number of children is $1/q$.

(b) Let X be the number of children.

Method 1.   Let $X_i = 1$ if the first i children are boys. Then $X = 6 + X_6 + X_7 + \cdots$, $EX = 6 + EX_6 + EX_7 + \cdots$.

$$EX_i = P(\text{first } i \text{ children are boys}) = p^i$$

$$EX = 6 + \underbrace{p^6 + p^7 + p^8 + \cdots}_{\text{geometric series}} = 6 + \frac{p^6}{1-p} = 6 + \frac{p^6}{q}$$

Method 2.   Condition on whether the first 6 are all boys or not. If the first 6 are all boys, then the game now consists of trying to get a girl. The number of trials to get a girl has a geometric distribution with parameter $P(\text{girl}) = q$, so in the *rest of the game* the expected number of trials to get a girl is $1/q$. We've already had 6 trials so

$$E(X|6\text{B}) = 6 + \frac{1}{q}$$

If at least 1 of the first 6 is a girl, then we stop after 6, so

$$E(X|\text{at least 1 girl in the first 6}) = 6$$

By the theorem of total expectation,

$$EX = p^6\left(6 + \frac{1}{q}\right) + (1 - p^6)6 = 6 + \frac{p^6}{q}$$

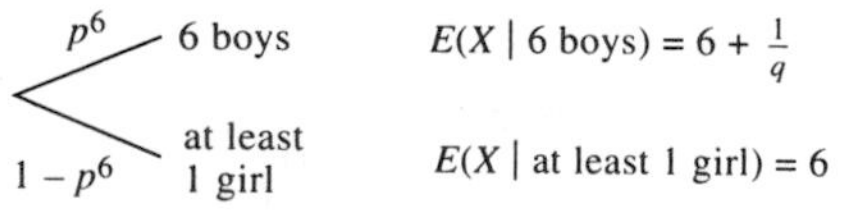

Figure P.16b   Method 2

17. $E(\text{winnings}) = \$1,000,000 P(\text{win}) + (-.25)P(\text{lose})$

$$= 1,000,000\frac{1}{10,000,001} - \frac{1}{4}\frac{10,000,000}{10,000,001} = -15 \text{ cents}$$

I wouldn't bother entering a contest that offered an average *loss* of 15 cents.

Solutions Section 4-1

1. Need $\displaystyle\int_{-\infty}^{\infty} f(x)\,dx = 1, \int_0^1 cx\,dx = 1, c = 1/\int_0^1 x\,dx) = 2.$

Then $\displaystyle P(X < .3) = \int_{-\infty}^{.3} f(x)\,dx = \int_0^{.3} 2x\,dx = .09$

2. **(a)** $1 - |x| = \begin{cases} 1 - x & \text{if } x \geq 0 \\ 1 + x & \text{if } x \leq 0 \end{cases}$

(b) By inspection, $f(x) \geq 0$. You can see from the diagram that the area under the graph is 1. Alternatively,

$$\int_{-1}^{1} (1 - |x|) \, dx = \int_{-1}^{0} (1 + x) \, dx + \int_{0}^{1} (1 - x) \, dx = \frac{1}{2} + \frac{1}{2} = 1$$

(c) $P\left(-\frac{1}{2} \leq X \leq \frac{1}{3}\right) = \int_{-1/2}^{1/3} (1 - |x|) \, dx$

$$= \int_{-1/2}^{0} (1 + x) \, dx + \int_{0}^{1/3} (1 - x) \, dx = \frac{47}{72}$$

Or just look at the areas in the diagram to get

$$P\left(-\frac{1}{2} \leq X \leq \frac{1}{3}\right) = \text{I} + \text{II} = 1 - (\text{III} + \text{IV}) = \frac{47}{72}$$

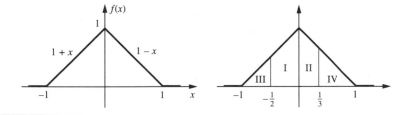

Figure P.2

3. **(a)** $P(X \geq 6) = \int_{6}^{\infty} e^{-x} \, dx = e^{-6}$

(b) Let X be the number of hours the first component lasts.
Let Y be the number of hours the second component lasts.
Let Z be the number of hours the third component lasts.

$$P(X \geq 6 \text{ and } Y \geq 6 \text{ and } Z \geq 6) = P(X \geq 6)P(Y \geq 6)P(Z \geq 6)$$
$$\text{(independence)}$$

$$= (e^{-6})^3 = e^{-18}$$

4. (a)

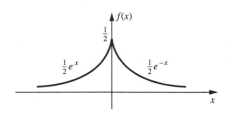

Figure P.4a

(b) $P(|X| < 4) = \int_{-4}^{4} \frac{1}{2} e^{-|x|} \, dx = \int_{-4}^{0} \frac{1}{2} e^{x} \, dx + \int_{0}^{4} \frac{1}{2} e^{-x} \, dx = 1 - e^{-4}$

or better still, by symmetry, $P(|X| < 4) = 2 \int_{0}^{4} \frac{1}{2} e^{-x} \, dx$

(c) Solution to $X^2 + X = 0$ is $X = -1, 0$. Solution to $X^2 + X > 0$ is $X < -1$ or $X > 0$.

$$P(X^2 + X \geq 0) = \int_{-\infty}^{-1} \frac{1}{2} e^{x} \, dx + \int_{0}^{\infty} \frac{1}{2} e^{-x} \, dx = \frac{1}{2e} + \frac{1}{2}$$

5. Both false. Suppose b is one of the values that X can assume. Then $P(X = b) = 0$ and $P(X \neq b) = 1 - P(X = b) = 1$, but $X = b$ is possible and $X \neq b$ is not sure.

6. Let X be uniformly distributed on [2,7]. Then X has density $f(x) = 1/5$ for $2 \leq x \leq 7$.
(a) $P(X = 4) = 0$
(b) $P(X \approx 4) = f(x) \, dx = \frac{1}{5} \, dx$

Solutions Section 4-2

1. (a) $P(X = 0) = \frac{1}{4}, P(X = 1) = \binom{2}{1} \left(\frac{1}{2}\right)^2 = \frac{1}{2}, P(X = 2) = \frac{1}{4}$, so

$$p(x) = \begin{cases} \dfrac{1}{4} & \text{if } x = 0 \\[2mm] \dfrac{1}{2} & \text{if } x = 1 \\[2mm] \dfrac{1}{4} & \text{if } x = 2 \end{cases}$$

(b) $F(x) = \begin{cases} 0 & \text{if } x \leq 0 \\ 1/4 & \text{if } 0 \leq x \leq 1 \\ 3/4 & \text{if } 1 \leq x \leq 2 \\ 1 & \text{if } x \geq 2 \end{cases}$

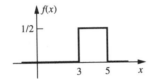

Figure P.1b

2. $f(x) = 1/2$, for $3 \leq x \leq 5$.

$$F(x) = \begin{cases} 0 & \text{if } x \leq 3 \\ \dfrac{1}{2}(x - 3) & \text{if } 3 \leq x \leq 5 \\ 1 & \text{if } x \geq 5 \end{cases}$$

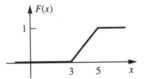

Figure P.2

3. $F(x) = \begin{cases} \dfrac{1}{3}x^2 & \text{if } 0 \leq x \leq 1 \\ \dfrac{1}{3} & \text{if } 1 \leq x \leq 3 \\ \dfrac{1}{4}(x - 1) & \text{if } 3 \leq x \leq 4 \end{cases}$

(a) 0

(b) $P(X = 3) = $ jump in F at $3 = \dfrac{1}{2} - \dfrac{1}{3} = \dfrac{1}{6}$

(c) $F\left(\dfrac{1}{2}\right) = \dfrac{1}{12}$

(d) lower $F(3) = \dfrac{1}{3}$

(e) upper $F(4) = 1$

(f) upper $F(4) - $ lower $F(3) = 1 - \dfrac{1}{3} = \dfrac{2}{3}$

(g) lower $F(4) - F(2) = \dfrac{3}{4} - \dfrac{1}{3} = \dfrac{5}{12}$

(h) $P(X = 3) + P(3\frac{1}{2} < X < 4)$

$$= \text{jump at } 3 + \text{lower } F(4) - F\left(3\frac{1}{2}\right) = \frac{1}{6} + \frac{3}{4} - \frac{5}{8}$$

(i) $\dfrac{PX \ge 3 \text{ and } X > 1/2)}{P(X > 1/2)} = \dfrac{P(X \ge 3)}{P(X > 1/2)} = \dfrac{1 - \text{lower } F(3)}{1 - F(1/2)}$

$$= \frac{1 - 1/3}{1 - 1/12} = \frac{8}{11}$$

4. (a) F   **(b)** T   **(c)** F   **(d)** T   **(e)** T

  (f) T   **(g)** T   **(h)** F   **(i)** F

  (j) T (Yes, some distribution functions can jump but *not* the distribution function of a *continuous* random variable, a random variable with a density.)

5. $F(-\infty) = 0, F(\infty) = 1$, can't predict $F(0)$.

  $f(-\infty) = 0, f(\infty) = 0$, can't predict $f(0)$.

  $P(X = 7) = 0$ because X is a continuous random variable.

6. (a) For $2 \le x \le 6$, $f(x) = 0$ and $F(x)$ stays constant; that is, the graph of $F(x)$ is horizontal (but $F(x)$ is not necessarily 0 since prob may have accumulated before $x = 2$).

  (b) $F(x) = 0$ for $x \le 2$ (no prob has accumulated yet).
  $F(x) = 1$ for $x \ge 6$ (all the prob has accumulated by $x = 6$).
  $f(x) = 0$ for $x \le 2$ and for $x \ge 6$.

  (c) For $x \le 3$, $F(x) = 0$ and $f(x) = 0$.

  (d) For $x \ge 4$, $F(x) = 1$ and $f(x) = 0$.

7. To get $f(x)$, put 1/3 unit of prob *at* $x = 5$. Spread out the other 2/3 unit of prob evenly over $[3, 6]$, which makes the density 2/9 over that interval $[3, 6]$ (see the diagram). All in all,

$$f(x) = \tfrac{2}{9} \text{ for } 3 \le x \le 6 \text{ (partial density)}$$

$$P(X = 5) = \tfrac{1}{3} \text{ (partial probability function)}$$

If you're familiar with delta functions, you can write the pseudo-density as

$$\frac{2}{9} + \frac{1}{3}\delta(x - 5) \text{ for } 3 \le x \le 6$$

To get $F(x)$, look at the cumulative area under $F(x)$. The area is 0 until $x = 3$. The area is $4/9$ when x reaches 5. At $x = 5$, suddenly area $1/3$ feeds in, so $F(x)$ jumps by $1/3$ at $x = 5$.

Then the area increases steadily until it reaches 1 at $x = 6$ and stays there for $x \geq 6$. Find the equations of the lines in the diagram to get

$$F(x) = \begin{cases} 0 & \text{if } x \leq 3 \\[2mm] \dfrac{2}{9}x - \dfrac{2}{3} & \text{if } x \leq 3 \leq 5 \\[2mm] \dfrac{2}{9}x - \dfrac{1}{3} & \text{if } 5 \leq x \leq 6 \\[2mm] 1 & \text{if } x \geq 6 \end{cases}$$

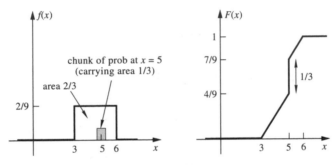

Figure P.7

8. **(a)** $P(\max \leq 7)$
    $$= P(X \leq 7 \text{ and } Y \leq 7 \text{ and } Z \leq 7)$$
    $$= P(X \leq 7)\, P(Y \leq 7)\, P(Z \leq 7) \quad \text{(by independence)}$$
    $$= [F(7)]^3$$

   (b) $P(\min \geq 5 \text{ and } \max \leq 7)$
    $$= P(5 \leq X, Y, Z \leq 7)$$
    $$= P(5 \leq X \leq 7)\, P(5 \leq Y \leq 7)\, P(5 \leq Z \leq 7)$$
    $$\text{(by independence)}$$
    $$= [F(7) - F(5)]^3$$

   (c) $P(\min \leq 5 \text{ and } \max \leq 7)$
    $$= P(\max \leq 7) - P(\min > 5 \text{ and } \max \leq 7)$$
    $$= [F(7)]^3 - [F(7) - F(5)]^3$$

9. (a) F has no jumps, so there is a density.

$$f(x) = F'(x) = \begin{cases} \dfrac{1}{2}e^x & \text{if } x \le 0 \\ 0 & \text{if } 0 \le x \le 2 \\ \dfrac{1}{6} & \text{if } 2 \le x \le 5 \\ 0 & \text{if } x \ge 5 \end{cases}$$

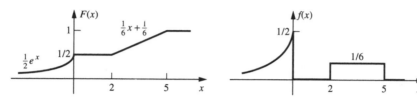

Figure P.9a

(b) F doesn't jump so there is a density

$$f(x) = F'(x) = \begin{cases} \dfrac{1}{2}e^x & \text{if } x \le 0 \\ \dfrac{1}{2}e^{-x} & \text{if } x \ge 0 \end{cases} = \dfrac{1}{2}e^{-|x|}$$

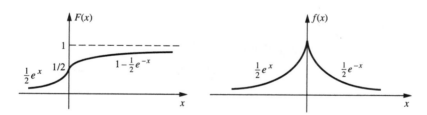

Figure P.9b

(c)   lower $F(0) = \dfrac{1}{4}e^{2x}\Big|_{x=0} = \dfrac{1}{4}$,   upper $F(0) = 1$.

So F jumps by $3/4$ at $x = 0$. No legal density. Here's the pseudo-density.

$$f(x) = F'(x) = \dfrac{1}{2}e^{2x} \text{ for } x \le 0$$

$$P(X = 0) = \dfrac{3}{4}(\text{ there's a chunk of prob of size } \dfrac{3}{4} \text{ at } x = 0)$$

The compact expression for the density is $\frac{1}{2}e^{2x} + \frac{3}{4}\delta(x)$ for $x \leq 0$.

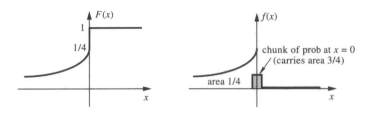

Figure P.9c

10. **(a)** If $x \leq 100$, then $F(x) = 0$. If $x \geq 100$, then

$$F(x) = \int_{-\infty}^{x} f(x)\, dx = \int_{100}^{x} \frac{100}{x^2}\, dx = 1 - \frac{100}{x}$$

(b) $f(x) = \begin{cases} 1+x & \text{if } -1 \leq x \leq 0 \\ 1-x & \text{if } 0 \leq x \leq 1 \end{cases}$

If $x \leq -1$, then $F(x) = 0$.

If $-1 \leq x \leq 0$, then $F(x) = \int_{-1}^{x} (1+x)dx = \frac{1}{2}x^2 + x + \frac{1}{2}$.

If $0 \leq x \leq 1$, then
$$F(x) = \int_{-1}^{0} (1+x)\, dx + \int_{0}^{x} (1-x)\, dx = -\frac{1}{2}x^2 + x + \frac{1}{2}.$$

If $x \geq 1$, then $F(x) = 1$.

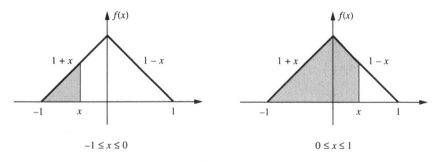

Figure P.10b

(c) $F(x) = \int_{-\infty}^{x} f(x)\,dx.$

If $x \le -.1$, then $F(x) = 0$.

If $-.1 \le x \le .1$, then $F(x) = $ area $A = 4(x + .1)$.

If $.1 \le x \le .5$, then $F(x) = $ area $B + $ area $C = \frac{4}{5} + \frac{1}{2}(x - .1)$.

If $x \ge .5$, then $F(x) = 1$.

All in all,

$$F(x) = \begin{cases} 0 & \text{if } x \le -.1 \\ 4(x + .1) & \text{if } -.1 \le x \le .1 \\ \dfrac{1}{2}x + \dfrac{3}{4} & \text{if } .1 \le x \le .5 \\ 1 & \text{if } x \ge .5 \end{cases}$$

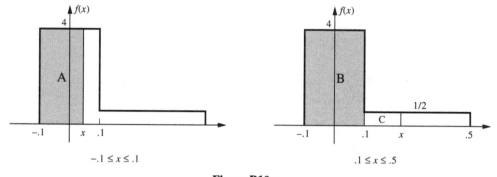

Figure P.10c

11. Look at the cumulative area under $f(x)$ to get this picture of $F(x)$:

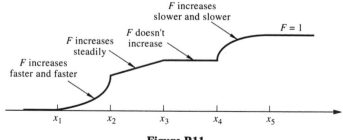

Figure P.11

Solutions Section 4-3

1. We need $f(x) \geq 0$ for all x and $\int_{-\infty}^{\infty} f(x)\, dx = 1$. The first part is clear and for the second part,

$$\int_{-\infty}^{\infty} f(x)\, dx = \int_0^{\infty} \lambda e^{-\lambda x}\, dx = -e^{-\lambda x}\, |_0^{\infty} = 1$$

2. Let X be the waiting time between particles. Then X is exponentially distributed with $\lambda = 3$.

(a) $P(X \leq 2) = \int_0^2 3e^{-3x}\, dx = 1 - e^{-6}$.

(b) $P(2 \leq X \leq 5) = \int_2^5 3e^{-3x}\, dx = e^{-6} - e^{-15}$.

(c) *Method 1.*   Use our exponentially distributed X. The particles arrive independently, so it is irrelevant that it has been at least 6 seconds since the last particle.

$$P(\text{nothing arrives in next 6 seconds})$$
$$= P(\text{waiting time for an arrival is } \geq 6 \text{ seconds})$$
$$= P(X \geq 6) = \int_6^{\infty} 3e^{-3x}\, dx = e^{-18}$$

Method 2.

$P(\text{nothing arrives in next 6 seconds}) = P(0 \text{ arrivals in 6 seconds})$. The number of arrival in 6 seconds has a Poisson distribution with $\lambda = 18$, so

$$P(0 \text{ arrivals in 6 seconds}) = \frac{18^0\, e^{-18}}{0!} = e^{-18}$$

(d) We want the prob of having to wait less than another 13 seconds for an arrival after we've already waited 7 seconds. The particles arrive independently, so this is the same as having to wait less than 13 seconds from scratch; that is,

$$P(X \leq 20 | X \geq 7) = P(X \leq 13) = \int_0^{13} 3e^{-3x}\, dx = 1 - e^{-39}$$

(e) The *number* of particles arriving in a second is a Poisson random variable with $\lambda = 3$. Call it Y. Then

$$P(1 \leq Y \leq 5) = P(Y = 1) + P(Y = 2) + P(Y = 3)$$
$$+ P(Y = 4) + P(Y = 5)$$
$$= e^{-3}\left(3 + \frac{3^2}{2!} + \frac{3^3}{3!} + \frac{3^4}{4!} + \frac{3^5}{5!}\right)$$

3. Let X be the duration (lifetime) of a call. Assume that X is exponential with

$$\lambda = \frac{1}{\text{average lifetime)}} = \frac{1}{3}$$

(a) $P(X < 2) = \int_0^2 \frac{1}{3} e^{-x/3} \, dx = 1 - e^{-2/3}$

(b) $P(X > 3) = \int_3^\infty \frac{1}{3} e^{-x/3} \, dx = 1/e$

(c) By the memoryless feature, the answer is the same as in part (a). The process does not care that the call has already lasted 4 minutes.

(d) We want the probability that the call lasts at least 1 *more* minute. By the memoryless feature, this is the same as the prob that a new call lasts at least 1 minute, which is

$$P(X \geq 1) = \int_1^\infty \frac{1}{3} e^{-x/3} \, dx = e^{-1/3}$$

(e) Must assume that the odds on a call's ending are the same no matter how long it has lasted already. In other words, a hangup is just as likely to occur in a new call as a call that has been going on for a while.

4. Let X be the lifetime of a gidget. Then X is exponential with $\lambda = 1/50,000$.

$P(\text{gidget dies on vacation}) = P(X \leq 5000) = \lambda e^{\lambda x} \, dx = 1 - e^{-1/10}$.

5. (a) A volley lasts $\frac{1}{2}$ minute on the average so $\lambda = 1/\text{av lifetime} = 2$.
   (b) A volley lasts 3 minutes on the average so $\lambda = 1/\text{av lifetime} = \frac{1}{3}$.
   (c) $\lambda = 1/2$.
   (d) Must assume that a long-lasting volley is just as likely to end as a volley that has just started.

6. Let X be the processing time (lifetime) of a customer where the average lifetime is 5 minutes. Or you can think of the customers arriving at the teller's window (like particles) at the average rate of 1 per 5 minutes, 1/5 per minute, and let X be the waiting time between customers.

In any case, X is exponential with $\lambda = 1/5$.

(a) $P(\text{the customer ahead of you takes between 5 and 10 minutes})$

$$= P(5 \leq X \leq 10) = \int_5^{10} \frac{1}{5} e^{-x/5} \, dx = e^{-1} - e^{-2}$$

(b) P(the customer who has been there for 7 minutes
 already takes another 20 minutes)

$$= P(X \leq 20) = \int_0^{20} \frac{1}{5} e^{-x/5} \, dx = 1 - e^{-4}$$

(c) Let Y be the waiting time for the 3rd customer (that's you). Then Y has a gamma distribution with $n = 3, \lambda = 1/5$, and

$$P(Y \geq 10) = \frac{\left(\frac{1}{5}\right)^3}{2!} \int_{10}^{\infty} y^2 \, e^{-y/5} \, dy$$

Either use integral tables to finish up, or use the gamma distribution function in (5) to get

$$P(Y \geq 10) = 1 - F_Y(10) = e^{-2} \left(1 + 2 + \frac{2^2}{2!} \right)$$

(d) Let Y be the number of customers reaching the teller in 3 minutes. The customers arrive at the rate of $1/5$ per minute, $3/5$ per 3 minutes. So Y has a Poisson distribution with $\lambda = 3/5$ and

$$P(Y \leq 2) = P(Y = 0) + P(Y = 1) + P(Y = 2)$$

$$= e^{-3/5} \left(1 + \frac{3}{5} + \frac{(3/5)^2}{2!} \right)$$

7. (a) and **(b)** Let X be the number of arrivals in a 10-minute interval. Then X is Poisson with $\lambda = 20$ and

P(4 arrive between 4:30 and 4:40)

$\quad = P$(4 arrive between 4:40 and 4:50)

$$= P(X = 4) = \frac{e^{-20} \, 20^4}{4!}$$

(c) Let X be the waiting time for the next arrival. Then X is exponential with $\lambda = 2$, and

$\quad P$(the next arrival is between 4:30 and 4:40)

$\qquad = P$(waiting time is between 10 and 20 minutes)

$$= P(10 \leq X \leq 20) = \int_{10}^{20} 2e^{-2x} \, dx = e^{-20} - e^{-40}$$

(d) Use same X as in part (c), but now you have to wait between 20 and 30 minutes, so

P(the next arrival is between 4:40 and 4:50)

$$= P(20 \leq X \leq 30) = \int_{20}^{30} 2e^{-2x} \, dx = e^{-40} - e^{-60}$$

(e) Let X be the waiting time for the next arrival. Then X is exp with $\lambda = 2$, and

$$P(X \geq 5) = \int_{5}^{\infty} 2e^{-2x} \, dx = -e^{-2x} \Big|_{5}^{\infty} = e^{-10}$$

(f) Same answer as (a) and (b).

(g) The particles arrive independently, so the number arriving between 6:30 and 6:45 is independent of the number arriving between 6:45 and 6:50.

We want

P(none in a 15-minute interval)P(3 in a 5-minute interval).

For the first prob, use the Poisson with $\lambda = 30$. For the second prob, use the Poisson with $\lambda = 10$. Answer is

$$e^{-30} \frac{e^{-10} \, 10^3}{3!}$$

(h) It's irrelevant that we've waited 3 days since particles arrive independently.

$$P(\text{waiting time } \leq 3) = \int_{0}^{3} 2e^{-2x} \, dx = -e^{-2x} \Big|_{0}^{3} = 1 - e^{-6}$$

8. Breakages arrive on the average once every 3 years, at the rate of $1/3$ per year. They arrive independently (when a glass breaks doesn't depend on when the last breakage occurred, that is, doesn't depend on when you started using the glass). So X is the waiting time for the 6th breakage to arrive and has a gamma distribution with $\lambda = 1/3$ and $n = 6$.

Solutions Section 4-4

1. (a) $F^*(.5) = .691$
   (b) 0
   (c) same as (a)

(d) $1 - P(X \leq .5) = 1 - F^*(.5) = .309$

(e) $P(-.5 \leq X^* \leq .5) = F^*(.5) - F^*(-.5) = .382$

(f) $F^*(3) - F^*(-2) = .976$

(g) $P(X^* > 1) + P(X^* < -1) = 2 F^*(-1)$ (by symmetry) $= .318$

2. **(a)** $F^*(b) = .75, b = .7$

   (b) $P(X^* \leq b) = .7, F^*(b) = .7, b = .5$

   (c) $F^*(b) = .9, b = 1.28$ (look at the diagram)

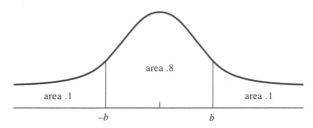

area .8

area .1

area .1

$-b$

b

Figure P.2c

3. $F^*(-.3) = P(X^* \leq -.3) = P(X^* \geq .3)$ (symm) $= 1 - P(X^* \leq .3)$,

   so $F^*(-3) = 1 - F^*(.3) = 1 - .618 = .382$.

4. **(a)** .5, since half the area is to the right of $x = \mu$ and half to the left.

   (b) $P(X \geq 11) + P(X \leq 9) = 2P(X \leq 9)$ by symmetry

   $$= 2P \left(\frac{x - 10}{3} \leq \frac{9 - 10}{3} \right) = 2P \left(X^* \leq -\frac{1}{3} \right) = 2F^* \left(-\frac{1}{3} \right)$$

   $$= 2(.382) = .764$$

5. **(a)** $P(\text{reject}) = P(X \geq 2.05) + P(X \leq 1.95) = 2P(X \leq 1.95)$

   $$= 2P \left(\frac{X - 2}{.08} \leq \frac{1.95 - 2}{.08} \right) = 2P(X^* \leq -.625)$$

   $$= 2F^*(-.625) = .548$$

   (b) $P(\text{reject}) = .2, \quad 2P(X \leq 1.95) = .2$ as in part (a),

   $$2P \left(\frac{X - 2}{\sigma} \leq \frac{1.95 - 2}{\sigma} \right) = .2, \ 2F^* \left(-\frac{.05}{\sigma} \right) = .2,$$

   $$F^* \left(-\frac{.05}{\sigma} \right) = .1, -\frac{.05}{\sigma} = -.1.28, \quad \sigma = .039$$

6. **(a)** Since half the area under the normal density is to the right of the mean and half is to the left of the mean, we must have $\mu = 110$. Then

$$P(X \geq 150) = .15, \quad P\left(\frac{X-110}{\sigma} \geq \frac{150-110}{\sigma}\right) = .15,$$

$$P\left(X^* \geq \frac{40}{\sigma}\right) = .15, \quad F^*\left(\frac{40}{\sigma}\right) = .85, \quad \frac{40}{\sigma} = 1, \quad \sigma = 40$$

(b) $P(X \geq 170 | X \geq 150) = \dfrac{P(X \geq 170 \text{ and } X \geq 150)}{P(X \geq 150)}$

$$= \frac{P(X \geq 170)}{P(X \geq 150)} = \frac{P(X^* \geq 1.5)}{.15}$$

$$= \frac{1 - F^*(1.5)}{.15} = \frac{.067}{.15} \approx .45$$

7. (a) We want b so that

$$P(X \geq b) = .25, \quad P\left(\frac{X-\mu}{\sigma} \geq \frac{b-\mu}{\sigma}\right) = .25,$$

$$P\left(X^* \geq \frac{b-\mu}{\sigma}\right) = .25, \quad F^*\left(\frac{b-\mu}{\sigma}\right) = .75, \quad \frac{b-\mu}{\sigma} = .7,$$

$$b = .7\sigma + \mu$$

The cutoff is .7 standard deviations above the mean.

(b) We know that the 50th percentile cutoff is $x = \mu$ since half the area under a normal density lies to the right of the mean and half lies to the left. And by symmetry from part (a), the cutoff of the lowest quartile is $\mu - .7\sigma$.

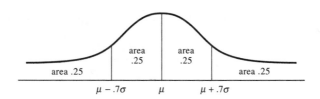

$$\mu - .7\sigma \qquad \mu \qquad \mu + .7\sigma$$

Figure P.7

8. $P(|X - \mu| \leq .5\sigma) = P(\mu - .5\sigma \leq X \leq \mu + .5\sigma)$

$$= P(-.5 \leq X^* \leq .5) \quad \text{(by (3))}$$

$$= F^*(.5) - F^*(-.5) = .382$$

9. Let X be the test score of a student. By (3),

$$P(X \geq \mu + \sigma) = P(X^* \geq 1) = 1 - F^*(1) = .159$$

$$P(\mu \leq X \leq \mu + \sigma) = P(0 \leq X^* \leq 1) = F^*(1) - F^*(0) = .341$$

$$P(\mu - \sigma \le X \le \mu) = P(-1 \le X^* \le 0) = .341 \text{ (by symmetry)}$$
$$P(\mu - 2\sigma \le X \le \mu - \sigma) = P(-2 \le X^* \le -1)$$
$$= F^*(-1) - F^*(-2) = .136$$
$$P(X \le \mu - 2\sigma) = P(X^* \le -2) = F^*(-2) = .023$$

So expect about 16% A's, 34% B's, 34% C's, 14% D's, 2% E's.

10. Choose $\mu = 60$. To choose σ, start with $P(X \ge 73) = .3$. Then

$$P\left(\frac{X - 60}{\sigma} \ge \frac{73 - 60}{\sigma}\right) = .3, \quad P\left(X^* \ge \frac{13}{\sigma}\right) = .3,$$

$$F^*\left(\frac{13}{\sigma}\right) = .7, \quad \frac{13}{\sigma} = .5, \quad \sigma = 26$$

11. Let X be the number of heads in 1000 tosses. Then X is binomial with $n = 1000, p = 1/2$. So X is approximately normal with $\mu = np = 500$, $\sigma^2 = npq = 250$.

$$P(X \le 510) = P\left(\frac{X - \mu}{\sigma} \le \frac{510 - \mu}{\sigma}\right) = P\left(X^* \le \frac{10}{\sqrt{250}}\right)$$

$$= P(X^* \le .632) = F^*(.632) = .726$$
$$\text{(all very approximately)}$$

12. First, find the prob of no drought in *one* summer.

$$P(\text{no drought in a summer}) = P(X \ge 4) = P\left(\frac{X - 10}{4} \ge \frac{4 - 10}{4}\right)$$

$$= P(X^* \ge -1.5)$$

$$= 1 - F^*(-1.5) = .933$$

Then, assuming that summers are independent of one another,

$$P(\text{no drought in 10 summers}) = (.933)^{10}$$

13. Let X be the number of Z supporters in the sample; X has a binomial distribution with $n = 400, p = .45$. So X is approximately normal with $\mu = 400 \times .45, \sigma^2 = 400 \times .45 \times .55$, and

$$P(X \ge 200) = P\left(X^* \ge \frac{200 - \mu}{\sigma}\right) = P(X^* \ge 2.01)$$

$$= 1 - F^*(2.01) = .023$$

14. Let X be the number of ticket buyers who show up for the plane. Each ticket buyer is a Bernoulli trial where $P(\text{show}) = .8$. Let n be the number of trials (the number of tickets sold). We want to find n so that $P(X \leq 300) = .95$.

X is approximately normal with $\mu = np = .8n$, $\sigma^2 = npq = .16n$. We want n so that

$$P\left(\frac{X - .8n}{.4\sqrt{n}} \leq \frac{300 - .8n}{.4\sqrt{n}}\right) = .95$$

$$P\left(X^* \leq \frac{300 - .8n}{.4\sqrt{n}}\right) = .95$$

$$\frac{300 - .8n}{.4\sqrt{n}} = 1.64$$

$$n \sim 359$$

15. Let X be the number of heads. Then X is approximately normal with $\mu = 300$, $\sigma^2 = 210$. We want b so that

$$P(X \geq b) = .9$$

$$P\left(\frac{X - \mu}{\sigma} \geq \frac{b - \mu}{\sigma}\right) = .9$$

$$P\left(X^* \geq \frac{b - 300}{\sqrt{210}}\right) = .9$$

$$P\left(X^* \leq \frac{b - 300}{\sqrt{210}}\right) = .1$$

$$\frac{b - 300}{\sqrt{210}} = -1.28$$

$$b = 300 - 1.28\sqrt{210} \sim 281.$$

16. Let X be the number who arrive in the first hour.

For any one person, $P(\text{arrive in the first hour}) = 1/5$. The experiment is like tossing a coin 1000 times where $P(\text{H}) = 1/5$ and X is the number of heads. So X has a binomial distribution with $n = 1000$, $p = 1/5$ and is approximately normal with $\mu = np = 200$, $\sigma^2 = npq = 160$. So

$$P(X \geq 175) = P\left(X^* \geq \frac{175 - 200}{\sqrt{160}}\right) = P(X^* \geq -1.98)$$

$$= P(X^* \leq 1.98) \quad \text{(by symmetry)} = .975$$

17. If X is normal, then $aX + b$ is normal with parameters $\mu = a\mu_X + b$, $\sigma^2 = a^2\sigma_X^2$. So

   (a) $5X - 2$ is normal with $\mu = 5\mu_X - 2$, $\sigma^2 = 25\sigma_X^2$.

   (b) $X - \mu_X$ is normal with $\mu = \mu_X - \mu_X = 0$, $\sigma^2 = \sigma_X^2$.

Solutions Section 4-6

1. $f(x) = 1/5$ for $0 \le x \le 5$.

Dist Function Method.

By inspection, $0 \le X \le 5$, so $0 \le Y \le 125$, so

$$F(y) = \begin{cases} 0 & \text{if } y \le 0 \\ 1 & \text{if } y \ge 125 \end{cases}$$

Let $0 \le y \le 125$ so that $0 \le \sqrt[3]{y} \le 5$. In the diagram, you can see that Y is below the y mark iff X is to the left of the $\sqrt[3]{y}$ mark, so

$$F(y) = P(Y \le y) = P(X \le \sqrt[3]{y})$$

$$= \text{ area under } f(x) \text{ up to } \sqrt[3]{y} = \frac{1}{5}\sqrt[3]{y}$$

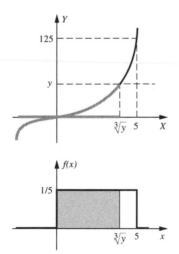

Figure P.1   Dist function method

All in all,

$$F(y) = \begin{cases} 0 & \text{if } y \leq 0 \\ \dfrac{1}{5}\sqrt[3]{y} & \text{if } 0 \leq y \leq 125 \\ 1 & \text{if } y \geq 125 \end{cases}$$

$$f(y) = F'(y) = \frac{1}{15y^{2/3}} \quad \text{if } 0 \leq y \leq 125$$

Density Function Method.   As before, Y is always between 0 and 125, so $f(y) = 0$ if $y \leq 0$ or $y \geq 125$.

Let $0 \leq y \leq 125$. If $y = x^3$, then

$$x = \sqrt[3]{y}, \quad \frac{dx}{dy} = \frac{1}{3y^{2/3}}$$

Since $0 \leq x \leq 5$, we have $f(x) = 1/5$.

$$f(y) = f(x)\,|\frac{dx}{dy}| = \frac{1}{5}\,\frac{1}{3y^{2/3}}$$

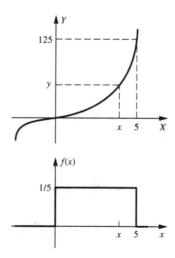

Figure P.1 (continued)   Density method

2. Dist Function Method.

By inspection, $0 \leq X \leq R$, so $0 \leq Y \leq 1$ and

$$F(y) = \begin{cases} 0 & \text{if } y \leq 0 \\ 1 & \text{if } y \geq 1 \end{cases}$$

Let $0 \le y \le 1$. In the diagram, Y is below the y mark iff X is left of the Ry mark, so

$$F(y) = P(Y \le y) = P\left(\frac{X}{R} \le y\right) = P(X \le Ry)$$

$$= \text{indicated area} = \int_{x=0}^{Ry} \frac{2x}{R^2}\, dx = y^2$$

So

$$F(y) = \begin{cases} 0 & \text{if } y \le 0 \\ y^2 & \text{if } 0 \le y \le 1 \\ 1 & \text{if } y \ge 1 \end{cases}$$

$$f(y) = F'(y) = 2y \text{ for } 0 \le y \le 1$$

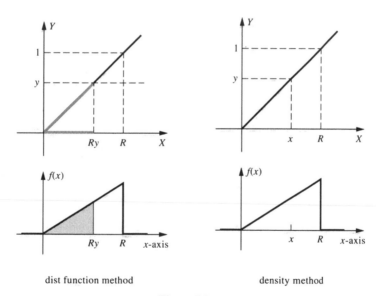

dist function method   density method

Figure P.2

Density Method.

By inspection, $0 \le Y \le 1$, so $f(y) = 0$ for $y \le 0$ or $y \le 1$.
Let $0 \le y \le 1$. Then

$$f(y) = f(x)\left|\frac{dx}{dy}\right| \text{ where } x = Ry,\ \frac{dx}{dy} = R,\ f(x) = \frac{2x}{R^2}$$

$$f(y) = \frac{2x}{R^2} \cdot R = \frac{2Ry}{R^2} \cdot R = 2y$$

So $f(y) = 2y$ for $0 \le y \le 1$.

3. **Dist Function Method.**

    By inspection, $X \geq 1$, so $Y \geq 2$ and $F(y) = 0$ if $y \leq 2$.

    Case 1. $2 \leq y \leq 4$

    In the diagram, Y is below the y mark iff X is to the left of the $y/2$ mark.

    $$F(y) = P(Y \leq y) = P(2X \leq y) = P\left(X \leq \frac{y}{2}\right)$$

    $$= \int_{x=1}^{y/2} \frac{1}{x^2}\, dx = 1 - \frac{2}{y}$$

    Case 2. $y \geq 4$

    In the diagram, Y is below the y mark iff X is to the left of the $\sqrt{y}$ mark.

    $$F(y) = P(Y \leq y) = P(X^2 \leq y)$$

    $$= P(X \leq \sqrt{y}) = \int_{1}^{\sqrt{y}} \frac{1}{x^2}\, dx = 1 - \frac{1}{\sqrt{y}}$$

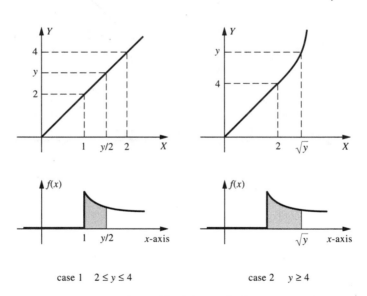

    case 1 $2 \leq y \leq 4$ case 2 $y \geq 4$

Figure P.3   Dist method

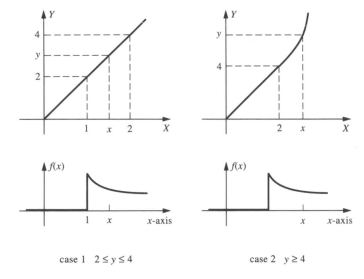

case 1 $2 \leq y \leq 4$ case 2 $y \geq 4$

Figure P.3 (continued) Density Method

All in all,

$$F(y) = \begin{cases} 0 & \text{if } y \leq 2 \\ 1 - \dfrac{2}{y} & \text{if } 2 \leq y \leq 4 \\ 1 - \dfrac{1}{\sqrt{y}} & \text{if } y \geq 4 \end{cases}$$

$$f(y) = F'(y) = \begin{cases} \dfrac{2}{y^2} & \text{if } 2 \leq y \leq 4 \\ \dfrac{1}{2y^{3/2}} & \text{if } y \geq 4 \end{cases}$$

Density method.

By inspection $X \geq 1$, Y is never ≤ 2, so $f(y) = 0$ if $y \leq 2$.

Case 1. $2 \leq y \leq 4$

Then $1 \leq x \leq 2$, $y = 2x$, $x = \frac{1}{2}y$, $dx/dy = \frac{1}{2}$.

$$f(y) = f(x) \left| \frac{dx}{dy} \right| = \frac{1}{x^2} \cdot \frac{1}{2} = \frac{1}{\frac{1}{4}y^2} \cdot \frac{1}{2} = \frac{2}{y^2}$$

Case 2. $y \geq 4$

Then $x \geq 2$, so $y = x^2$, $x = \sqrt{y}$, $dx/dy = 1/(2\sqrt{y})$.

$$f(y) = f(x) \left| \frac{dx}{dy} \right| = \frac{1}{x^2} \frac{1}{2\sqrt{y}} = \frac{1}{(\sqrt{y})^2} \frac{1}{2\sqrt{y}} = \frac{1}{2y^{3/2}}$$

4. Dist Method.

$$F(y) = P(Y \leq y) = P\left(\frac{1}{X} \leq y \right)$$

Solving the inequality $1/X \leq y$ is tricky. Look at the graph of $Y = 1/X$.

If $y \geq 0$, then Y is below y iff X is to the right of $1/y$ or to the left of 0, so

$$P(Y \leq y) = P(X \leq 0) + P\left(X \geq \frac{1}{y} \right)$$

If $y \leq 0$, then Y is below y iff X is between $1/y$ and 0, so

$$P(Y \leq y) = P\left(\frac{1}{y} \leq X \leq 0 \right)$$

So we need at least two cases. Furthermore we need more cases since $P(X \geq 1/y)$ and $P(1/y \leq X \leq 0)$ depend on where $1/y$ is in the $f(x)$ picture.

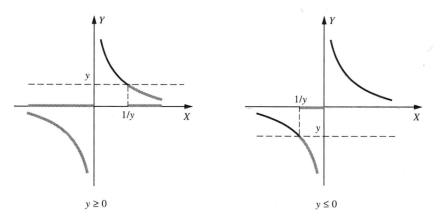

Figure P.4   Solving the inequality $Y \leq y$

Case 1. $y \le -1$

$$F(y) = P(Y \le y) = P\left(\frac{1}{y} \le X \le 0\right) = \frac{1}{2} \cdot -\frac{1}{y} = -\frac{1}{2y}$$

Case 2. $-1 \le y \le 0$

$$F(y) = P(Y \le y) = P\left(\frac{1}{y} \le X \le 0\right) = \frac{1}{2}$$

Case 3. $0 \le y \le 1$

$$F(y) = P(Y \le y) = P(X \le 0) + P\left(X \ge \frac{1}{y}\right) = \frac{1}{2} + 0 = \frac{1}{2}$$

Case 4. $y \ge 1$

$$F(y) = P(Y \le y) = P(X \le 0) + P\left(X \ge \frac{1}{y}\right) = \frac{1}{2} + \frac{1}{2}\left(1 - \frac{1}{y}\right)$$

$$= 1 - \frac{1}{2y}$$

All in all,

$$F(y) = \begin{cases} -\dfrac{1}{2y} & \text{if } y \le -1 \\[2mm] \dfrac{1}{2} & \text{if } -1 \le y \le 1 \\[2mm] 1 - \dfrac{1}{2y} & \text{if } y \ge 1 \end{cases}$$

$$f(y) = F'(y) = \begin{cases} \dfrac{1}{2y^2} & \text{if } y \le -1 \\[2mm] 0 & \text{if } -1 \le y \le 1 \\[2mm] \dfrac{1}{2y^2} & \text{if } y \ge 1 \end{cases}$$

Density Method. $f(y) = f(x)\,|dx/dy|$. We always have $y = 1/x$, $x = 1/y$, $dx/dy = -1/y^2$. But $f(x)$ changes formulas, so we need cases.

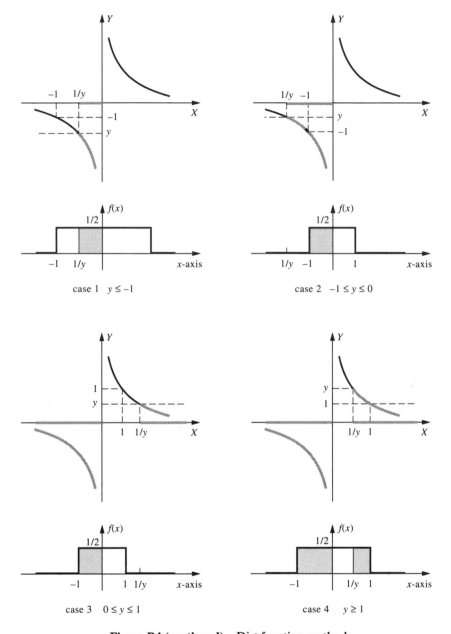

Figure P.4 (continued) Dist function method

Case 1. If $y \leq -1$ then $-1 \leq x \leq 0$, $f(x) = 1/2$, $f(y) = (1/2)(1/y^2)$.

Case 2. If $-1 \leq y \leq 0$ then $x \leq -1$, $f(x) = 0$, $f(y) = 0$.

Case 3.   If $0 \leq y \leq 1$ then $x \geq 1$, so $f(x) = 0$, $f(y) = 0$.

Case 4.   If $y \geq 1$ then $-1 \leq x \leq 0$, $f(x) = 1/2$, $f(y) = (1/2) (1/y^2)$.
All in all, $f(y) = 1/2y^2$ if $y \leq -1$ or $y \geq 1$.

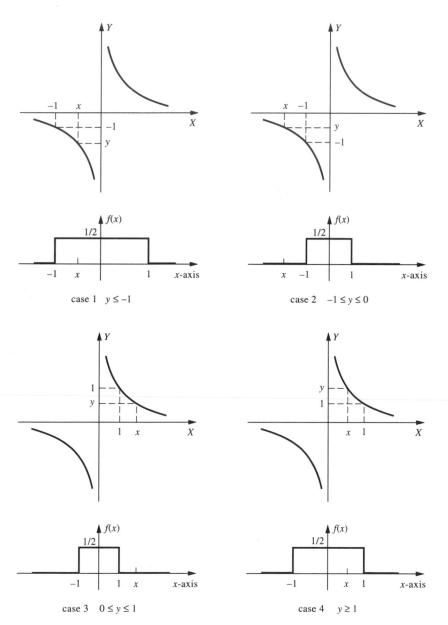

Figure P.4 (continued)   Density method

5. **Dist Method.**   By inspection, $Y \geq 0$, so $F(y) = 0$ if $y \leq 0$.

Case 1. $0 \leq y \leq 1$

Look at the graph of $Y = |X|$ to see that Y is below the y mark iff X is between $-y$ and y. So

$$F(y) = P(Y \leq y) = P(-y \leq X \leq y) = \text{ indicated areas}$$

$$= \frac{1}{2}y + \int_0^y \frac{1}{2}e^{-x}\,dx = \frac{1}{2}y + \frac{1}{2} - \frac{1}{2}e^{-y}$$

Case 2. $y \geq 1$

$$F(y) = P(Y \leq y) = P(-y \leq X \leq y) = \text{ indicated area}$$

$$= \frac{1}{2} + \int_0^y \frac{1}{2}e^{-x}\,dx$$

Better still,

$$F(y) = 1 - \text{ unindicated area } = 1 - \int_y^\infty \frac{1}{2}e^{-x}\,dx = 1 - \frac{1}{2}e^{-y}$$

$$f(y) = F'(y) = \begin{cases} \dfrac{1}{2} + \dfrac{1}{2}e^{-y} & \text{if } 0 \leq y \leq 1 \\[2mm] \dfrac{1}{2}e^{-y} & \text{if } y \geq 1 \end{cases}$$

Density Method.

By inspection, Y is never neg, so $f(y) = 0$ for $y \leq 0$.

Consider $y \geq 0$. If $y = |x|$, then there are two corresponding x's for each y.

$$x_1 = -y, \quad x_2 = y, \quad \frac{dx_1}{dy} = -1, \quad \frac{dx_2}{dy} = 1$$

$$f(y) = f(x_1)\left|\frac{dx_1}{dy}\right| + f(x_2)\left|\frac{dx_2}{dy}\right|$$

We always have $x_2 \geq 0$, $f(x_2) = \frac{1}{2}e^{-x_2}$. But $f(x_1)$ depends on where x_1 is, so we need cases.

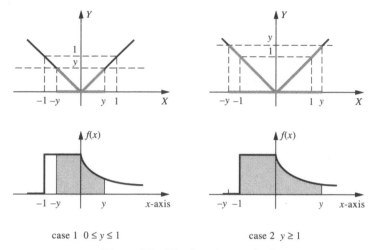

case 1 $0 \le y \le 1$ case 2 $y \ge 1$

Figure P.5   Dist function method

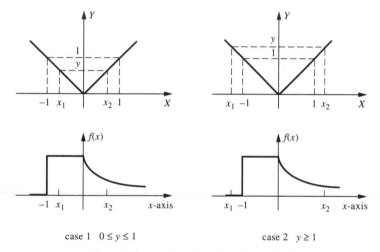

case 1 $0 \le y \le 1$ case 2 $y \ge 1$

Figure P.5 (continued)   Density method

Case 1. $0 \le y \le 1$

Then $-1 \le x_1 \le 0$, $x_2 \ge 0$, $f(x_1) = 1/2$.

$$f(y) = \frac{1}{2} \cdot 1 + \frac{1}{2}e^{-x_2} \cdot 1 = \frac{1}{2} + \frac{1}{2}e^{-y}$$

Case 2. $y \ge 1$

Then $x_1 \le -1$, $f(x_1) = 0$.

$$f(y) = \frac{1}{2}e^{-x_2} \cdot 1 = \frac{1}{2}e^{-y}$$

6. Dist Method.   We know $X \geq 0$, so $0 \leq Y \leq 1$.

Consider $0 \leq y \leq 1$. Look at the picture of Y versus X to see that Y is below the y mark iff X is to the left of y or to the right of $1/y$. So

$$F(y) = P(Y \leq y) = P(X \leq y) + P(X \geq 1/y)$$

$$= \int_0^y e^{-x}\, dx + \int_{1/y}^{\infty} e^{-x}\, dx = 1 - e^{-y} + e^{-1/y}$$

All in all,

$$F(y) = \begin{cases} 0 & \text{if } y \leq 0 \\ 1 - e^{-y} + e^{-1/y} & \text{if } 0 \leq y \leq 1 \\ 1 & \text{if } y > 1 \end{cases}$$

$$f(y) = F'(y) = e^{-y} + \frac{1}{y^2} e^{-1/y} \quad \text{if } 0 \leq y \leq 1$$

Density Method.

We know $X > 0$, so $0 \leq Y \leq 1$, so $f(y) = 0$ if y is outside $[0, 1]$.
Suppose $0 \leq y \leq 1$. Then

$$f(y) = f(x_1)|\frac{dx_1}{dy}| + f(x_2)|\frac{dx_2}{dy}|$$

where

$$x_1 = y, \quad x_2 = \frac{1}{y}, \quad \frac{dx_1}{dy} = 1, \quad \frac{dx_2}{dy} = -\frac{1}{y^2}$$

So

$$f(y) = e^{-x_1} \cdot 1 + e^{-x_2} \cdot \frac{1}{y^2}$$

Replace x's by y's to get the final answer

$$f(y) = e^{-y} + \frac{1}{y^2} e^{-1/y} \quad \text{for } 0 \leq y \leq 1$$

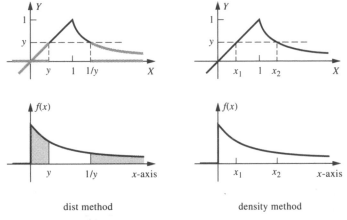

dist method   density method

Figure P.6

7. (a) Y is not discrete because it takes on all the values between 2 and 6. But Y is not continuous because

$$P(Y = 6) = P(X \geq 3) = \int_3^\infty \frac{1}{x^2}\, dx = \frac{1}{3}$$

(b) By inspection, $X \geq 1$, so $2 \leq Y \leq 6$.

We know from part (a) that $F(y)$ jumps by $1/3$ at $y = 6$.

Consider $2 \leq y \leq 6$. The diagram shows that Y is under the y mark iff X is to the left of the $y/2$ mark. So

$$F(y) = P(Y \leq y) = P\left(X \leq \frac{1}{2}y\right) = \int_1^{y/2} \frac{1}{x^2}\, dx = 1 - \frac{2}{y}$$

All in all,

$$F(y) = \begin{cases} 0 & \text{if } y \leq 2 \\ 1 - \dfrac{2}{y} & \text{if } 2 \leq y \leq 6 \\ 1 & \text{if } y \geq 6 \end{cases}$$

(c) $f(y) = F'(y) = \dfrac{2}{y^2} \quad \text{for } 2 \leq y < 6$

$P(Y = 6) = \frac{1}{3}$

Overall,

$$f(y) = \frac{2}{y^2} + \frac{1}{3}\delta(y - 6) \quad \text{for } 2 \leq y \leq 6$$

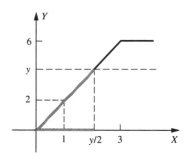

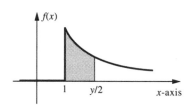

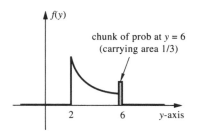

Figure P.7b   Dist function method   **Figure P.7c**   Pseudo-density

8. Dist Function Method.

By inspection, Y is never ≤ 0, so $F(y) = 0$ for $y \leq 0$.

Consider $y \geq 0$. Look at the graph of $Y = X^2$ to see that Y is below y iff X is between $-\sqrt{y}$ and $\sqrt{y}$. So

$$F(y) = P(Y \leq y) = P(-\sqrt{y} \leq X \leq \sqrt{y}) = \int_{-\sqrt{y}}^{\sqrt{y}} f(x)\, dx$$
$$= \text{area under } f(x) \text{ between } -\sqrt{y} \text{ and } \sqrt{y}$$

Case 1.   Let $0 \leq y \leq 1$. Then $-1 \leq -\sqrt{y} \leq 0$ and

$$F(y) = \int_{-\sqrt{y}}^{0} \frac{1}{2}\, dx + \int_{0}^{\sqrt{y}} \frac{1}{2} e^{-x}\, dx = \frac{1}{2}\sqrt{y} + \frac{1}{2} - \frac{1}{2} e^{-\sqrt{y}}$$

Case 2.   Let $y \geq 1$. Then $-\sqrt{y} \leq -1$ and

$$F(y) = \int_{-1}^{0} \frac{1}{2}\, dx + \int_{0}^{\sqrt{y}} \frac{1}{2} e^{-x}\, dx = 1 - \frac{1}{2} e^{-\sqrt{y}}$$

Density method.

By inspection, Y is never ≤ 0, so $f(y) = 0$ for $y \leq 0$.

For any $y \geq 0$, there are *two* corresponding x's, so

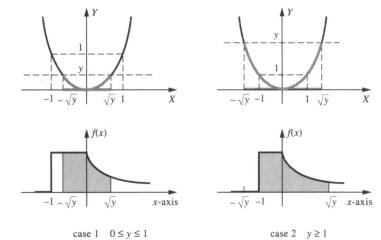

case 1 $0 \le y \le 1$ case 2 $y \ge 1$

Figure P.8   Dist function method

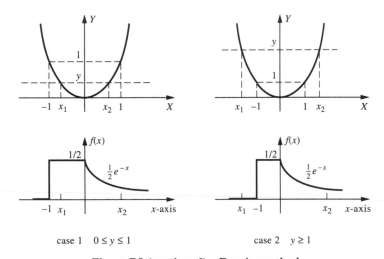

case 1 $0 \le y \le 1$ case 2 $y \ge 1$

Figure P.8 (continued)   Density method

$$f(y) = f(x_1) \left| \frac{dx_1}{dy} \right| + f(x_2) \left| \frac{dx_2}{dy} \right|$$

where

$$x_1 = -\sqrt{y}, \quad x_2 = \sqrt{y}, \quad \frac{dx_1}{dy} = -\frac{1}{2\sqrt{y}}, \quad \frac{dx_2}{dy} = \frac{1}{2\sqrt{y}}$$

Case 1.   Let $0 \le y \le 1$.

Then $-1 \le x_1 \le 0$, $f(x_1) = 1/2$, $f(x_2) = \frac{1}{2}e^{-x_2}$,

$$f(y) = \frac{1}{2}\frac{1}{2\sqrt{y}} + \frac{1}{2}e^{-x_2}\frac{1}{2\sqrt{y}} = \frac{1 + e^{-\sqrt{y}}}{4\sqrt{y}}$$

Case 2.   Let $y \geq 1$. Then $x_1 \leq -1$, $f(x_1) = 0$,

$$f(y) = \frac{1}{2}e^{-x_2}\frac{1}{2\sqrt{y}} = \frac{e^{-\sqrt{y}}}{4\sqrt{y}}$$

Solutions Section 4-7

1. Y has distribution function $F(y) = 1 - e^{-\lambda y}, y \geq 0$.

    Let $X = 1 - e^{-\lambda Y}$. Then $e^{-\lambda Y} = 1 - X$, $-\lambda Y = \ln(1 - X)$, so
    $Y = \frac{1}{\lambda}\ln(1 - X)$.

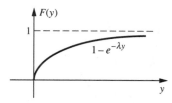 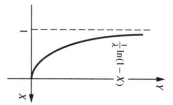

Figure P.1

2. **(a)** If $0 \leq X \leq 1/3$, let $X = \frac{1}{3}Y^2$, $Y = \sqrt{3X}$.

    If $1/3 \leq X \leq 1$, let $X = \frac{2}{3}Y - \frac{5}{3}$, $Y = (3X + 5)/2$.

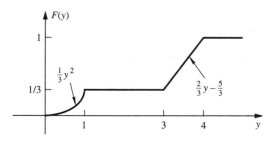

Figure P.2a

(b) Let X be uniform on $[0,1]$. Let

$$Y = \begin{cases} \sqrt{3X} & \text{if } 0 \leq X \leq \dfrac{1}{3} \\[2mm] \dfrac{3X + 5}{2} & \text{if } \dfrac{1}{3} \leq X \leq 1 \end{cases}$$

Case 1. $0 \leq y \leq 1$

$$F_Y(y) = P(Y \leq y) = P\left(0 \leq X \leq \frac{1}{3}y^2\right) = \frac{1}{3}y^2$$

Case 2. $1 \leq y \leq 3$

$$F_Y(y) = P(Y \leq y) = P\left(0 \leq X \leq \frac{1}{3}\right) = \frac{1}{3}$$

Case 3. $3 \leq y \leq 4$

$$F_Y(y) = P(Y \leq y) = P\left(0 \leq X \leq \frac{2}{3}y - \frac{5}{3}\right) = \frac{2}{3}y - \frac{5}{3}$$

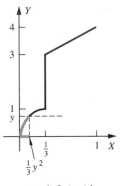

case 1 $0 \leq y \leq 1$

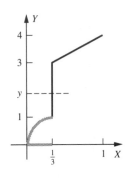

case 2 $1 \leq y \leq 3$

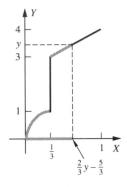

case 3 $3 \leq y \leq 4$

Figure P.2b

3. If $0 \leq X \leq \frac{1}{2}$, let $X = 1 - \frac{2}{Y}, Y = \frac{2}{1-X}$.

If $\frac{1}{2} \leq X \leq 1$, let $X = 1 - \frac{1}{\sqrt{Y}}, Y = \frac{1}{(1-X)^2}$

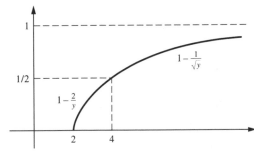

Figure P.3

4. (a) One possibility is to let

$$Y = \begin{cases} 1 & \text{if } 0 \le X \le .1 \\ 2 & \text{if } .1 \le X \le .5 \\ e & \text{if } .5 \le X \le .8 \\ -3 & \text{if } .8 \le X \le 1 \end{cases}$$

(b) Here's one possibility.

$$Y = \begin{cases} 1 & \text{if } 0 \le X \le \dfrac{1}{2} \\[2mm] 2 & \text{if } \dfrac{1}{2} \le X \le \dfrac{3}{4} \\[2mm] 3 & \text{if } \dfrac{3}{4} \le X \le \dfrac{7}{8} \\[2mm] \vdots \end{cases}$$

In general,

$$Y = n \quad \text{if } \frac{2^{n-1} - 1}{2^{n-1}} \le X \le \frac{2^n - 1}{2^n}$$

(c) To give Y a Poisson distribution you have to make $P(Y = k)$ come out to be $e^{-\lambda}\lambda^k/k!$. One way is to choose

$$Y = \begin{cases} 0 & \text{if } 0 \le X \le e^{-\lambda} \\ 1 & \text{if } e^{-\lambda} \le X \le e^{-\lambda} + \lambda e^{-\lambda} \\ 2 & \text{if } e^{-\lambda} + \lambda e^{-\lambda} \le X \le e^{-\lambda} + \lambda e^{-\lambda} + \dfrac{e^{-\lambda}\lambda^2}{2!} \\ \vdots \end{cases}$$

Here's a check that it works:

$$P(Y = 2) = P(e^{-\lambda} + \lambda e^{-\lambda} \le X \le e^{-\lambda} + \lambda e^{-\lambda} + \frac{e^{-\lambda}\lambda^2}{2!})$$

$$= \text{length of } X\text{-interval} \quad (\text{because } X \text{ is uniform on } [0, 1])$$

$$= \frac{e^{-\lambda}\lambda^2}{2!}$$

5. (a) *Case 1.*

If $0 \le X \le 1/100$, take $Y = -10$.

So $P(Y = -10) = P(0 \le X \le 1/100) = \text{jump} = 1/100$.

Case 2.

If $1/100 \le X \le 1/4$, let $X = 1/Y^2$, $\quad Y = \sqrt{1/X}$.

Case 3.

If $1/4 \le X \le 1/2$, let $X = \frac{1}{4}Y - \frac{1}{2}$, $\quad Y = 4X + 2$.

Case 4.

If $1/2 \le X \le 1$, take $Y = 4$ (where there is a jump).

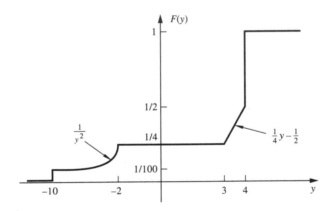

Figure P.5a

(b) If $2 \le y \le 3$, $F(y) = \frac{1}{4}y$. So for $1/2 \le X \le 3/4$, let $X = \frac{1}{4}Y$, $Y = 4X$. So let

$$
Y = \begin{cases}
-1 & \text{if } 0 \le X \le \dfrac{1}{10} \\[2mm]
1 & \text{if } \dfrac{1}{10} \le X \le \dfrac{1}{4} \\[2mm]
2 & \text{if } \dfrac{1}{4} \le X \le \dfrac{1}{2} \\[2mm]
4X & \text{if } \dfrac{1}{2} \le X \le \dfrac{3}{4} \\[2mm]
3 & \text{if } \dfrac{3}{4} \le X \le 1
\end{cases}
$$

6. *Method 1.*   First find $F(y)$ (see the diagram).

Then let $X = \frac{1}{4}Y - \frac{3}{4}, Y = 4X + 3$.

Method 2.   If X is uniform on $[0,1]$ and $Y = aX + b$, then (by inspection?) Y is uniform on $[b, a + b]$. So we can get Y uniform on $[3,7]$ by setting $b = 3, a + b = 7, a = 4$. So choose $Y = 4X + 3$.

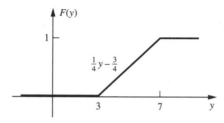

Figure P.6

Solutions Section 5-1

1. (a) $\displaystyle\int_{\text{plane}} f(x, y)\, dA = 1, \quad K = \dfrac{1}{\int_{x=0}^{\infty} \int_{y=x}^{\infty} e^{-(x+y)}\, dy\, dx}$

$$\text{inner} = -e^{(x+y)}\Big|_{y=x}^{\infty} = e^{-2x}, \quad \text{outer} = -\frac{1}{2} e^{-2x}\Big|_{x=0}^{\infty} = \frac{1}{2}$$

so $K = 2$.

(b) $P(Y < 2X) = \displaystyle\int_{x=0}^{\infty} \int_{y=x}^{2x} 2e^{-(x+y)}\, dy\, dx$

$$= \int_{0}^{\infty} (2e^{-2x} - 2e^{-3x})\, dx = \frac{1}{3}$$

(c) $P(\max \le 4) = \displaystyle\int_{\text{fav}} f(x, y)\, dA = \int_{x=0}^{4} \int_{y=x}^{4} 2e^{-(x+y)}\, dy\, dx$

$$= 1 + e^{-8} - 2e^{-4}$$

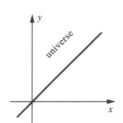

Figure P.1a

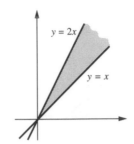

Figure P.1b

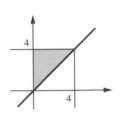

Figure P.1c

2. (a) $P(Y \geq X^2) = \int_{\text{fav}} \frac{6}{7}x \, dA = \int_{\text{I}} + \int_{\text{II}}$

$$= \int_{x=0}^{(-1+\sqrt{5})/2} \int_{y=1-x}^{2-x} \frac{6}{7}x \, dy \, dx$$

$$+ \int_{x=(-1+\sqrt{5})/2}^{1} \int_{y=x^2}^{2-x} \frac{6}{7}x \, dy \, dx$$

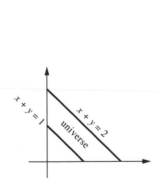

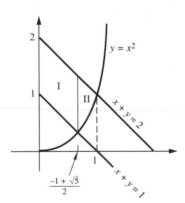

Figure P.2a

(b) $P(\text{max} > 1) = 1 - \int_{\text{unfav}} \frac{6}{7}x \, dA = 1 - \int_{x=0}^{1} \int_{y=1-x}^{1} \frac{6}{7}x \, dy \, dx$

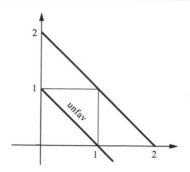

Figure P.2b

3. (a) $P\left(X < \dfrac{1}{2} \,\&\, Y > \dfrac{1}{2}\right) = P(\text{I}) + P(\text{II}) = \dfrac{3}{2}\ \text{area I} + \dfrac{1}{2}\ \text{area II} = \dfrac{1}{4}$

(b) $P\left(X < \dfrac{1}{2} \Big| Y > \dfrac{1}{2}\right) = \dfrac{P(X < 1/2 \,\&\, Y > 1/2)}{P(Y > 1/2)}$

From part (a), the numerator is $1/4$.

$P\left(Y > \dfrac{1}{2}\right) = P(\text{III}) + P(\text{IV}) = \dfrac{3}{2}\ \text{area III} + \dfrac{1}{2}\ \text{area IV} = \dfrac{3}{8}.$

Answer is $\dfrac{1/4}{3/8} = \dfrac{2}{3}.$

(c) $P\left(\min < \dfrac{2}{3}\right) = 1 - P(\text{V}) = 1 - \dfrac{1}{2}\ \text{area V} = \dfrac{17}{18}$

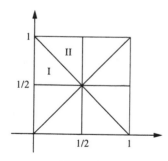

Figure P.3a

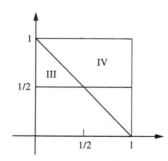

Figure P.3b

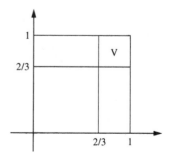

Figure P.3c

4. (a) (See inequality review if you didn't get the graph of $y/x \le -3/2$).

$$P\left(\dfrac{Y}{X} \le -\dfrac{3}{2}\right) = \dfrac{\text{fav}}{\text{total}}$$

$$\text{total area} = \int_{-1}^{1} (1 - x^2)\,dx = \frac{4}{3},$$

$$\text{fav area} = \int_{x=-1/2}^{0} \left(1 - x^2 - \frac{3}{2}x\right)\,dx = \frac{13}{48}$$

Answer is $13/64$.

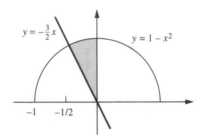

Figure P.4a

(b) $f(x, y) = \dfrac{1}{\text{total area}} = \dfrac{3}{4}$ for (x, y) in the given universe

5. $f(x) = \dfrac{1}{2}$ for $0 \le x \le 2$, $f(y) = e^{-y}$ for $y \ge 0$.

$$f(x, y) = f(x)f(y) = \frac{1}{2}e^{-y} \text{ for } 0 \le x \le 2, y \ge 0.$$

$$P(Y - X \ge 3) = \int_{x=0}^{2}\int_{y=x+3}^{\infty} \frac{1}{2}e^{-y}\,dy\,dx = \frac{1}{2}e^{-3} - \frac{1}{2}e^{-5}.$$

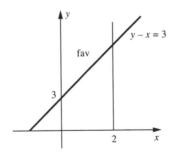

Figure P.5

6. X and Y are jointly uniform on the square $-1 \le x \le 1, -1 \le y \le 1$.

(a) $P\left(XY \le \dfrac{1}{4}\right) = \dfrac{\text{fav}}{\text{total}}$

total $= 4$,

fav $= 4 -$ unfav $= 4 - 2$ area I

$$= 4 - 2\int_{1/4}^{1}\left(1 - \frac{1}{4}\frac{1}{x}\right)dx = 4 - 2\left(\frac{3}{4} - \frac{1}{4}\ln 4\right)$$

Answer is $\dfrac{5}{8} + \dfrac{1}{8}\ln 4$

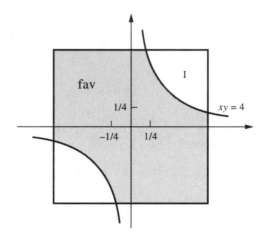

Figure P.6a

(b) $P\left(\dfrac{Y}{X} \le \dfrac{1}{2}\right) = \dfrac{\text{fav}}{\text{total}} = \dfrac{2(1 + \frac{1}{2}\cdot\frac{1}{2}\cdot 1)}{4} = \dfrac{5}{8}$

(c) The graph of $|x| + |y| = 1$ consists of four segments:

$x + y = 1$ for $x > 0, y > 0$,
$x - y = 1$ for $x > 0, y < 0$,
$-x + y = 1$ for $x < 0, y > 0$,
$-x - y = 1$ for $x < 0, y < 0$.

$$P(|X| + |Y| \le 1) = \frac{\text{fav}}{\text{total}} = \frac{\sqrt{2}\sqrt{2}}{4} = \frac{1}{2}$$

(d) $f(x, y) = \dfrac{1}{\text{total}} = \dfrac{1}{4}$ for $-1 \le x \le 1, -1 \le y \le 1$

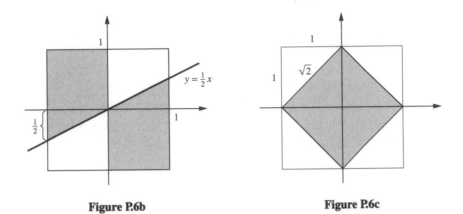

Figure P.6b   **Figure P.6c**

7. Let X be Mary's arrival time. Let Y be John's arrival time. Then X is uniform on [15,45], Y is uniform on [0,60], and point (X,Y) is uniformly distributed in a rectangle.

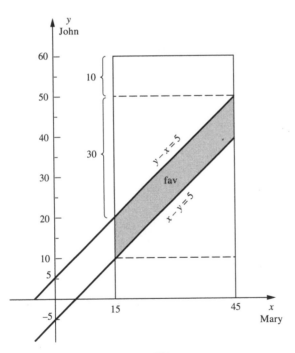

Figure P.7a

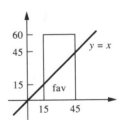

Figure P.7b

(a) $P(\text{meet}) = P(|X - Y| \le 5) = \text{fav/total}.$

total $= 1800$, unfav $= 2\left(10 \cdot 30 + \dfrac{1}{2} \cdot 900\right) = 1500,$

fav $= 300$, answer $= \dfrac{1}{6}$

(b) Just look at the diagram to see that half the area is favorable, so prob is $1/2$.

8. (a) $p(2, 4) = P(D_1 = 2, \max = 4) = P(D_1 = 2, D_2 = 4) = \dfrac{1}{36}$

$p(4, 2) = P(D_1 = 4, \max = 1) = 0$

$p(4, 4) = P(D_1 = 4, \max = 4)$

$\qquad = P(D_1 = 4, D_2 = 1 \text{ or } 2 \text{ or } 3 \text{ or } 4) = \dfrac{4}{36}$

and so on

(b) $F(2, 5) = P(X \le 2, Y \le 5)$

$=$ sum of indicated entries in the northwest corner $= 10/36$

$y\backslash^x$	1	2	3	4	5	6
1	$\frac{1}{36}$	0	0	0	0	0
2	$\frac{1}{36}$	$\frac{2}{36}$	0	0	0	0
3	$\frac{1}{36}$	$\frac{1}{36}$	$\frac{3}{36}$	0	0	0
4	$\frac{1}{36}$	$\frac{1}{36}$	$\frac{1}{36}$	$\frac{4}{36}$	0	0
5	$\frac{1}{36}$	$\frac{1}{36}$	$\frac{1}{36}$	$\frac{1}{36}$	$\frac{5}{36}$	0
6	$\frac{1}{36}$	$\frac{1}{36}$	$\frac{1}{36}$	$\frac{1}{36}$	$\frac{1}{36}$	$\frac{6}{36}$

Figure P.8

9. The joint density of ind random variables is the product of the individual densities

(a) $f(x_1, \ldots, x_4) = e^{-x_1} e^{-x_2} e^{-x_3} e^{-x_4}$

$\qquad = e^{-(x_1 + x_2 + x_3 + x_4)}$ for $x_1, x_2, x_3, x_4 \ge 0$

(b) $f(x_1, \ldots, x_4) = \dfrac{1}{256}\, x_1 x_2 x_3 x_4$ for $1 \le x_1, x_2, x_3, x_4 \le 3$

(c) $f(x_1, \ldots, x_4) = g(x_1)\, g(x_2)\, g(x_3)\, g(x_4)$

Solutions Section 5-2

1. total area $= \int_{-1}^{1} (1 - x^2)\, dx = \dfrac{4}{3}$,

$f(x, y) = \dfrac{1}{\text{area}} = \dfrac{3}{4}$ for (x, y) in the region

$f(x) = \displaystyle\int_{y=-\infty}^{\infty} f(x, y)\, dy = \int_{y=0}^{1-x^2} \dfrac{3}{4}\, dy = \dfrac{3}{4}(1 - x^2)$ for $0 \le x \le 1$

$f(y) = \displaystyle\int_{x=-\sqrt{1-y}}^{\sqrt{1-y}} \dfrac{3}{4}\, dy = \dfrac{3}{2}\sqrt{1 - y}$ for $0 \le y \le 1$

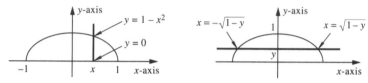

Figure P.1

2. (a) *Case 1.* $0 \le x \le 1$

$$f(x) = \int_{y=1-x}^{2-x} \dfrac{6}{7}x\, dy = \dfrac{6}{7}x$$

Case 2. $1 \le x \le 2$

$$f(x) = \int_{y=0}^{2-x} \dfrac{6}{7}x\, dy = \dfrac{6}{7}x(2 - x)$$

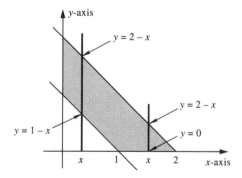

Figure P.2a

(b) *Case 1.* $0 \le y \le 1$

$$f(y) = \int_{x=1-y}^{2-y} \frac{6}{7}x\,dx = \frac{3}{7}(3 - 2y)$$

Case 2. $1 \le y \le 2$

$$f(y) = \int_{x=0}^{2-y} \frac{6}{7}x\,dx = \frac{3}{7}(2 - y)^2$$

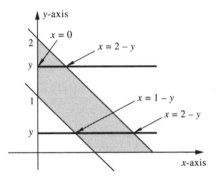

Figure P.2b

3. $f(x) = \displaystyle\int_{y=-\infty}^{\infty} f(x,y)\,dy$

$$= \int_{y=0}^{1-x} \frac{3}{2}\,dy + \int_{y=1-x}^{1} \frac{1}{2}\,dy$$

$$= \frac{3}{2} - x \quad \text{for } 0 \le x \le 1$$

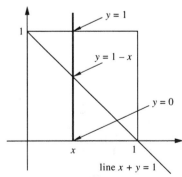

Figure P.3

4. (a) $f(x) = \int_{y=-\infty}^{\infty} f(x,y)\, dy$

$$= \int_{y=0}^{1} \frac{x+y}{4}\, dy + \int_{y=1}^{x/2+1} \frac{1}{4}\, dy = \frac{3}{8}x + \frac{1}{8}$$

$$\text{for } 0 \le x \le 2$$

(b) *Case 1.* $0 \le y \le 1$

$$f(y) = \int_{x=0}^{2} \frac{x+y}{4}\, dx = \frac{1}{2} + \frac{1}{2}y$$

Case 2. $1 \le y \le 2$

$$f(y) = \int_{x=2y-2}^{2} \frac{1}{4}\, dx = 1 - \frac{1}{2}y$$

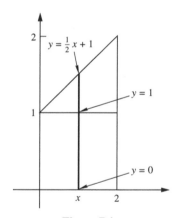

Figure P.4a

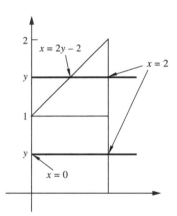

Figure P.4b

5. (a) Not ind   (universe is not a rectangle)

(b) Ind

$f(x) = Cx$ for $0 \le x \le 1$,

$f(y) = Kye^{-y^2/4}$ for $y \ge 0$ where $CK = 1$

$\int_{0}^{1} Cx\, dx = 1, C = 2, K = \frac{1}{2}$

(c) Not ind   (universe is not a rectangle).

(d) Not ind since $2(x+y)$ doesn't separate.

(e) Ind; X is uniform on $[1,2]$, Y is uniform on $[0,3]$.

$$f(x) = 1 \text{ for } 1 \le x \le 2, f(y) = \frac{1}{3} \text{ for } 0 \le y \le 3$$

6. (a) $f(y) = Ce^{-2y}$ for $y \ge 0$,

$f(x) = K$ for $0 \le x \le 4$ where $CK = 1/2$.

By inspection, Y must be exponential with $\lambda = 2$, so $C = 2$.

And X is uniform on $[0,4]$, so $K = 1/4$.

(b) $f(x) = \displaystyle\int_{y=0}^{\infty} \frac{1}{2} e^{-2y} \, dy = \frac{1}{4}$ for $0 \le x \le 4$

$f(y) = \displaystyle\int_{x=0}^{4} \frac{1}{2} e^{-2y} \, dx = 2e^{-2y}$ for $y \le 0$

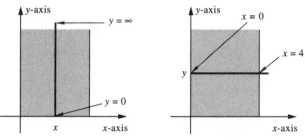

Figure P.6b

7. $p(1,1) = P(\min = 1, \max = 1) = P(D_1 = D_2 = 1) = \dfrac{1}{36}$

$p(1,2) = P(\min = 1, \max = 2)$

$$= P(D_1 = 1, D_2 = 2 \text{ or vice versa}) = \frac{2}{36}$$

$p(2,1) = P(\min = 2, \max = 1) = 0,$

and so on.

Clearly not ind, since knowing the min affects the odds on the max. For example,

$$P(\max = 5) = P(Y = 5) = \frac{9}{36} \quad \text{but} \quad P(\max = 5| \min = 6) = 0$$

$y\backslash^x$	1	2	3	4	5	6	$p(y)$
1	$\frac{1}{36}$	0	0	0	0	0	$\frac{1}{36}$
2	$\frac{2}{36}$	$\frac{1}{36}$	0	0	0	0	$\frac{3}{36}$
3	$\frac{2}{36}$	$\frac{2}{36}$	$\frac{1}{36}$	0	0	0	$\frac{5}{36}$
4	$\frac{2}{36}$	$\frac{2}{36}$	$\frac{2}{36}$	$\frac{1}{36}$	0	0	$\frac{7}{36}$
5	$\frac{2}{36}$	$\frac{2}{36}$	$\frac{2}{36}$	$\frac{2}{36}$	$\frac{1}{36}$	0	$\frac{9}{36}$
6	$\frac{2}{36}$	$\frac{2}{36}$	$\frac{2}{36}$	$\frac{2}{36}$	$\frac{2}{36}$	$\frac{1}{36}$	$\frac{11}{36}$
$p(x)$	$\frac{11}{36}$	$\frac{9}{36}$	$\frac{7}{36}$	$\frac{5}{36}$	$\frac{3}{36}$	$\frac{1}{36}$	

Figure P.7

8. $f(x, y) = \dfrac{1}{\text{total area}} = \dfrac{1}{4\pi}$ for (x, y) in circle

Circle has equ $x^2 + y^2 = 4$, so

$$f(x) = \int_{y=-\sqrt{4-x^2}}^{\sqrt{4-x^2}} \frac{1}{4\pi}\, dy = \frac{2}{4\pi}\sqrt{4 - x^2} \text{ for } -2 \leq x \leq 2$$

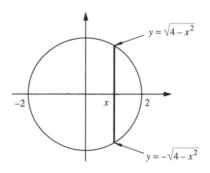

Figure P.8

9. (a) F   (would be true if X and Y were independent).

It may be possible for X to be .6, and it may be possible for Y to be .7, but it may not be possible to get $X = .6$ and $Y = .7$ simultaneously.

Look at problem 1 where the 2-dim universe is the non-rectangle $0 \leq y \leq 1 - x^2$ but the X universe is [0,1] and the Y universe is [0,1].

(b) F   (would be true if X and Y were independent).

(c) F   (would be true only if X and Y were jointly uniform on a rectangle).

(d) T   (the joint density is not separable).

(e) F   (would be true if the universe were a rectangle).

(f) T

(g) F

Solutions Section 5-3

1. $f(x) = 1$ for $0 \leq x \leq 1$, $f(y) = e^{-y}$ for $y \geq 0$

$f(x, y) = e^{-y}$ in the strip $0 \leq x \leq 1, y \geq 0$

$$F_Z(z) = P(Z \leq z) = P(X + Y \leq z) = \int_{\text{fav}} f(x, y) \, dA$$

Case 1.   If $z \leq 0$, then $F(z) = 0$ because $X + Y$ is never ≤ 0.

Case 2.   Let $0 \leq z \leq 1$.

$$F(z) = \int_{x=0}^{z} \int_{y=0}^{z-x} e^{-y} \, dy \, dx = \int_{x=0}^{z} (1 - e^{x-z}) \, dx$$

$$= (x - e^{x-z})\big|_{x=0}^{z} = z - 1 + e^{-z}$$

Case 3.   Let $z \geq 1$.

$$F(z) = \int_{x=0}^{1} \int_{y=0}^{z-x} e^{-y} \, dy \, dx = 1 - e^{1-z} + e^{-z}$$

$$f(z) = F'(z) = \begin{cases} 1 - e^{-z} & \text{if } 0 \leq z \leq 1 \\ e^{1-z} - e^{-z} & \text{if } z \geq 1 \end{cases}$$

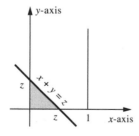

Figure P.1a $0 \leq z \leq 1$

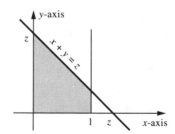

Figure P.1b $z \geq 1$

2. $f(x, y) = 12xy(1 - x)$ in a unit square. Z is always between 0 and 1, so

$$F(z) = \begin{cases} 0 & \text{if } z \leq 0 \\ 1 & \text{if } z \geq 1 \end{cases}$$

If $0 \leq z \leq 1$, then

$$F(z) = P(Z \leq z) = P(I^2 R \leq z) = 1 - \int_{\text{unfav}} 12xy(1-x)\, dA$$

$$= 1 - \int_{x=\sqrt{z}}^{1} \int_{y=z/x^2}^{1} 12xy(1-x)\, dy\, dx$$

$$= 1 - (1 - 3z^2 - 6z + 8z^{3/2})$$

$$f(z) = F'(z) = 6z + 6 - 12\sqrt{z} \quad \text{for } 0 \leq z \leq 1$$

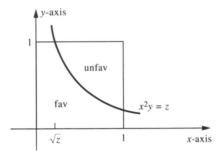

Figure P.2

3. **(a)** (X, Y) is uniform in a square. The max is always between 0 and 1, so

$$F_{\max}(z) = \begin{cases} 0 & \text{if } z \leq 0 \\ 1 & \text{if } z \geq 1 \end{cases}$$

Let $0 \leq z \leq 1$.

Method 1.

$$\begin{aligned} F_{\max}(z) &= P(\max \leq z) = P(X \leq z \text{ and } Y \leq z) \\ &= P(X \leq z)\, P(Y \leq z) \text{ (by independence)} \\ &= z \cdot z \quad \text{(since } X \text{ and } Y \text{ are uniform on } [0,1]) \\ &= z^2 \end{aligned}$$

Method 2.

$$F_{\max}(z) = P(\max \le z) = \frac{\text{fav area}}{\text{total}} \; (\text{ see the diagram}) = z^2$$

Then $f_{\max}(z) = F'_{\max}(z) = 2z$ if $0 \le z \le 1$.

(b) Let $Z = XY$. Then $0 \le Z \le 1$, so $f(z) = 0$ if $z \le 0$ or $z \ge 1$.
Let $0 \le z \le 1$. Then $F(z) = P(Z \le z) = \text{fav/total}$.

$$\text{total area} = 1,$$

$$\text{fav area} = \text{I} + \text{II} = z + \int_{x=z}^{1} \frac{z}{x}\, dx = z - z\ln z$$

$F(z) = z - z\ln z$ for $0 \le z \le 1$

$f(z) = F'(z) = 1 - (z \cdot \frac{1}{z} + \ln z) = -\ln z$ if $0 \le z \le 1$

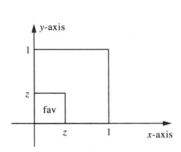

Figure P.3a   Method 2   **Figure P.3b**

4. $F(z) = P(Z \le z) = \text{fav/total}$, where total area $= 1$.
For $0 \le z \le 1$,

$$\text{fav area} = \text{II} + (\text{ACD} - \text{I} - \text{III})$$

$$= \frac{1}{2}z^2 + \left[\frac{1}{2} - \frac{1}{4}(1-z)^2 - \frac{1}{4}(1-z^2)\right] = z$$

$$F(z) = \begin{cases} 0 & \text{if } z \le 0 \\ z & \text{if } 0 \le z \le 1 \\ 1 & \text{if } z \ge 1 \end{cases}$$

$f(z) = F'(z) = 1$ if $0 \le z \le 1$, so Z is uniform on $[0,1]$.

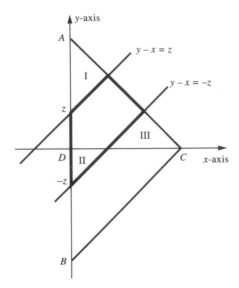

Figure P.4

5. $f(x, y) = f(x)f(y) = (1/2\pi) \cdot e^{-(x^2+y^2)/2}$

$T \geq 0$, so $F_T(t) = 0$ for $t \leq 0$.

Suppose $t \geq 0$. Then

$$F(t) = P(T \leq t) = P\left(\frac{1}{\sqrt{X^2 + Y^2}} \leq t\right) = \int_{\text{fav}} f(x, y)\, dA$$

The graph of $1/\sqrt{x^2 + y^2} = t$ is the circle $x^2 + y^2 = 1/t^2$; radius is $1/t$.
The fav region is outside.

$$F(t) = \int_{\theta=0}^{2\pi} \int_{r=1/t}^{\infty} \frac{1}{2\pi} e^{-r^2/2} r\, dr\, d\theta = e^{-1/2t^2}$$

$$f(t) = F'(t) = \frac{1}{t^3} e^{-1/2t^2} \quad \text{if } t \geq 0$$

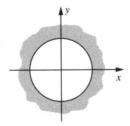

Figure P.5

6. *Case 1.* $z \leq 0$

$$F(z) = P\left(\frac{Y}{X} \leq z\right) = \int_{\text{fav}} f(x,y) \, dA = \int_{\text{I}} + \int_{\text{II}}$$

$$= \int_{x=-\infty}^{0} \int_{y=zx}^{\infty} f(x)f(y) \, dy \, dx + \int_{x=0}^{\infty} \int_{y=-\infty}^{zx} f(x)f(y) \, dy \, dx$$

Case 2. $z \geq 0$

$$F(z) = \int_{\text{III}} + \int_{\text{IV}} = \int_{x=-\infty}^{0} \int_{y=zx}^{\infty} f(x)f(y) \, dy \, dx$$

$$+ \int_{x=0}^{\infty} \int_{y=-\infty}^{zx} f(x)f(y) \, dy \, dx$$

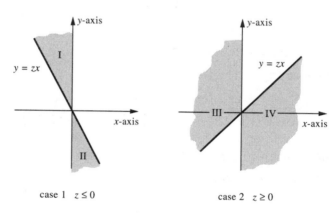

case 1 $z \leq 0$ case 2 $z \geq 0$

Figure P.6

7. (a) $F_{\max}(z) = P(\max \leq z) = \int_{\text{fav}} f(x,y) \, dA$

$$= \int_{x=0}^{z} \int_{y=x}^{z} 2e^{-(x+y)} \, dy \, dx$$

$$= e^{-2z} - 2e^{-z} + 1 \quad \text{for } z \geq 0$$

$$f_{\max}(z) = F'_{\max} = -2e^{-2z} + 2e^{-z} \quad \text{for } z \geq 0$$

(b) $F_{\min}(z) = P(\min \leq z) = \int_{\text{fav}} f(x,y) \, dA$

$$= \int_{x=0}^{z} \int_{y=x}^{\infty} 2e^{-(x+y)} \, dy \, dx = 1 - e^{-2z} \quad \text{for } z \geq 0$$

$$f_{\min}(z) = F'_{\min}(z) = 2e^{-2z} \text{ for } z \geq 0$$

(The min has an exponential distribution with $\lambda = 2$.)

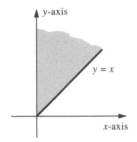

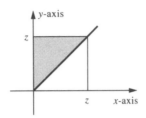

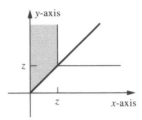

Figure P.7   Universe   **Figure P.7a**   **Figure P.7b**

Solutions Review Problems for Chapters 4 and 5

1. $f(x) = 0$ for $2 \leq x \leq 8$.

$F(x)$ is constant for $2 \leq x \leq 8$ (but not necessarily 0). Probability may accumulate until $x = 2$. As x goes from 2 to 8, no new probability accumulates, $F(x)$ stays constant, and the graph of $F(x)$ is horizontal.

2. (a) $m = (a + b)/2$, the midpoint of $[a, b]$, so that half the length is to the left and half to the right.

(b) $m = \mu$. The normal density is symmetric around $x = \mu$. Half the probability is to the left, half to the right.

(c) X has density $f(x) = \lambda e^{-\lambda x}$ for $x \geq 0$, so

$$P(X \leq m) = \int_0^m \lambda e^{-\lambda x}\, dx = 1 - e^{-\lambda m}$$

We need $1 - e^{-\lambda m} = 1/2,\, e^{-\lambda m} = 1/2,\, -\lambda m = \ln 1/2,$
$m = (1/\lambda) \ln 2$.

3. Dist Function Method Y is always between 0 and 2.

Case 1. $0 \leq y \leq 1/2$

$$F(y) = P(Y \leq y) = P\left(X \leq \frac{1}{2}y\right) + P\left(X \geq \sqrt{\frac{2}{y}}\right)$$

$$= \frac{1}{2}\left(\frac{1}{2}y\right)^2 = \frac{1}{8}y^2$$

Case 2. $1/2 \le y \le 2$

$$F(y) = P(Y \le y) = P\left(X \le \frac{1}{2}y\right) + P\left(X \ge \sqrt{\frac{2}{y}}\right)$$

$$= \frac{1}{2}\left(\frac{1}{2}y\right)^2 + \frac{1}{2}\left(2 - \sqrt{\frac{2}{y}}\right)^2$$

So

$$F(y) = \begin{cases} 0 & \text{if } y \le 0 \\[2mm] \dfrac{1}{8}y^2 & \text{if } 0 \le y \le \dfrac{1}{2} \\[3mm] \dfrac{1}{8}y^2 + \dfrac{1}{2}\left(2\sqrt{\dfrac{2}{y}}\right)^2 & \text{if } \dfrac{1}{2} \le y \le 2 \\[3mm] 1 & \text{if } y \ge 2 \end{cases}$$

$$f(y) = F'(y)$$

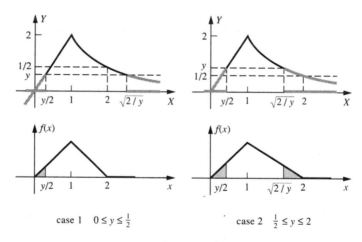

case 1 $0 \le y \le \frac{1}{2}$ case 2 $\frac{1}{2} \le y \le 2$

Figure P.3   Distribution function method

Density Function Method

First of all, $0 \le Y \le 2$, so $f(y) = 0$ for $y \le 0$ or $y \ge 2$.
For $0 \le y \le 2$,

$$f(y) = f(x_1)\left|\frac{dx_1}{dy}\right| + f(x_2)\left|\frac{dx_2}{dy}\right|$$

where

$$x_1 = \frac{1}{2}y, \quad \frac{dx_1}{dy} = \frac{1}{2}, \quad x_2 = \sqrt{\frac{2}{y}}, \quad \frac{dx_2}{dy} = -\frac{1}{2}\sqrt{2}\,y^{-3/2}$$

We always have $f(x_1) = x_1$.

Case 1. $0 \le y \le 1/2$

Then $x_2 \ge 2$, $f(x_2) = 0$ and $f(y) = x_1 \cdot \frac{1}{2} = \frac{1}{4}y$.

Case 2. $1/2 \le y \le 2$

Then $1 \le x_2 \le 2$, $f(x_2) = 2 - x_2$, and

$$f(y) = x_1 \cdot \frac{1}{2} + (2 - x_2)\frac{1}{2}\sqrt{2}\,y^{-3/2}$$

$$= \frac{1}{4}y + \left(2 - \sqrt{\frac{2}{y}}\right)\frac{1}{2}\sqrt{2}\,y^{-3/2}$$

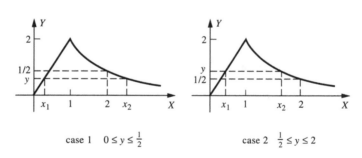

case 1 $0 \le y \le \frac{1}{2}$ case 2 $\frac{1}{2} \le y \le 2$

Figure P.3 (continued)   Density method

4. (a) $P(X + Y \ge 1) = \dfrac{\text{fav area}}{\text{total}}$

$$\text{fav area} = \int_{x=0}^{1} (y_{\text{upper}} - y_{\text{lower}})\,dx$$

$$= \int_{0}^{1} (1 - x^2 - (1 - x))\,dx = \frac{1}{6}$$

$$\text{total area} = 2\int_{0}^{1} (1 - x^2)\,dx = \frac{4}{3}$$

$$P(X + Y \ge 1) = \frac{1/6}{4/3} = \frac{1}{8}$$

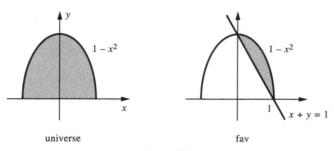

universe   fav

Figure P.4

(b) Total area $= 4/3$, $f(x, y) = 1/$total area $= 3/4$ for $0 \le y \le 1 - x^2$.

5. Let X be the number of defectives in the sample. The sampling is without replacement from a large population, so we can think of it as sampling with replacement. Then X has a binomial dist with $n = 100$, $p = .2$ and X is approx normal with $\mu = np = 20$, $\sigma^2 = npq = 16$.

$$P(\text{batch is rejected}) = P(15\% \text{ or more of sample is defective})$$
$$= P(X \ge 15) = P\left(X^* \ge \frac{15 - \mu}{\sigma}\right)$$
$$= P(X^* \ge -1.25)$$
$$= 1 - P(X^* \le -1.25) = 1 - F^*(-1.25) = .9$$

6. $f(x, y) = 1/$total area $= 1$ for (x, y) in the triangle.

$$f(y) = \int_{x=y-1}^{1-y} f(x, y)\, dx = \int_{x=y-1}^{1-y} 1\, dx = 2 - 2y \quad \text{for } 0 \le y \le 1$$

For $f(x)$ we need cases.

If $0 \le x \le 1$, then $f(x) = \int_{y=0}^{1-x} 1\, dy = 1 - x$.

If $-1 \le x \le 0$, then $f(x) = \int_{y=0}^{x+1} 1\, dy = 1 + x$.

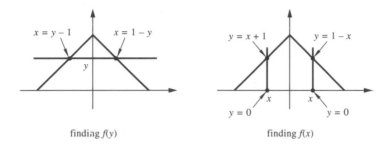

finding $f(y)$ finding $f(x)$

Figure P.6

7. Y takes on only the values $-1, 0, 1$ (Y is *discrete*).

$$p(-1) = P(Y = -1) = P(X \le -1)$$

$$= P\left(X^* \le \frac{-1 - \mu}{\sigma}\right) = P(X^* \le -1) = F^*(-1) = .159$$

$$p(0) = P(Y = 0) = P(-1 \le X \le 2)$$

$$= P\left(\frac{-1 - \mu}{\sigma} \le X^* \le \frac{2 - \mu}{\sigma}\right) = P\left(-1 \le X^* \le \frac{1}{2}\right)$$

$$= F^*\left(\frac{1}{2}\right) - F^*(-1) = .532$$

$$p(1) = P(Y = 1) = P(X \ge 2) = P\left(X^* \ge \frac{2 - \mu}{\sigma}\right)$$

$$= 1 - F^*\left(\frac{1}{2}\right) = .309$$

Or better still, $p(1) = 1 - p(-1) - p(0)$.

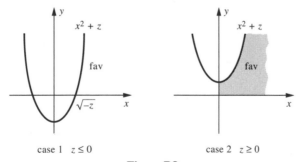

case 1 $z \le 0$ case 2 $z \ge 0$

Figure P.8

8. Let $Z = Y - X^2$. We'll get $F(z)$.

X has density $f_X(x) = g(x)$ for $x \ge 0$ (because X is non-negative).

Y has density $f_Y(y) = g(y)$ for $y \ge 0$ (*same* g but with letter y instead of x).

$$f(x,y) = f_X(x)f_Y(y) = g(x)g(y) \text{ for } x \geq 0, y \geq 0$$

$$F(z) = P(Z \leq z) = P(Y - X^2 \leq z) = \int_{\text{fav}} f(x,y)\, dA$$

Case 1.   If $z \leq 0$, then $F(z) = \int_{x=\sqrt{-z}}^{\infty} \int_{y=0}^{x^2+z} g(x)g(y)\, dy\, dx$.

Case 2.   If $z \geq 0$, then $F(z) = \int_{x=0}^{\infty} \int_{y=0}^{x^2+z} g(x)g(y)\, dy\, dx$.

9. We want c so that $P(X \geq c) = 10^{-22}$. For $0 \leq x \leq 1$,

$$P(X \geq c) = \int_{x=c}^{\infty} f(x)\, dx = \int_{x=c}^{1} 11(1-x)^{10}\, dx$$

$$= -(1-x)^{11}\Big|_{x=c}^{1} = (1-c)^{11}$$

So $(1-c)^{11} = 10^{-22} = (10^{-2})^{11}, 1-c = 10^{-2}, c = 1 - 10^{-2} = .99$.

10. **(a)** Point (X,Y) is uniformly distributed in a rectangle. Total area $= 6$, fav $= 9/4$,

$$P(Y \geq 2X) = \frac{\text{fav}}{\text{total}} = \frac{3}{8}$$

(b) $f(x,y) = 1/\text{total area} = \frac{1}{6}$ for $0 \leq x \leq 2, 0 \leq y \leq 3$

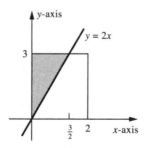

Figure P.10

11. If $0 \leq r \leq a$, then the fav region is a sphere of radius R inside the universe of radius a so

$$F(r) = P(R \leq r) = \frac{\text{fav vol}}{\text{total vol}} = \frac{\frac{4}{3}\pi r^3}{\frac{4}{3}\pi a^3} = \left(\frac{r}{a}\right)^3$$

$$F(r) = \begin{cases} 0 & \text{if } r \leq 0 \\ \left(\dfrac{r}{a}\right)^3 & \text{if } 0 \leq r \leq a \\ 1 & \text{if } r \geq a \end{cases}$$

$$f(r) = \frac{3r^2}{a^2} \quad \text{if } 0 \leq r \leq a$$

12. **(a)** $P(X = 2) = $ jump at $2 = 1/4$

(b) $P(X > 2) = 1 - P(X \leq 2) = 1 - $ upper $F(2) = 1 - 3/4 = 1/4$

(c) $P(X < 2) = $ lower $F(2) = 1/2$

(d) $P(X = 3) = 0$ (no jump at 3)

(e) The segment between $x = 2$ and $x = 3$ is part of line $y = \frac{1}{8}x + \frac{1}{2}$, so $F(3) = 7/8$ and

$$P(1 < X < 3) = F(3) - \text{upper } F(1) = \frac{7}{8} - \frac{1}{2} = \frac{3}{8}$$

(f) $P(X \geq 2 | X > 0) = \dfrac{P(X \geq 2 \text{ and } X > 0)}{P(X > 0)}$

$$= \frac{P(X \geq 2)}{P(X > 0)} = \frac{1 - \text{lower } F(2)}{1 - \text{upper } F(0)} = \frac{4}{7}$$

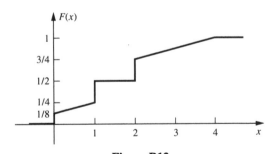

Figure P.12

13. **(a)** $f(x) \geq 0$

$$\int_{-\infty}^{\infty} f(x)\, dx = \int_{-1}^{1} \frac{1}{4}\, dx + \int_{1}^{\infty} \frac{1}{2x^2}\, dx = \frac{1}{2} + \frac{1}{2} = 1$$

(b) $\displaystyle\int_0^1 \frac{1}{4}\, dx + \int_1^2 \frac{1}{2x^2}\, dx = \frac{1}{4} + \frac{1}{4} = \frac{1}{2}$

(c) $F(x) = \int_{-\infty}^{x} f(x)\, dx$

If $x \leq -1$, then $F(x) = 0$.

If $-1 \leq x \leq 1$, then $F(x) = \int_{-1}^{x} \frac{1}{4}\, dx = \frac{1}{4}(x+1)$.

If $x \geq 1$, then $F(x) = \int_{-1}^{1} \frac{1}{4}\, dx + \int_{1}^{x} \frac{1}{2x^2}\, dx = 1 - \frac{1}{2x}$.

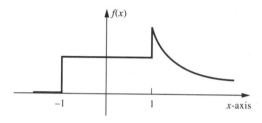

Figure P.13

14. Let $Y = X^3$. Then
$$F_Y(y) = P(Y \leq y) = P(X^3 \leq y) = P(X \leq \sqrt[3]{y}) = F_X(\sqrt[3]{y})$$

15. **(a)** Let X be the lifetime of an extraphasor. Then X has an exponential distribution with $\lambda = 1/10$, and

P(lasts at least another 5 round trips|has lasted at least 5 RT already)

$\quad = P(X \geq 5) \quad$ (by the memoryless property of exponential)

$\quad = \int_{5}^{\infty} \frac{1}{10} e^{-x/10}\, dx = e^{-1/2}$

(b) Breakdowns arrive independently (because the age of an extraphasor has nothing to do with when it dies) at the rate of $1/10$ per round trip. So the number of breakdowns in *two* round trips is Poisson with $\lambda = 2/10$ and

$$P(3 \text{ breakdowns in 2 trips}) = \frac{e^{-.2}(.2)^3}{3!}$$

(c) The number of breakdowns in 1 round trip is Poisson with $\lambda = 1/10$. So

P(1 breakdown in each of two trips)

$$= [P(1 \text{ breakdown in one trip})]^2$$

$$= \left[\frac{e^{-.1} \times .1}{1!} \right]^2$$

16. $-x^2 - 3x = -\left(x + \frac{3}{2}\right)^2 + \frac{9}{4}$, so

$$f(x) = Ke^{-(x+3/2)^2+9/4} = Ke^{9/4}\,e^{-(x+3/2)^2}$$

This is the normal density with

$$\mu = -\frac{3}{2}, \quad 2\sigma^2 = 1, \quad \sigma^2 = \frac{1}{2}, \quad Ke^{9/4} = \frac{1}{\sigma\sqrt{2\pi}}, \quad K = \frac{e^{-9/4}}{\sqrt{\pi}}$$

17. Think of X as the waiting time for the first red arrival, where red particles arrive independently at the rate of λ per hour.

Think of Y as the waiting time for the first blue arrival, where blue particles arrive independently at the rate of λ per hour.

Since X and Y are independent, the reds arrive independently of the blues.

So we have a mixed stream of particles arriving independently at the rate of 2λ per hour. The min of X and Y can be thought of as the waiting time for the first of the mixed streams to arrive, so the min is exponential with parameter 2λ.

18. (a) Z is the minimum of X and Y.

Method 1.

$$\begin{aligned} F_{\min}(z) &= P(\min \leq z) = 1 - P(X \geq z \text{ and } Y \geq z) \\ &= 1 - P(X \geq z)P(Y \geq z) \\ &= 1 - \int_{x=z}^{\infty} e^{-x}\,dx \int_{y=z}^{\infty} e^{-y}\,dy \\ &= 1 - e^{-z}e^{-z} = 1 - e^{-2z} \text{ for } z \geq 0 \end{aligned}$$

$$f_{\min}(z) = 2e^{-2z} \quad \text{for } z \geq 0$$

So the min has an exponential distribution with $\lambda = 2$.

Method 2.   By problem 17, the min of two ind exponentials with common parameter λ is exponential with parameter 2λ.

Method 3.

$$F_{\min}(z) = P(\min \leq z)$$
$$= P(X \leq z \text{ or } Y \leq z)$$
$$= P(X \leq z) + P(Y \leq z) - P(X \leq z \text{ and } Y \leq z)$$
$$= P(X \leq z) + P(Y \leq z) - P(X \leq z)P(Y \leq z)$$
$$= 2 \int_{x=0}^{z} e^{-x} \, dx - \left[\int_{x=0}^{z} e^{-x} \, dx \right]^2$$

Method 4. (see diagram)

The joint density of X and Y is $f(x,y) = e^{-x}e^{-y}$ for $x \geq 0, y \geq 0$.

$$F_{\min}(z) = P(\min \leq z) = \int_{\text{fav}} f(x,y) \, dA$$

$$= 1 - \int_{\text{unfav}} f(x,y) \, dA$$

$$= 1 - \int_{y=z}^{\infty} \int_{x=z}^{\infty} e^{-x}e^{-y} \, dx \, dy \quad \text{for } z \geq 0$$

(b) (see diagram) $Z = X + Y$

$$F(z) = \int_{\text{fav}} f(x,y) \, dA = \int_{y=0}^{z} \int_{x=0}^{z-y} e^{-x}e^{-y} \, dx \, dy$$

$$= 1 - ze^{-z} - e^{-z} \text{ for } z \geq 0$$

$$f(z) = ze^{-z} \text{ for } z \geq 0$$

(c) Z is the max of X and Y.

Method 1.

$$F_{\max}(z) = P(\max \leq z) = P(X \leq z \text{ and } Y \leq z)$$
$$= P(X \leq z)P(Y \leq z)$$
$$= \left[\int_{0}^{z} e^{-x} \, dx \right]^2 = (1 - e^{-z})^2 \text{ for } z \geq 0$$

Method 2. (see diagram)

$$F_{\max}(z) = P(\max \le z) = \int_{\text{fav}} f(x,y)\, dA$$

$$= \int_{y=0}^{z} \int_{x=0}^{z} e^{-x} e^{-y}\, dx\, dy \text{ for } z \ge 0$$

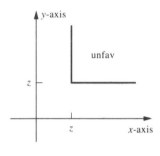

method 4

Figure P.18a

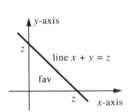

Figure P.18b

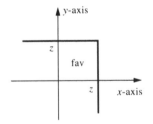

Figure P.18c

Solutions Section 6-1

1. **(a)** Binomial with $n = 30$, $p = 1/3$.
   (b) Can't do by inspection.
   (c) Poisson with parameter $\lambda = 6$.
   (d) Poisson with parameter $\lambda = 7$.
   (e) Gamma with $n = 2$, $\lambda = 3$.
   (f) Can't do by inspection.
   (g) Normal with $\mu = 5$, $\sigma^2 = 11$.

2. Toss a coin where $P(H) = p$. Then X_1 can be thought of as the number of tosses to get the first head, X_2 can be thought of as the number of tosses to get the second head after the first head, and so on. So

$X_1 + \cdots + X_{10}$ is the number of tosses to get the 10th head; it has a negative binomial distribution with parameters p, $r = 10$. In other words,

$$P(X_1 + \cdots + X_{10} = k) = \binom{k-1}{9} p^9 (1-p)^{k-10} p$$

for $k = 10, 11, 12, \ldots$ (see Section 2.2)

3. Imagine particles arriving independently at the average rate of 2 per minute. Then X can be thought of as the waiting time for the 5th arrival, Y can be thought of as the waiting time from the 5th arrival to the 12th arrival, and Z can be thought of as the waiting time from the 12th arrival to the 20th. So $X + Y + Z$ is the waiting time for the 20th arrival; it has a gamma distribution with parameters $n = 20$, $\lambda = 2$.

4. The machine's lifetime is the sum of the lifetimes of the 10 components. The sum of iid exponentials is gamma, so the machine's lifetime has a gamma distribution with parameters $n = 10$, λ.

Solutions Section 6-2

1. (a) $F_{5\text{th}}(x) = P(\text{5th largest is } \leq x)$

$= P(\text{at most 4 are } \geq x)$

$= P(\text{13 are } \leq x, 4 \text{ are } \geq x)$

$\qquad + \cdots + P(\text{17 are } \leq x, \text{ none is } \geq x)$

$= \binom{17}{13} [F(x)]^{13} [1 - F(x)]^4 + \binom{17}{14} [F(x)]^{14} [1 - F(x)]^3$

$\qquad + \binom{17}{15} [F(x)]^{15} [1 - F(x)]^2$

$\qquad + 17[F(x)]^{16} [1 - F(x)] + [F(x)]^{17}$

$f_{5\text{th}}(x)\, dx = P(1 \text{ is } \approx x, 4 \text{ are } \geq x, 12 \text{ are } \leq x)$

$= \dfrac{17!}{4!\, 12!} f(x)\, dx\, [1 - F(x)]^4 [F(x)]^{12}$

$f_{5\text{th}}(x) = \dfrac{17!}{4!\, 12!} f(x) [1 - F(x)]^4 [F(x)]^{12}$

(b) $F_{10\text{th}}(x) = P(\text{10th largest is } \leq x) = P(\text{at most 9 are } \geq x)$

$\qquad = P(8 \text{ are } \leq x, 9 \text{ are } \geq x) + P(9 \text{ are } \leq x, 8 \text{ are } \geq x)$

$\qquad\qquad\qquad\qquad + \cdots + P(17 \text{ are } \leq x)$

$\qquad = \begin{pmatrix} 17 \\ 8 \end{pmatrix} [F(x)]^8 [1 - F(x)]^9 + \begin{pmatrix} 17 \\ 9 \end{pmatrix} [F(x)]^9 [1 - F(x)]^8$

$\qquad\qquad\qquad\qquad + \cdots + [F(x)]^{17}$

$f_{10\text{th}}(x)\, dx = P(1 \text{ is } \approx x, 9 \text{ are } \geq x, 7 \text{ are } \leq x)$

$f_{10\text{th}}(x) = \dfrac{17!}{9!\,7!}\, f(x)\, [1 - F(x)]^9\, [F(x)]^7$

(c) $F_{13\text{th}}(x) = P(\text{13th largest is } \leq x) = P(\text{at most 12 are } \geq x)$

$\qquad = P(5 \text{ are } \leq x, 12 \text{ are } \geq x)$

$\qquad\qquad + P(6 \text{ are } \leq x, 11 \text{ are } \geq x) + \cdots + P(\text{ all are } \leq x)$

$\qquad = 1 - P(\text{none or 1 or 2 or 3 or 4 are } \leq x)$

$\qquad = 1 - [1 - F(x)]^{17} - 17\, F(x)[1 - F(x)]^{16}$

$\qquad\qquad - \begin{pmatrix} 17 \\ 2 \end{pmatrix} [F(x)]^2 [1 - F(x)]^{15}$

$\qquad\qquad - \begin{pmatrix} 17 \\ 3 \end{pmatrix} [F(x)]^3 [1 - F(x)]^{14}$

$\qquad\qquad - \begin{pmatrix} 17 \\ 4 \end{pmatrix} [F(x)]^4 [1 - F(x)]^{13}$

$f_{13\text{th}}(x)\, dx = P(\text{13th is } \approx x)$

$\qquad\qquad = P(1 \text{ is } \approx x, 12 \text{ are } \geq x, 4 \text{ are } \leq x)$

$\qquad\qquad = \dfrac{17!}{12!\,4!\,1!} f(x)\, dx\, [F(x)]^4 [1 - F(x)]^{12}$

$f_{13\text{th}}(x) \qquad = \dfrac{17!}{12!\,4!}\, f(x)\, [F(x)]^4 [1 - F(x)]^{12}$

(d) $F_{\min}(x) = P(\min \leq x) = P(\text{at least 1 is } \leq x)$

$\qquad\qquad = 1 - P(\text{all are } \geq x)$

$\qquad\qquad = 1 - [1 - F(x)]^{17}$

$$f_{\min}(x)\,dx = P(1 \text{ is} \approx x, \text{ others are } \geq x)$$

$$= \frac{17!}{16!\,1!}\,f(x)\,dx\,[1 - F(x)]^{16}$$

$$f_{\min}(x) = 17\,f(x)\,[1 - F(x)]^{16}$$

(easy to see this is the deriv of $F_{\min}(x)$)

2. (a) $F_{4\text{th}}(x) = P(\text{4th largest is } \leq x) = P(\text{at most 3 are } \geq x)$

$$= P(3 \text{ are } \geq x, 29 \text{ are } \leq x)$$

$$+ P(2 \text{ are } \geq x, 30 \text{ are } \leq x) + \cdots + P(32 \text{ are } \leq x)$$

$$= \binom{32}{29}[F(x)]^{29}\,[1 - F(x)]^3 + \binom{32}{30}[F(x)]^{30}\,[1 - F(x)]^2$$

$$+ 32[F(x)]^{31}\,[1 - F(x)] + [F(x)]^{32}$$

$$f_{4\text{th}}(x)\,dx = P(\text{4th largest is } \approx x)$$

$$= P(1 \text{ is } \approx x, 3 \text{ are } \geq x, 28 \text{ are } \leq x)$$

$$f_{4\text{th}}(x) = \frac{32!}{3!\,28!}f(x)\,[1 - F(x)]^3\,[F(x)]^{28}$$

$$P(\text{4th largest is between 7 and 8}) = F_{4\text{th}}(8) - F_{4\text{th}}(7)$$

(b) $F_{\max}(x) = P(\max \leq x) = P(\text{all are } \leq x) = [F(x)]^{32}$

$$f_{\max}(x)\,dx = P(\max \approx x) = P(1 \text{ is } \approx x, \text{ others } \leq x)$$

$$= \frac{32!}{31!\,1!}f(x)\,dx\,[F(x)]^{31}$$

$$f_{\max}(x) = 32f(x)\,[F(x)]^{31} \quad (\text{the derivative of } F_{\max}(x))$$

(c) $F_{27\text{th}}(x) = P(\text{27th largest is } \leq x) = P(\text{at most 26 are } \geq x)$

$$= P(26 \geq x, 6 \leq x) + P(25 \geq x, 7 \leq x)$$

$$+ \cdots + P(\text{all are } \leq x)$$

$$= 1 - P(27 \geq x, 5 \leq x) - P(28 \geq x, 4 \leq x)$$

$$- \cdots - P(\text{ all are } \geq x)$$

$$= 1 - \binom{32}{5} [F(x)]^5 [1 - F(x)]^{27}$$

$$- \binom{32}{4} [F(x)]^4 [1 - F(x)]^{28} - \cdots - [1 - F(x)]^{32}$$

$f_{27\text{th}}(x)\, dx = P(27\text{th is} \approx x)$

$$= P(1 \text{ is} \approx x, 26 \text{ are} \geq x, 5 \text{ are} \leq x)$$

$f_{27\text{th}}(x) \quad = \dfrac{32}{26!\, 5!}\, f(x) [1 - F(x)]^{26} [F(x)]^5$

(d) (same as 30th largest)

$F_{3\text{rd smallest}}(x) = P(3\text{rd smallest is} \leq x) = P(\text{at least 3 are} \leq x)$

$$= 1 - P(\text{ none is} \leq x)$$

$$- P(1 \text{ is} \leq x) - P(2 \text{ are} \leq x)$$

$$= 1 - [1 - F(x)]^{32} - 32[F(x)] [1 - F(x)]^{31}$$

$$- \binom{32}{2} [F(x)]^2 [1 - F(x)]^{30}$$

$f_{3\text{rd smallest}}(x)\, dx = P(3\text{rd smallest is} \approx x)$

$$= P(1 \text{ is} \approx x, 2 \text{ are} \leq x, \text{ others are} \geq x)$$

$f_{3\text{rd smallest}}(x) \quad = \dfrac{32!}{1!\, 2!\, 29!}\, f(x) [F(x)]^2 [1 - F(x)]^{29}$

3. (a) $P(\text{next to smallest is} \geq 4) = P(\text{at least 8 are} \geq 4)$

$$= P(8 \text{ are} \geq 4) + P(9 \text{ are} \geq 4)$$

Think of each X_i as the result of a coin toss where

$$P(\text{success}) = P(X_i \geq 4) = \frac{\text{fav length}}{\text{total}} = \frac{8}{11}$$

Then

$$P(\text{at least 8 are} \geq 4) = 9 \left(\frac{8}{11}\right)^8 \cdot \frac{3}{11} + \left(\frac{8}{11}\right)^9$$

(b) The 7th largest is between 1 and 12, so

$$F(x) = \begin{cases} 0 & \text{if } x \leq 1 \\ 1 & \text{if } x \geq 12 \end{cases}$$

Now consider $1 \leq x \leq 12$.

$$F_{7\text{th}}(x) = P(\text{7th largest is } \leq x) = P(\text{at most 6 are } \geq x)$$

$$= P(6 \geq x, 3 \leq x) + P(5 \geq x, 4 \leq x) + \cdots + P(\text{all } \leq x)$$

$$= 1 - P(\text{none } \leq x) - P(\text{1 is } \leq x) - P(\text{2 are } \leq x)$$

Think of each X_i as the result of a coin toss where

(*)
$$P(\text{result is } \leq x) = P(X_i \leq x) = \frac{\text{fav length}}{\text{total}} = \frac{x-1}{11}$$

$$P(\text{result is } \geq x) = P(X_i \geq x) = \frac{\text{fav length}}{\text{total}} = \frac{12-x}{11}$$

Use the binomial distribution to get

$$F_{7\text{th}}(x) = 1 - \left(\frac{12-x}{11}\right)^9 - 9\left(\frac{x-1}{11}\right)\left(\frac{12-x}{11}\right)^8$$

$$- \binom{9}{2}\left(\frac{x-1}{11}\right)^2\left(\frac{12-x}{11}\right)^7 \qquad \text{for } 1 \leq x \leq 12$$

(c) Seventh largest is between 1 and 12, so $f_{7\text{th}}(x) = 0$ for $x \leq 1$ and for $x \geq 12$.

Consider $1 \leq x \leq 12$.

$$f_{7\text{th}}(x)\, dx = P(\text{7th largest is } \approx x)$$

$$= P(\text{1 is } \approx x, 2 \text{ are } \leq x, 6 \text{ are } \geq x)$$

Think of each X_i as the result of tossing a 3-sided die where in addition to the two possibilities above in (*) we have the third possibility

$$P(\text{result } \approx x) = P(X_i \approx x) = f_{X_i}(x)\, dx = \frac{1}{11}\, dx$$

Then

$$f_{7\text{th}}(x)\, dx = \frac{9!}{1!\, 6!\, 2!}\, \frac{1}{11}\, dx \left(\frac{x-1}{11}\right)^2\left(\frac{12-x}{11}\right)^6 \qquad \text{(multinomial)}$$

Cancel dx's to get

$$f_{7\text{th}}(x) = \frac{9!}{1!\, 6!\, 2!}\, \frac{1}{11} \left(\frac{x-1}{11}\right)^2\left(\frac{12-x}{11}\right)^6 \qquad \text{for } 1 \leq x \leq 12$$

4. (a) $P(\max \leq 9) = P(\text{all} \leq 9) = [F(9)]^n$

(b) $F_{\max}(x) = P(\max \leq x) = P(\text{all} \leq x) = [F(x)]^n$

(c) (i) $f_{\max}(x) = F'_{\max}(x) = n[F(x)]^{n-1}F'(x) = n[F(x)]^{n-1}f(x)$

(ii) $f_{\max}(x)\,dx = P(\max \approx x) = P(1 \text{ is } \approx x, \text{ others } \leq x)$

$$= \frac{n!}{1!\,(n-1)!}\,f(x)\,dx\,[F(x)]^{n-1} \quad \text{(multinomial)}$$

So

$$f_{\max}(x) = nf(x)[F(x)]^{n-1}$$

(d) $P(\min \geq 2) = P(\text{each } X_i \text{ is } \geq 2) = [1 - F(2)]^n$

(e) $F_{\min}(x) = P(\min \leq x) = P(\text{at least 1 of the } X_i\text{'s is } \leq x)$

$$= 1 - P(\text{all } X_i\text{'s are } \geq x) = 1 - [1 - F(x)]^n$$

(f) *Method 1.*

$$f_{\min}(x) = F'_{\min}(x) = -n[1 - F(x)]^{n-1} \cdot -F'(x)$$
$$= n[1 - F(x)]^{n-1}\,f(x)$$

Method 2.

$$f_{\min}(x)\,dx = P(\min \approx x)$$
$$= P(1 \text{ of the } X_i\text{'s is } \approx x, \text{ others are } \geq x)$$
$$= \frac{n!}{1!\,(n-1)!}\,f(x)\,dx[1 - F(x)]^{n-1} \quad \text{(multinomial)}$$

So

$$f_{\min}(x) = nf(x)\,[1 - F(x)]^{n-1}$$

5. (a) $P(M \text{ lasts} \leq 5 \text{ hours}) = P(\text{each } X_i \text{ is} \leq 5) = [F(5)]^7$

(b) $P(M \text{ lasts} \leq 5 \text{ hours}) = P(\text{at least 1 of the } X_i\text{'s is} \leq 5)$

$$= 1 - P(\text{all are} \geq 5) = 1 - [1 - F(5)]^7$$

(c) $P(M \text{ lasts} \leq 5 \text{ hours}) = P(6\ X_i\text{'s are} \leq 5) + P(\text{all } X_i\text{'s are} \leq 5)$

$$= [F(5)]^7 + \binom{7}{6}[F(5)]^6\,[1 - F(5)]$$

6. (a) The machine's lifetime is the 4th largest of the X_i's, so

$$F_{\text{machine}}(x) = F_{4\text{th}}(x)$$
$$= P(\text{at most 3 of the } X_i\text{'s are } \geq x)$$
$$= P(3 \geq x, 3 \leq x) + P(2 \geq x, 4 \leq x)$$
$$+ P(1 \geq x, 5 \leq x) + P(\text{all } \leq x)$$

Since X_i is exponential, if $x \geq 0$, then $P(X_i \leq x) = \int_0^x \lambda e^{-\lambda x} \, dx = 1 - e^{-\lambda x}$. So

$$F_{4\text{th}}(x) = \binom{6}{3}[1 - e^{-\lambda x}]^3 \, [e^{-\lambda x}]^3 + \binom{6}{4}[1 - e^{-\lambda x}]^4 \, [e^{-\lambda x}]^2$$

$$+ 6 \, [1 - e^{-\lambda x}]^5 \, e^{-\lambda x} + [1 - e^{-\lambda x}]^6 \quad \text{for } x \geq 0$$

(b) $f_{4\text{th}}(x) \, dx = P(\text{4th largest is} \approx x)$

$$= P(1 \text{ of the } X_i\text{'s is} \approx x, 2 \text{ are } \leq x, 3 \text{ are } \geq x)$$

For $x \geq 0$, the density of each X_i is $\lambda e^{-\lambda x}$ and $P(X_i \approx x) = \lambda e^{-\lambda x} \, dx$, so

$$P(1 \text{ is} \approx x, 2 \text{ are } \leq x, 3 \text{ are } \geq x)$$

$$= \frac{6!}{1! \, 2! \, 3!} \, \lambda e^{-\lambda x} \, dx [1 - e^{-\lambda x}]^2 \, [e^{-\lambda x}]^3 \quad \text{(multinomial)}$$

Finally,

$$f_{4\text{th}}(x) = \frac{6!}{1! \, 2! \, 3!} \, \lambda e^{-4\lambda x} \, [1 - e^{-\lambda x}]^2 \quad \text{for } x \geq 0$$

7. It's as if there are 7 independent trials of an experiment where each trial has 8 possible outcomes (toss an 8-sided die 7 times).

$P(\text{outcome is } [0,1]) = P(0 \leq X_i \leq 1) = F(1) - F(0)$
$P(\text{outcome is } [1,2]) = P(1 \leq X_i \leq 2) = F(2) - F(1)$

$\vdots$

$P(\text{outcome is } [6,7]) = P(6 \leq X_i \leq 7) = F(7) - F(6)$
$P(\text{outcome is } [7, \infty) = P(X_i \geq 7) = 1 - F(7)$

Use the multinomial formula to get

$P(1 \text{ each of } [0, 1], [1, 2], \ldots, [6, 7])$

$$= \frac{7!}{1! \ldots 1!} \, [F(1) - F(0)] \, [F(2) - F(1)] \ldots [F(7) - F(6)]$$

But

$$F(k+1) - F(k) = \frac{k+1}{k+2} - \frac{k}{k+1} = \frac{1}{(k+1)(k+2)}$$

So the answer simplifies to

$$7! \, \frac{1}{1 \cdot 2} \, \frac{1}{2 \cdot 3} \cdots \frac{1}{7 \cdot 8} = 7! \, \frac{1}{7! \, 8!} = \frac{1}{8!}$$

8. The X_i's can be thought of as the results of 7 Bernoulli trials with parameter p where

$$p = P(X_i \geq 5) = \int_5^\infty f(x) \, dx$$

The expected number of successes is $np = 7 \int_5^\infty f(x) \, dx$.

9. (a) $P(\max \geq 9, \text{next} \leq 5) = P(1 \text{ of the } X_i\text{'s is} \geq 9, \text{others are} \leq 5)$.
It's like tossing a 3-sided die 7 times where the 3 outcomes are $\geq 9, \leq 5$, and between 5 and 9. Answer is

$$\frac{7!}{1! \, 6!} \, [1 - F(9)] \, [F(5)]^6$$

(b) $P(\max \leq 9, \text{ next } \leq 5)$

$$= P(\text{all } \leq 5) + P(6 \text{ are } \leq 5, 1 \text{ is between 5 and 9})$$

$$= [F(5)]^7 + \frac{7!}{1! \, 6!} \, [F(5)]^6 \, [F(9) - F(5)]$$

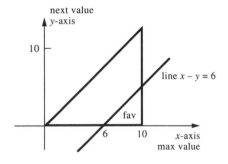

Figure P.9

10. First we'll find the joint density of the functions $\max(x)$ and $\text{next}(y)$. If x is a possible max value and y is a possible next value, then we must have $y \leq x$ so the universe is $0 \leq y \leq x \leq 10$ and

$f(x, y) \, dx \, dy$

$\quad = P(\max \approx x, \text{ next } \approx y)$

$\quad = P(1 \text{ of the } X_i\text{'s is } \approx x, 1 \text{ is } \approx y, \text{ others are between } 0 \text{ and } y)$

Think of the 8 X_i's as the results of tossing a 5-sided die 8 times, where

$$P(x \leq X_i \leq 10) = \frac{\text{fav}}{\text{total}} = \frac{10 - x}{10}$$

$$P(X_i \approx x) = f(x) \, dx = \frac{1}{10} \, dx \quad (f \text{ is the common density of the } X_i\text{'s})$$

$$P(y \leq X_i \leq x) = \frac{\text{fav}}{\text{total}} = \frac{x - y}{10}$$

$$P(X_i \approx y) = f(y) \, dy = \frac{1}{10} \, dy$$

$$P(0 \leq X_i \leq y) = \frac{\text{fav}}{\text{total}} = \frac{y}{10}$$

Then

$$P(1 \text{ is } \approx x, 1 \text{ is } \approx y, \text{ others between } 0 \text{ and } y)$$

$$= \frac{8!}{1! \, 1! \, 6!} \frac{1}{10} \, dx \, \frac{1}{10} \, dy \left(\frac{y}{10} \right)^6$$

So

$$f(x, y) = \frac{56}{10^8} y^6 \text{ for } 0 \leq y \leq x \leq 10$$

$$P(\max - \text{next} \geq 6) = \int_{\text{fav}} f(x, y) \, dA$$

$$= \int_{x=6}^{10} \int_{y=0}^{y=x-6} \frac{56}{10^8} y^6 \, dy \, dx$$

$$= \left(\frac{4}{10} \right)^8$$

11. **(a)** You can't think of $X_1, \ldots, X_n$ as the results of n coin (or die) tosses. So you can't use the binomial (or the multinomial) distribution to compute probabilities.

   (b) You *can* think of $X_1, \ldots, X_n$ as the results of coin (or die) tosses, but however success might be defined, its probability would be different for each coin. So you wouldn't be tossing the *same* coin n times, so you couldn't use the binomial distribution to compute probs.

12. (a) If X and Y are independent, then $f(x,y) = f_X(x)f_Y(y)$. So

$$f(x_1, \ldots, x_7) = f(x_1)f(x_2)\ldots f(x_7)$$

where f is the common density of the X_i's.

(b) The universe is the set of points $(x_1, \ldots, x_7)$ where $x_1 \geq x_2 \geq x_3 \geq \ldots \geq x_7$ (the universe is restricted because the value of largest must be $\geq$ value of 2nd largest, etc.). Then

$$g(x_1, \ldots, x_7)\, dx_1 dx_2 \ldots dx_7$$
$$= P(\text{largest} \approx x_1, \text{2nd largest} \approx x_2, \ldots, \text{7th largest} \approx x_7)$$
$$= P(1 \text{ of the } X_i\text{'s is } \approx x_1, 1 \text{ is } \approx x_2, \ldots, 1 \text{ is } \approx x_7)$$

Think of tossing an 8-sided die 7 times where

$$P(\text{result} \approx x_1) = P(X_i \approx x_1) = f(x_1)\, dx_1$$
$$P(\text{result} \approx x_2) = P(X_i \approx x_2) = f(x_2)\, dx_2$$
$$\vdots$$
$$P(\text{result} \approx x_7) = P(X_i \approx x_7) = f(x_7)\, dx_7$$
$$P(\text{other}) = \text{who cares}$$

Then

$$g(x_1, \ldots, x_7)\, dx_1 dx_2 \ldots dx_7 = \frac{7!}{1! \ldots 1!} f(x_1)\, dx_1 f(x_2)\, dx_2 \ldots f(x_7)\, dx_7$$

$$g(x_1, \ldots, x_7) = 7!\, f(x_1)f(x_2)\ldots f(x_7) \text{ for } x_1 \geq x_2 \geq x_3 \ldots \geq x_7$$

Solutions Section 7-1

1. $EX = \displaystyle\int_{x=0}^{\infty} x \cdot \frac{1}{4}xe^{-x/2}\, dx = \frac{1}{4}\int_0^{\infty} x^2 e^{-x/2}\, dx = \frac{1}{4}\frac{2!}{(1/2)^3} = 4$

2. $EX = \displaystyle\int_{-1}^{1} x(1 - |x|)\, dx = \int_{-1}^{0} x(1 - -x)\, dx + \int_0^1 x(1 - x)\, dx$

$$= \left(\frac{x^2}{2} + \frac{x^3}{3}\right)\Bigg|_{-1}^{0} + \left(\frac{x^2}{2} - \frac{x^3}{3}\right)\Bigg|_0^1 = 0$$

(By inspection, $EX = 0$ because the density is symmetric w.r.t. the y-axis, so $xf(x)$ is an odd function.)

3. $E(X^3) = \displaystyle\int_{x=-\infty}^{\infty} x^3 f(x)\, dx = \int_{x=0}^{\infty} x^3 \frac{1}{4}xe^{-x/2}\, dx = \frac{1}{4}\frac{4!}{(1/2)^5}$

4. $E(|X|) = \int_{-\infty}^{\infty} |x| \frac{1}{\sqrt{2\pi}} e^{-x^2/2} \, dx$

$$= \frac{1}{\sqrt{2\pi}} \int_{-\infty}^{0} -xe^{-x^2/2} \, dx + \frac{1}{\sqrt{2\pi}} \int_{0}^{\infty} xe^{-x^2/2} \, dx$$

$$= \frac{2}{\sqrt{2\pi}} \int_{0}^{\infty} xe^{-x^2/2} \, dx \text{ (by symmetry)} = \frac{2}{\sqrt{2\pi}}$$

5. $E\left(\sin \frac{\pi X}{2}\right) = \sum_{k=0}^{4} \sin \frac{k\pi}{2} \, P(X = k)$

$$= \sum_{k=0}^{4} \sin \frac{k\pi}{2} \binom{4}{k} p^k q^{n-k} = 4pq^3 - 4p^3 q$$

6. *Method 1.* (law of the uncon stat)

$$E(\text{Int } X) = \int_{x=-\infty}^{\infty} \text{Int } x \, f(x) \, dx$$

$$= \int_{0}^{1} 0 \, e^{-x} \, dx + \int_{1}^{2} 1 \, e^{-x} \, dx + \int_{2}^{3} 2 e^{-x} \, dx + \cdots$$

$$= \frac{1}{e} + \frac{1}{e^2} + \frac{1}{e^3} + \cdots \left(\text{geometric series with } r = \frac{1}{e}, a = \frac{1}{e}\right)$$

$$= \frac{1/e}{1 - 1/e} = \frac{1}{e - 1}$$

Method 2. (directly)   Let $Y = \text{Int } X$. Then Y is *discrete* and

$$EY = 1P(Y = 1) + 2P(Y = 2) + 3P(Y = 3) + \cdots$$

$$= 1P(1 \le X < 2) + 2P(2 \le X < 3) + 3P(3 \le X < 4) + \cdots$$

$$= 1 \int_{0}^{1} e^{-x} \, dx + 2 \int_{2}^{3} e^{-x} \, dx + 3 \int_{3}^{4} e^{-x} \, dx + \cdots$$

(same as method 1 now)

7. (a) $E(X - Y) = EX - EY = 1/\lambda - 1/\lambda = 0$

(b) $f(x, y) = f(x) f(y) = e^{-x} e^{-y}$ for $x \ge 0, y \ge 0$ because X and Y are ind

$$|x - y| = \begin{cases} x - y & \text{if } x \ge y \\ y - x & \text{if } y \ge x \end{cases}$$

$$E|X - Y| = \int_{\text{plane}} |x - y| e^{-x} e^{-y} \, dA$$

$$= \int_I (x - y) e^{-x} e^{-y} \, dA + \int_{II} (y - x) e^{-x} e^{-y} \, dA$$

$$= 2 \int_I (x - y) e^{-x} e^{-y} \, dA \quad \text{(by symmetry)}$$

$$= 2 \int_{x=0}^{\infty} \int_{y=0}^{x} x e^{-x} e^{-y} \, dy \, dx$$

$$- 2 \int_{y=0}^{\infty} \int_{x=y}^{\infty} y e^{-x} e^{-y} \, dx \, dy$$

$$= 1 \text{ eventually}$$

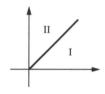

Figure P.7b

8. **(a)** Use a coord system with origin at the center of the circle. Then

$$f(x,y) = \frac{1}{\text{total area}} = \frac{1}{\pi a^2} \text{ for } (x,y) \text{ in the circle}$$

$$E\left(\sqrt{X^2 + Y^2}\right) = \int_{\text{circle}} \sqrt{x^2 + y^2} \, \frac{1}{\pi a^2} \, dA$$

$$= \frac{1}{\pi a^2} \int_{\theta=0}^{2\pi} \int_{r=0}^{a} r \, r \, dr \, d\theta = \frac{2}{3} a$$

(b) Let R be the distance to the center.

$$F_R(r) = P(R \le r) = \begin{cases} 0 & \text{if } r \le 0 \\ \dfrac{\text{fav}}{\text{total}} = \dfrac{\pi r^2}{\pi a^2} & \text{if } 0 \le r \le a \\ 1 & \text{if } r \ge a \end{cases}$$

$$f_R(r) = F_R'(r) = 2r/a^2 \text{ for } 0 \le r \le a$$

$$ER = \int_{r=0}^{a} r \cdot \frac{2r}{a^2} \, dr = \frac{2}{a^2} \int_{r=0}^{a} \cdot r^2 \, dr = \frac{2a}{3}$$

9. $f(x, y) = \dfrac{1}{\text{area}} = \dfrac{1}{\frac{1}{2}\pi R^2}$ for (x, y) in the semicircle

$$EY = \int_{\text{semi}} yf(x, y)\, dA = \int_{\theta=0}^{\pi} \int_{r=0}^{R} r\sin\theta \, \frac{2}{\pi R^2} \, r\, dr\, d\theta = \frac{4R}{3\pi}$$

10. $Y = g(X)$ where $g(X) = \begin{cases} X^2 & \text{if } X \le 6 \\ 12 & \text{if } X > 6 \end{cases}$

$$EY = \int_{x=-\infty}^{\infty} g(x)\, f(x)\, dx = \int_{x=0}^{6} x^2 \, \frac{1}{10} \, dx + \int_{6}^{10} 12 \cdot \frac{1}{10} \, dx = 12$$

11. (a) X and Y are not ind since $X^2 + Y^2 = 1$.

     Θ has density $f(\theta) = 1/2\pi$ for $0 \le \theta \le 2\pi$.

$$EX = E(\cos\Theta) = \int_{\theta=0}^{2\pi} \cos\theta \cdot \frac{1}{2\pi} \, d\theta = 0 \text{ and similarly } EY = 0.$$

$$E(XY) = E(\cos\Theta\sin\Theta) = \int_{\theta=0}^{2\pi} \cos\theta\sin\theta \frac{1}{2\pi} d\theta$$

$$= \frac{1}{2\pi} \frac{1}{2} \sin^2\theta \Big|_{\theta=0}^{2\pi} = 0$$

     So $(EX)(EY)$ happens to equal $E(XY)$ (both are 0).

(b) $EX = 1$. X has density $f(x) = 1/2$ for $0 \le x \le 2$.

$$EY = E(X^2) = \int_{x=0}^{2} x^2 \frac{1}{2} dx = \frac{4}{3}$$

$$E(XY) = E(X^3) = \int_{x=0}^{2} x^3 \frac{1}{2} dx = 2$$

     So $(EX)(EY) = 1 \cdot 4/3 = 4/3$ and $E(XY) = 2$. They aren't equal.

12. $\max(x, y) = \begin{cases} x & \text{in regions } C\&D \\ y & \text{in regions } A\&B \end{cases}$ (see the diagram)

$$E(\max) = \int_{\text{plane}} \max(x,y) f(x,y) dA = \int_A + \int_B + \int_C + \int_D$$

$$= \int_{y=1/2}^{1} \int_{x=1-y}^{y} \frac{1}{2} y \, dx \, dy + \int_{x=0}^{1/2} \int_{y=x}^{1-x} \frac{3}{2} y \, dy \, dx$$

$$+ \int_{y=0}^{1/2} \int_{x=y}^{1-y} \frac{3}{2} x \, dx \, dy + \int_{x=1/2}^{1} \int_{y=1-x}^{x} \frac{1}{2} x \, dy \, dx$$

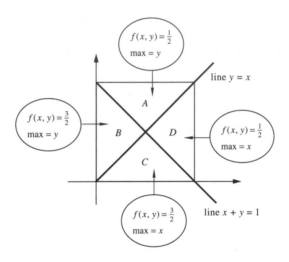

Figure P.12

13. *Method 1.*

$$\text{score} = \min(X,5) = \begin{cases} X & \text{if } X \le 5 \\ 5 & \text{if } X > 5 \end{cases}$$

where $f_X(x) = \dfrac{1}{8}$ for $0 \le x \le 8$

$$E(\text{score}) = \int_{x=0}^{5} x \cdot \frac{1}{8} \, dx + \int_{x=5}^{8} 5 \cdot \frac{1}{8} \, dx = \frac{55}{16}$$

Method 2.   Score is a mixed random variable. The pseudo-density has a chunk of probability of size 3/8 at $x = 5$ because

$$P(\text{score} = 5) = P(5 \le X \le 8) = \frac{3}{8}$$

We have

$$E(\text{score}) = \sum_x x \, P(\text{score} = x)$$

The sum is part ordinary sum (the $x = 5$ term) and part integral.

$$E(\text{score}) = 5P(\text{score} = 5) + \int_{x=0}^{5} x f_X(x)\, dx$$

$$= 5 \cdot \frac{3}{8} + \int_{x=0}^{5} x \cdot \frac{1}{8}\, dx = \frac{55}{16}$$

Method 3.   Use the theorem of total expectation.

$$E(\text{score}) = P(X \le 5)\, E(\text{score}|X \le 5) + P(X > 5)\, E(\text{score}|X > 5)$$

If $X \le 5$, then your score is uniform on [0,5] and $E(\text{score}|X \le 5) = 5/2$. So

$$E(\text{score}) = \frac{5}{8} \cdot \frac{5}{2} + \frac{3}{8} \cdot 5 = \frac{55}{16}$$

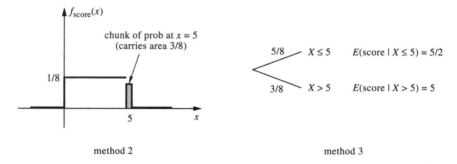

method 2   method 3

Figure P.13

14. (a) T   **(b)** F   **(c)** F   **(d)** T

15. Let Y be the replacement price. Then $Y = 2e^{.1T}$, where $f_T(t) = e^{-t}$, $t \ge 0$, and

$$EY = \int_{t=0}^{\infty} 2e^{.1t} f(t)\, dt = \int_{t=0}^{\infty} 2e^{.1t} e^{-t}\, dt = \int_{t=0}^{\infty} 2e^{-.9t}\, dt = \$2.22$$

Solutions Section 7-2

1. $EX = 1P(X = 1) + \cdots + 6P(X = 6) = \dfrac{1}{6}(1 + 2 + 3 + 4 + 5 + 6) = \dfrac{7}{2}$

$E(X^2) = 1^2 P(X = 1) + \cdots + 6^2 P(X = 6)$

(law of the unconscious stat)

$$= \frac{1}{6}(1^2 + \cdots + 6^2) = \frac{91}{6}$$

$$\text{Var } X = E(X^2) - (EX)^2 = \frac{91}{6} - \frac{49}{4} = \frac{35}{12}$$

2. $EX = \dfrac{a+b}{2}$

$$E(X^2) = \int_a^b x^2 f(x)\, dx = \int_a^b x^2 \frac{1}{b-a}\, dx = \frac{b^3 - a^3}{3} \cdot \frac{1}{b-a}$$

$$= \frac{1}{3}(b^2 + ab + a^2)$$

$$\text{Var } X = E(X^2) - (EX)^2 = \frac{(b-a)^2}{12}$$

3. (a) Let X be the number of boys.

$$EX = 1P(X=1) + 2P(X=2) + 3P(X=3)$$

$$= 1P(\text{BG}) + 2P(\text{BBG}) + 3P(\text{BBB})$$

$$= \frac{1}{4} + 2 \cdot \frac{1}{8} + 3 \cdot \frac{1}{8} = \frac{7}{8}$$

$$E(X^2) = 1^2 \cdot \frac{1}{4} + 2^2 \cdot \frac{1}{8} + 3^2 \cdot \frac{1}{8} = \frac{15}{8}$$

$$\text{Var } X = \frac{15}{8} - \frac{49}{64} = \frac{71}{64}$$

(b) Let Y be the number of girls. Then Y is either 0 or 1, and

$$EY = 1P(Y=1) = P(\text{G or BG or BBG}) = \frac{1}{2} + \frac{1}{4} + \frac{1}{8} = \frac{7}{8}$$

$$E(Y^2) = 1^2 P(Y=1) = \frac{7}{8}$$

$$\text{Var } Y = \frac{7}{8} - \frac{49}{64} = \frac{7}{64}$$

4. $f(x) = \begin{cases} 1 + x & \text{if } -1 \le x \le 0 \\ 1 - x & \text{if } 0 \le x \le 1 \end{cases}$

$$EX = \int_{-1}^1 x\, f(x)\, dx = \int_{-1}^0 x(1+x)\, dx + \int_0^1 x(1-x)\, dx = 0$$

(You can get $EX = 0$ by inspection since $f(x)$ is an even function and $xf(x)$ is an odd function.)

$$E(X^2) = \int_{-1}^{0} x^2(1+x)\,dx + \int_{0}^{1} x^2(1-x)\,dx = \frac{1}{6}$$

$$\text{Var } X = E(X^2) - (EX)^2 = \frac{1}{6}$$

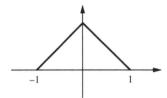

Figure P.4

5. $EY = \int_{x=1}^{2} x^3 \cdot \frac{1}{x^2}\,dx + \int_{2}^{\infty} 8 \cdot \frac{1}{x^2}\,dx = \frac{11}{2}$

$$EY^2 = \int_{1}^{2} x^6\,\frac{1}{x^2}\,dx + \int_{2}^{\infty} 64 \cdot \frac{1}{x^2}\,dx = \frac{191}{5}$$

$$\text{Var } Y = E(Y^2) - (EY)^2 = \frac{191}{5} - \frac{121}{4}$$

6. Let X have a Poisson distribution with parameter λ.

   (a) X almost has a binomial distribution with parameters n, p where $np = \lambda$. Guess that Var X is the limit of the binomial variance as $n \to \infty, p \to 0$, and np stays λ, so

$$\text{Var } X = \lim npq = \lim np(1-p) = \lim \lambda(1-p) = \lambda$$

   (b) We know that $EX = \lambda$ (Section 3.1).

$$E(X^2) = \sum_{k=0}^{\infty} k^2\,\frac{e^{-\lambda}\lambda^k}{k!}$$

$$= e^{-\lambda}\left[\lambda + \frac{2^2\lambda^2}{2!} + \frac{3^2\lambda^3}{3!} + \frac{4^2\lambda^4}{4!} + \cdots\right]$$

$$= e^{-\lambda}\left[\lambda + 2\lambda^2 + \frac{3\lambda^3}{2!} + \frac{4\lambda^4}{3!} + \cdots\right]$$

We know that

$$1 + \lambda + \frac{\lambda^2}{2!} + \frac{\lambda^3}{3!} + \cdots = e^{\lambda}$$

So

$$\lambda + \lambda^2 + \frac{\lambda^3}{2!} + \frac{\lambda^4}{3!} + \cdots = \lambda e^\lambda \quad \text{(multiply by } \lambda)$$

$$1 + 2\lambda + \frac{3\lambda^2}{2!} + \frac{4\lambda^3}{3!} + \cdots = \lambda e^\lambda + e^\lambda \quad \text{(differentiate w.r.t. } \lambda)$$

$$\lambda + 2\lambda^2 + \frac{3\lambda^3}{2!} + \frac{4\lambda^4}{3!} + \cdots = \lambda^2 e^\lambda + \lambda e^\lambda \quad \text{(multiply by } \lambda)$$

So

$$E(X^2) = e^{-\lambda} \left(\lambda^2 e^\lambda + \lambda e^\lambda \right) = \lambda^2 + \lambda$$

$$\text{Var } X = E(X^2) - (EX)^2 = \lambda^2 + \lambda - \lambda^2 = \lambda$$

7. (a) Var $10X = 100$ Var $X = 200$
   (b) Var$(10X + 3) = 100$ Var $X = 200$
   (c) $E(-X) = -EX = -10$
   (d) Var$(-X) = (-1)^2$ Var $X = 2$
   (e) Var $X = E(X^2) - (EX)^2$, so $E(X^2) = 2 + 100 = 102$
   (f) $\frac{4}{9}$ Var $X = \frac{8}{9}$

8. (a) False   Var$(X + Y)$ may or may not equal Var $X +$ Var Y.
   (b) True
   (c) True   If X and Y were ind, then Var$(X + Y)$ would equal Var X + Var Y, so they can't be ind. (Statements (b) and (c) are logically equivalent.)
   (d) False   Var$(X + Y)$ can equal Var X + Var Y even if X and Y are not ind.
   (e) True $E(X + Y) = EX + EY$ always, whether or not X and Y are ind.
   (f) False $E(X + Y)$ is always $EX + EY$.

9. (a) Var$(X - Y)$

$$= E\left[X - Y - E(X - Y)\right]^2$$

$$= E\left[(X - EX) + (Y - EY)\right]^2$$

$$= E\left[(X - EX)^2 + (Y - EY)^2 - 2(X - EX)(Y - EY)\right]$$

$$= E(X - EX)^2 + E(Y - EY)^2 - 2E\left[(X - EX)(Y - EY)\right]$$

$$= \text{Var } X + \text{Var } Y - 2\,\text{Cov}(X, Y)$$

(b) $\mathrm{Var}(X - Y) = E(X - Y)^2 - [E(X - Y)]^2$

$$= E(X^2 - 2XY + Y^2) - [EX - EY]^2$$
$$= E(X^2) - 2E(XY) + E(Y^2) - (EX)^2$$
$$-(EY)^2 + 2(EX)(EY)$$
$$= E(X^2) - (EX)^2 + E(Y^2) - (EY)^2$$
$$- 2[EXY - EX\,EY]$$
$$= \mathrm{Var}\,X + \mathrm{Var}\,Y - 2\,\mathrm{Cov}(X, Y)$$

(c) $\mathrm{Var}(X-Y) = \mathrm{Var}(X+-Y) = \mathrm{Var}\,X + \mathrm{Var}(-Y) + 2\mathrm{Cov}(X, -Y)$.
But $\mathrm{Var}(-Y) = \mathrm{Var}\,Y$, and

$$\mathrm{Cov}(X, -Y) = E(X \cdot -Y) - E(X)\,E(-Y)$$
$$= -E(XY) + (EX)(EY) = -\mathrm{Cov}(X, Y)$$

so

$$\mathrm{Var}(X - Y) = \mathrm{Var}\,X + \mathrm{Var}\,Y - 2\,\mathrm{Cov}(X, Y)$$

10. $E(aX^2 + bX + c) = aE(X^2) + b\,EX + c$

$$= a(\mathrm{Var}\,X + (EX)^2) + b\,EX + c$$
$$= a(\sigma^2 + \mu^2) + b\mu + c$$

11. $\mathrm{Var}(X_1 + \cdots + X_n) = \mathrm{Var}\,X_1 + \cdots + \mathrm{Var}\,X_n + 2\sum_{i<j} \mathrm{Cov}(X_i, X_j)$

$\mathrm{Var}\,X_i = E(X_i^2) - (EX_i)^2 = 11 - 9 = 2$

$\mathrm{Cov}(X_i, X_j) = E(X_iX_j) - (EX_i)(EX_j) = 5 - 9 = -4$

$\mathrm{Var}(X_1 + \cdots + X_n) = 2n + 2\binom{n}{2} \cdot -4$

12. (a) T   **(b)** F   **(c)** F   **(d)** T

13. (a) $EX = 1P(X = 1) + 2P(X = 2) + 3P(X = 3)$

$\qquad = \frac{1}{3}(1 + 2 + 3) = 2$

$\quad EY = 1P(Y = 1) + 2P(Y = 2) + 3P(Y = 3)$

Remember that, by symmetry, $P(Y = 1) = P(1$ on 2nd draw$)$ $= P(1$ on first draw$) = \frac{1}{3}$, so $EY = 2$ also.

$EXY = 2P(X = 1, Y = 2$ or vice versa$)$

$\qquad + 3P(X = 1, Y = 3$ or vv$) + 6P(X = 2, Y = 3$ or vv$)$

$\qquad = 2 \cdot 2 \cdot \dfrac{1}{3}\dfrac{1}{2} + 3 \cdot \dfrac{1}{3} + 6 \cdot \dfrac{1}{3} = \dfrac{11}{3}$

$$\text{Cov}(X, Y) = E(XY) - (EX)(EY) = \frac{11}{3} - 4 = -\frac{1}{3}$$

(b) If the drawing is with replacement, then X and Y are independent and $\text{Cov}(X, Y) = 0$.

14. The box is 60% reds and 40% black, so

$$EX = 60\% \times 2 = \frac{6}{5} \quad \text{and} \quad EY = 40\% \times 2 = \frac{4}{5}$$

$$EX^2 = 1^2 P(1R) + 2^2 P(2R) = \frac{3 \cdot 2}{\binom{5}{2}} + 4\frac{\binom{3}{2}}{\binom{5}{2}} = \frac{9}{5}$$

$$\text{Var } X = EX^2 - (EX)^2 = \frac{9}{5} - \frac{36}{25} = \frac{9}{25}$$

$$E(XY) = 1 P(X = 1, Y = 1) + 0 \times \text{ doesn't matter } = \frac{3 \cdot 2}{\binom{5}{2}} = \frac{3}{5}$$

$$\text{Cov}(X, Y) = E(XY) - (EX)(EY) = \frac{3}{5} - \frac{6}{5}\frac{4}{5} = -\frac{9}{25}$$

15. $EU = 3EX + 2EY = 10; \; EV = -2$

Var $U = 9 \text{ Var } X + 4 \text{ Var } Y = 13$

$$E(UV) = E(6X^2 - 5XY - 6Y^2) = 6E(X^2) - 5E(XY) - 6E(Y^2)$$
$$= 6E(X^2) - 5(EX)(EY) - 6E(Y^2) \quad \text{(since } X, Y \text{ are ind)}$$
$$= -20$$

$\text{Cov}(U, V) = E(UV) - (EU)(EV) = 0$

16. The 10 tosses are Bernoulli trials, so X has a binomial distribution with

$$n = 10, \quad p = \frac{1}{6}, \quad \text{and } EX = np = \frac{10}{6}, \quad \text{Var } X = npq = \frac{50}{36}$$

17. $X = X_1 + \cdots + X_n, \quad EX = EX_1 + \cdots + EX_n$
$EX_i = P(\text{match on } i\text{th}) = 1/n$
$EX = n \cdot 1/n = 1$

The X_i's are not independent, so

$$\text{Var } X = \text{Var } X_1 + \cdots + \text{Var } X_n + 2 \sum_{i<j} \text{Cov}(X_i, X_j)$$

$$\text{Var } X_i = E(X_i^2) - (EX_i)^2 = EX_i - (EX_i^2) \text{ (because } X_i^2 = X_i)$$

$$= \frac{1}{n} - \frac{1}{n^2} = \frac{n-1}{n^2}$$

$\mathrm{Cov}(X_i, X_j) = E(X_i X_j) - (EX_i) EX_j)$ where

$X_i X_j = 1$ if $X_i = 1$ and $X_j = 1$, that is, if ith and jth match, so

$$\mathrm{Cov}(X_i, X_j) = P(i\text{th and }j\text{th match}) - (EX_i)(EX_j)$$

$$= \frac{1}{n}\frac{1}{n-1} - \frac{1}{n}\frac{1}{n} = \frac{1}{n^2(n-1)}$$

Finally,

$$\mathrm{Var}\, X = n \cdot \frac{n-1}{n^2} + 2\binom{n}{2}\frac{1}{n^2(n-1)} = 1$$

18. $X = X_1 + \cdots + X_{10}, \quad EX = EX_1 + \cdots + EX_{10}$

$$EX_i = P(X_i = 1) = P(W_i \text{ is paired with a man}) = \frac{10}{19}$$

$$EX = \text{ sum of 10 } \frac{10}{19}\text{'s} = \frac{100}{19}$$

Now we want $\mathrm{Var}\, X$. The X_i's are not independent (if, say, W_1 gets a man then the chances of W_2's getting a man go down). So

$$\mathrm{Var}\, X = \mathrm{Var}\, X_1 + \cdots + \mathrm{Var}\, X_{10} + 2\sum_{i<j}\mathrm{Cov}(X_i, X_j)$$

Now we need $\mathrm{Var}\, X_i$ and $\mathrm{Cov}(X_i, X_j)$.

$\mathrm{Var}\, X_i = E(X_i^2) - (EX_i)^2$. But $X_i^2 = X_i$, so $\mathrm{Var}\, X_i = EX_i - (EX_i)^2$,

$EX_i = P(X_i = 1) = P(W_i \text{ is paired with a man}) = 10/19$

$$\mathrm{Var}\, X_i = \frac{10}{19} - \left(\frac{10}{19}\right)^2$$

$\mathrm{Cov}(X_i, X_j) = EX_i X_j - (EX_i)(EX_j)$

where $X_i X_j = \begin{cases} 1 & \text{if } X_i \text{ and } X_j \text{ are both 1, that is,} \\ & \quad \text{if } W_i \text{ and } W_j \text{ both get men} \\ 0 & \text{otherwise} \end{cases}$

So

$EX_i X_j = P(W_i \text{ gets man and } W_j \text{ gets man})$

$$= P(W_i \text{ gets man})P(W_j \text{ gets man}|W_i \text{ gets man}) = \frac{10}{19}\frac{9}{17}$$

$$\text{Cov}(X_i, X_j) = \frac{10}{19}\frac{9}{17} - \left(\frac{10}{19}\right)^2$$

Finally,

$$\text{Var } X = 10\left[\frac{10}{19} - \left(\frac{10}{19}\right)^2\right] + 2\binom{10}{2}\left[\frac{10}{19}\frac{9}{17} - \left(\frac{10}{19}\right)^2\right]$$

(There's a factor of $\binom{10}{2}$ because that's how many terms there are in the sum.)

19. (a) We know that $EX = 1/p$ (Section 3.2).

$$\begin{aligned}
E(X^2) &= 1^2 P(X = 1) + 2^2 P(X = 2) + 3^2 P(X = 3) + \cdots \\
&= 1^2 P(H) + 2^2 P(TH) + 3^2 P(TTH) + \cdots \\
&= p + 4qp + 9q^2p + 16q^3p + \cdots \\
&= p(1 + 4q + 9q^2 + 16q^3 + \cdots)
\end{aligned}$$

We know that

$$1 + x + x^2 + x^3 + x^4 + \cdots = \frac{1}{1 - x} \quad \text{for } -1 < x < 1$$

so

$$1 + 2x + 3x^2 + 4x^3 + \cdots = \frac{1}{(1-x)^2} \quad \text{(differentiate)}$$

$$x + 2x^2 + 3x^3 + 4x^4 + \cdots = \frac{x}{(1-x)^2} \quad \text{(multiply by } x\text{)}$$

$$1 + 4x + 9x^2 + 16x^3 + \cdots = \frac{1+x}{(1-x)^3} \quad \text{(differentiate)}$$

So

$$E(X^2) = p \cdot \frac{1+q}{(1-q)^3} = \frac{1+q}{p^2}$$

$$\text{Var } X = E(X^2) - (EX)^2 = \frac{1+q}{p^2} - \frac{1}{p^2} = \frac{q}{p^2}$$

(b) Let X_i be the number of trials to get the ith head after the $(i-1)$st head. Then $X = X_1 + \cdots + X_r$. Each X_i has a geometric distribution with parameter p. So

$$EX_i = 1/p, \quad \text{Var } X_i = q/p^2 \quad \text{(by part (a))}$$

$$EX = EX_1 + \cdots + EX_r = r \cdot \frac{1}{p}$$

The X_i's are independent, so

$$\text{Var } X = \text{Var } X_1 + \cdots + \text{Var } X_r = r \cdot \frac{q}{p^2}$$

Solutions Section 8-1

1. (a) $f(y) = \int_{x=2y}^{1} \frac{24}{5}(x+y)\,dx = \frac{24}{5}\left(\frac{1}{2} - 4y^2 + y\right)$ for $0 \le y \le \frac{1}{2}$

$$f(x|y) = \frac{f(x,y)}{f(y)} = \frac{2x+2y}{1-8y^2+2y}$$ for $0 \le y \le \frac{1}{2}$, $2y \le x \le 1$

Case 1. $0 \le y \le 1/4$

$$P\left(X \ge \frac{1}{2}\middle| Y = y\right) = \int_{x=1/2}^{1} \frac{2x+2y}{1-8y^2+2y}\,dx = \frac{y+3/4}{1-8y^2+2y}$$

Case 2. $1/4 \le y \le 1/2$

$$P\left(X \ge \frac{1}{2}\middle| Y = y\right) = 1$$

(everything in the conditional universe is fav)

(b) *Case 1.* $0 \le y \le .05$

$$P(.1 \le X \le .2|Y = y) = \int_{x=.1}^{.2} \frac{2x+2y}{1-8y^2+2y}\,dx$$

Case 2. $.05 \le y \le .1$

$$P(.1 \le X \le .2|Y = y) = \int_{x=2y}^{.2} \frac{2x+2y}{1-8y^2+2y}\,dx$$

Case 3. $.1 \le y \le 1/2$

$$P(.1 \le X \le .2|Y = y) = 0$$

(nothing in the conditional universe is fav)

(c) $E(X|Y = y) = \dfrac{1}{1-8y^2+2y} \displaystyle\int_{x=2y}^{1} x(2x+2y)\,dx$

$$= \frac{\frac{2}{3} + y - \frac{28}{3}y^3}{1-8y^2+2y} \quad \text{for } 0 \le y \le \frac{1}{2}$$

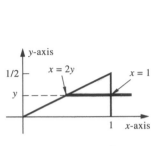

universe of $X \mid Y$ $P(X \geq \frac{1}{2} \mid Y = y)$

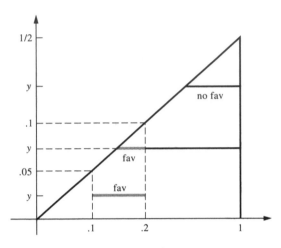

$P(.1 \leq X \leq .2 \mid Y = y)$

Figure P.1

2. (a) $f(x) = \int_{y=x}^{\infty} e^{-y}\, dy = e^{-x}$ for $x \geq 0$

$$f(y|x) = \frac{f(x,y)}{f(x)} = e^{x-y} \quad \text{for } x \geq 0, \ y \geq x$$

$$f(y) = \int_{x=0}^{y} e^{-y}\, dx = ye^{-y} \quad \text{for } y \geq 0$$

$$f(x|y) = \frac{e^{-y}}{ye^{-y}} = \frac{1}{y} \quad \text{for } y \geq 0, \ 0 \leq x \leq y$$

which means $X|Y$ is uniform on $[0, y]$

(b) *Case 1.*   If $0 \leq y \leq 2$, then $P(X \leq 2|Y = y) = 1$.

Case 2.   If $y \geq 2$, then $P(X \leq 2 | Y = y) = \int_{x=0}^{2} 1/y\, dx = 2/y$.

Or better still, since $X|Y$ is uniform on $[0, y]$,

$$P(X \leq 2 | Y = y) = \frac{\text{fav length}}{\text{total}} = \frac{2}{y}$$

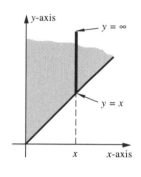

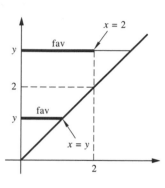

universe of $Y \mid X$ universe of $X \mid Y$ $P(X \leq 2 \mid Y = y)$

Figure P.2

(c) $E(Y|X = x) = \int_{y=x}^{\infty} ye^{x-y}\, dy = e^x \int_{y=x}^{\infty} ye^{-y}\, dy$ (use tables)

$$= 1 + x \text{ for } x \geq 0$$

(d) $X|Y$ is uniform in $[0, y]$ so $E(X|Y = y) = \frac{1}{2}y$ for $y \geq 0$.

3. (a) If $0 \leq x \leq 1$, then $Y|X$ is uniform on $[0, x]$.
    If $1 \leq x \leq 2$, then $Y|X$ is uniform on $[0, 1]$.
    If $2 \leq x \leq 3$, then $Y|X$ is uniform on $[x - 2, 1]$.
    If $Y = y$ where $0 \leq y \leq 1$, then $X|Y$ is uniform on $[y, y + 2]$.

$$f(y|x) = \begin{cases} \dfrac{1}{x} & \text{for } 0 \leq x \leq 1, 0 \leq y \leq x \\[2mm] 1 & \text{for } 1 \leq x \leq 2, 0 \leq y \leq 1 \\[2mm] \dfrac{1}{3-x} & \text{for } 2 \leq x \leq 3, x - 2 \leq y \leq 1 \end{cases}$$

$$f(x|y) = \frac{1}{2} \quad \text{for } 0 \leq y \leq 1, y \leq x \leq y + 2$$

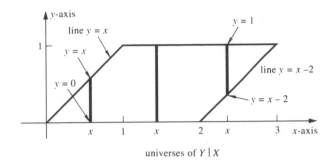

universes of $Y \mid X$

Figure P.3a

(b) If $Y = \frac{1}{2}$, $X|Y$ is uniform on $\left[\frac{1}{2}, 2\frac{1}{2}\right]$, so

$$P\left(1 \le X \le 2 \middle| Y = \frac{1}{2}\right) = \frac{\text{fav length}}{\text{total}} = \frac{1}{2}$$

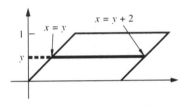

$X \mid Y$ is unif on $[y, y + 2]$

Figure P.3b

(c) *Case 0.* $0 \le x \le 1/2$

$$P\left(Y \le \frac{1}{2} \middle| X = x\right) = 1 \text{ (the whole universe is fav)}$$

Case 1. $1/2 \le x \le 1$

$Y|X$ is uniform on $[0, x]$.

$$P\left(Y \le \frac{1}{2} \middle| X = x\right) = \frac{\text{fav length}}{\text{total}} = \frac{1/2}{x} = \frac{1}{2x}$$

Case 2. $1 \le x \le 2$

$Y|X$ is uniform on $[0,1]$.

$$P\left(Y \le \frac{1}{2} \middle| X = x\right) = \frac{\text{fav length}}{\text{total}} = \frac{1/2}{1} = \frac{1}{2}$$

Case 3. $2 \leq x \leq 2\frac{1}{2}$

$Y|X$ is unif on $[x-2, 1]$.

$$P\left(Y \leq \frac{1}{2}\middle| X = x\right) = \frac{\text{fav length}}{\text{total}} = \frac{1/2 - (x-2)}{1 - (x-2)} = \frac{2\frac{1}{2} - x}{3 - x}$$

Case 4. $2\frac{1}{2} \leq x \leq 3$

$$P\left(Y \leq \frac{1}{2}\middle| X = x\right) = 0$$

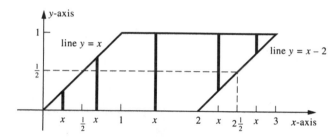

Figure P.3c

(d) If $0 \leq x \leq 1$, then $Y|X$ is unif on $[0, x]$.
   If $1 \leq x \leq 2$, then $Y|X$ is unif on $[0, 1]$.
   If $2 \leq x \leq 3$, then $Y|X$ is unif on $[x - 2, 1]$.

$$E(Y|X = x) = \begin{cases} \dfrac{1}{2}x & \text{if } 0 \leq x \leq 1 \\[2mm] \dfrac{1}{2} & \text{if } 1 \leq x \leq 2 \\[2mm] \dfrac{x-1}{2} & \text{if } 2 \leq x \leq 3 \end{cases}$$

(e) If $0 \leq y \leq 1$, then $X|Y$ is unif on $[y, y + 2]$.

$$E(X|Y = y) = \frac{2y + 2}{2} = y + 1 \qquad \text{if } 0 \leq y \leq 1$$

4. (a) $\int_{\text{plane}} f(x,y)\,dA = 1, \quad c = \dfrac{1}{\displaystyle\int_{x=1}^{5}\int_{y=0}^{2x} xy^2\,dy\,dx}$

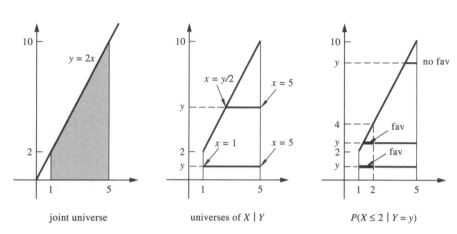

Figure P.4

(b) First, get $f(y)$.

Case 1. $0 \le y \le 2$

$$f(y) = \int_{x=1}^{5} cxy^2\,dx = 12cy^2$$

Case 2. $2 \le y \le 10$

$$F(y) = \int_{x=y/2}^{5} cxy^2\,dx = c\left(\frac{25}{2}y^2 - \frac{1}{8}y^4\right)$$

$$f(x|y) = \begin{cases} \dfrac{x}{12} & \text{for } 0 \le y \le 2, \quad 1 \le x \le 5 \\[2ex] \dfrac{8x}{100 - y^2} & \text{for } 2 \le y \le 10, \quad \dfrac{1}{2}y \le x \le 5 \end{cases}$$

Finally,

Case 1. $0 \le y \le 2$

$$P(X \le 2 | Y = y) = \int_{1}^{2} \frac{x}{12}\,dx = \frac{1}{8}$$

Case 2. $2 \le y \le 4$

$$P(X \le 2 | Y = y) = \int_{y/2}^{2} \frac{8x}{100 - y^2} \, dx = \frac{16 - y^2}{100 - y^2}$$

Case 3. $4 \le y \le 10$

$$P(X \le 2 | Y = y) = 0$$

(c) *Case 1.* $0 \le y \le 2$

$$E(X | Y = y) = \int_{x=1}^{5} x \, \frac{x}{12} \, dx = \frac{31}{9}$$

Case 2. $2 \le y \le 10$

$$E(X | Y = y) = \int_{x=y/2}^{5} x \, \frac{8x}{100 - y^2} \, dx = \frac{8}{100 - y^2} \frac{125 - \frac{1}{8} y^3}{3}$$

5. (a) For $-1 \le x \le 1$, $Y | X$ is uniform on $\left[-\sqrt{1 - x^2}, \ \sqrt{1 - x^2} \right]$, so

$$f(y | x) = \frac{1}{\text{total length}} = \frac{1}{2\sqrt{1 - x^2}}$$

$$\text{for } -1 \le x \le 1, \quad -\sqrt{1 - x^2} \le y \le \sqrt{1 - x^2}$$

(b) $f(x, y) = \dfrac{1}{\text{area}} = \dfrac{1}{\pi}$ for (x, y) in the circle

$$f(x) = \int_{y = -\sqrt{1-x^2}}^{\sqrt{1-x^2}} f(x, y) \, dy = \frac{1}{\pi} \, 2\sqrt{1 - x^2} \text{ for } -1 \le x \le 1$$

$$f(y | x) = \frac{f(x, y)}{f(x)} = \text{same answer as part (a)}$$

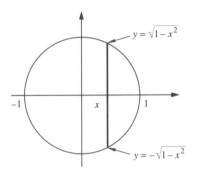

Figure P.5

Solutions Section 8-2

1. **(a)** $f(x) = 1/8$ for $0 \leq x \leq 8$.

$Y|X$ is uniform on $[0, x]$ so $f(y|x) = \dfrac{1}{x}$ for $0 \leq x \leq 8, 0 \leq y \leq x$

$$f(x, y) = f(x)f(y|x) = \frac{1}{8}\frac{1}{x} \text{ for } 0 \leq x \leq 8, 0 \leq y \leq x$$

$$f(y) = \int_{x=-\infty}^{\infty} f(x, y)\, dx = \int_{x=y}^{8} \frac{1}{8}\frac{1}{x}\, dx$$

$$= \frac{1}{8}\ln 8 - \frac{1}{8}\ln y \text{ for } 0 \leq y \leq 8$$

$$f(x|y) = \frac{f(x, y)}{f(y)} = \frac{1/x}{\ln 8 - \ln y} \text{ for } 0 \leq y \leq 8, y \leq x \leq 8$$

(b) *Method 1.* (theorem of total prob)
$Y|X$ is uniform on $[0, x]$, so

$$P(3 \leq Y \leq 5 | X = x) = \frac{\text{fav length}}{\text{total}} = \begin{cases} 0 & \text{if } 0 \leq x \leq 3 \\[2mm] \dfrac{x-3}{x} & \text{if } 3 \leq x \leq 5 \\[2mm] \dfrac{2}{x} & \text{if } 5 \leq x \leq 8 \end{cases}$$

$$P(3 \le Y \le 5) = \int_{x=0}^{8} P(3 \le Y \le 5 | X = x)\, f(x)\, dx$$

$$= \int_{3}^{5} \frac{x-3}{x} \frac{1}{8}\, dx + \int_{5}^{8} \frac{2}{x} \frac{1}{8}\, dx$$

$$= \frac{1}{8}\left(2 + 3\ln 3 - 5\ln 5 + 2\ln 8\right)$$

Method 2.

$$P(3 \le Y \le 5) = \int_{\text{fav}} f(x,y)\, dA = \int_{y=3}^{5} \int_{x=y}^{8} \frac{1}{8} \frac{1}{x}\, dx\, dy$$

(If you set up the double integral with inner y limits so that you have to split up the region, you'll get the sum of the same two integrals as with the theorem of total prob in method 1.)

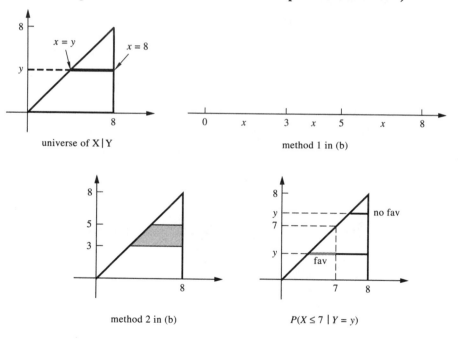

universe of $X \mid Y$ method 1 in (b)

method 2 in (b) $P(X \le 7 \mid Y = y)$

Figure P.1

(c) *Case 1.* $0 \le y \le 7$

$$P(X \le 7 | Y = y) = \int_{x=y}^{7} \frac{1/x}{\ln 8 - \ln y}\, dx = \frac{\ln 7 - \ln y}{\ln 8 - \ln y}$$

Case 2. $7 \le y \le 8$

$$P(X \le 7|Y = y) = 0$$

(d) *With* $Y|X$ is uniform on $[0, x]$, so $E(Y|X = x) = \frac{1}{2}x$.

$$EY = \int_{x=0}^{8} E(Y|X = x)f(x)\,dx = \int_{x=0}^{8} \frac{1}{2}x\,\frac{1}{8}\,dx = 2$$

Without

$$EY = \int_{\text{universe}} yf(x, y)dA = \int_{x=0}^{8} \int_{y=0}^{x} y\frac{1}{8}\frac{1}{x}dy\,dx = 2$$

(e) $E(X|Y = y) = \int_{x=y}^{8} xf(x|y)\,dx$

$$= \int_{x=y}^{8} x\,\frac{1/x}{\ln 8 - \ln y}\,dx = \frac{8 - y}{\ln 8 - \ln y}\ \text{for } 0 \le y \le 8$$

2. (a) $f(y|x) = \dfrac{x}{y^2}$ for $x \ge 0, y \ge x$

$$f(x, y) = f(x)\,f(y|x) = \frac{xe^{-x}}{y^2}\ \text{for } x \ge 0, y \ge x$$

$$P(Y \le 2) = \int_{\text{fav}} f(x, y)\,dA$$

$$= \int_{x=0}^{2} \int_{y=x}^{2} \frac{xe^{-x}}{y^2}\,dy\,dx$$

$$= \frac{1}{2} + \frac{1}{2}\,e^{-2}$$

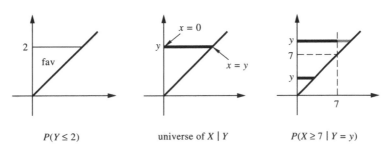

$P(Y \le 2)$ universe of $X \mid Y$ $P(X \ge 7 \mid Y = y)$

Figure P.2

(b) $f(y) = \int_{x=-\infty}^{\infty} f(x,y)\, dx = \int_{x=0}^{x=y} \frac{xe^{-x}}{y^2}\, dx$

$= \frac{1}{y^2}(1 - ye^{-y} - e^{-y})$ for $y \geq 0$

$f(x|y) = \frac{f(x,y)}{f(y)} = \frac{xe^{-x}}{1 - ye^{-y} - e^{-y}}$ for $y \geq 0, 0 \leq x \leq y$

Case 1. If $0 \leq y \leq 7$, then $P(X \geq 7|Y = y) = 0$.

Case 2. If $y \geq 7$, then

$$P(X \geq 7|Y = y) = \int_{x=7}^{y} \frac{xe^{-x}}{1 - ye^{-y} - e^{-y}}\, dx$$

(c) $EY = \int_{universe} y\, f(x,y)\, dA = \int_{x=0}^{\infty} \int_{y=x}^{\infty} y\, \frac{xe^{-x}}{y^2}\, dy\, dx = \infty$

(d) $E(X|Y = y) = \int_{x=0}^{y} x\, f(x|y)\, dx = \int_{x=0}^{y} x\, \frac{xe^{-x}}{1 - ye^{-y} - e^{-y}}\, dx$

for $y \geq 0$

3. (a) $Y|X$ is uniform on $[0, x]$.

$f(y|x) = \frac{1}{x}$ for $x \geq 1, 0 \leq y \leq x$

$f(x,y) = f(x)f(y|x) = \frac{1}{x^3}$ for $x \geq 1, 0 \leq y \leq x$

$P(Y \leq 5) = \int_{fav} f(x,y)\, dA = \int_{I} + \int_{II}$

$= \int_{x=1}^{5} \int_{y=0}^{x} \frac{1}{x^3}\, dy\, dx + \int_{x=5}^{\infty} \int_{y=0}^{5} \frac{1}{x^3}\, dy\, dx = \frac{9}{10}$

(b) $f(y) = \int_{x=-\infty}^{\infty} f(x,y)\, dx$

If $0 \leq y \leq 1$, then $f(y) = \int_{x=1}^{\infty} 1/x^3\, dx = 1/2$.

If $y \geq 1$, then $f(y) = \int_{x=y}^{\infty} 1/x^3\, dx = 1/2y^2$.

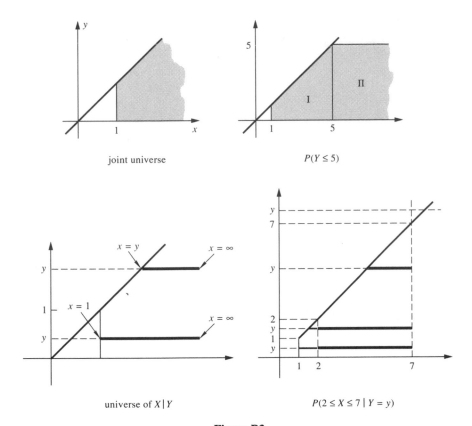

Figure P.3

So

$$
f(x|y) = \frac{f(x,y)}{f(y)} = \begin{cases} \dfrac{2y^2}{x^3} & \text{for } y \geq 1, x \geq y \\[2mm] \dfrac{2}{x^3} & \text{for } 0 \leq y \leq 1, x \geq 1 \end{cases}
$$

If $0 \leq y \leq 1$, $P(2 \leq X \leq 7|Y = y) = \int_2^7 2/x^3\, dx.$

If $1 \leq y \leq 2$, $P(2 \leq X \leq 7|Y = y) = \int_2^7 2y^2/x^3\, dx.$

If $2 \leq y \leq 7$, $P(2 \leq X \leq 7|Y = y) = \int_y^7 2y^2/x^3\, dx.$

If $y \geq 7$, $P(2 \leq X \leq 7|Y = y) = 0.$

(c) *Method 1.* (theorem of total exp)

$Y|X$ is unif on $[0, x]$, so $E(Y|X = x) = \frac{1}{2}x$ and

$$EY = \int_{x=1}^{\infty} \frac{1}{2}x \, \frac{1}{x^2} \, dx = \infty$$

Method 2.

$$EY = \int_{univ} y \, f(x, y) \, dA = \int_{x=1}^{\infty} \int_{y=0}^{x} y \, \frac{1}{x^3} \, dy \, dx = \infty$$

(d) $E(X|Y) = \int_{x=1}^{\infty} x \, f(x|y) \, dx$

$$= \begin{cases} \int_{x=1}^{\infty} x \, \frac{2}{x^3} \, dx \ = 2 & \text{if } 0 \le y \le 1 \\ \int_{x=y}^{\infty} x \, \frac{2y^2}{x^3} \, dx = 2y & \text{if } y \ge 1 \end{cases}$$

4. (a) $f(x) = \dfrac{1}{5}$ for $0 \le x \le 5$

$Y|X$ is uniform on $[x, 2x]$.

$$f(y|x) = \frac{1}{x} \text{ for } 0 \le x \le 5, x \le y \le 2x$$

$$f(x, y) = f(x)f(y|x) = \frac{1}{5}\frac{1}{x} \text{ for } 0 \le x \le 5, x \le y \le 2x$$

$$P(Y \le 2) = \int_{y=0}^{2} \int_{x=y/2}^{x=y} \frac{1}{5}\frac{1}{x} \, dx \, dy = \frac{2}{5} \ln 2$$

(b) $f(y) = \displaystyle\int_{x=y/2}^{y} \frac{1}{5}\frac{1}{x} \, dx = \frac{1}{5} \ln 2$ for $0 \le y \le 5$

$$f(x|y) = \frac{f(x, y)}{f(y)} = \frac{1}{\ln 2}\frac{1}{x} \text{ for } 0 \le y \le 5, \frac{1}{2}y \le x \le y$$

$$P(1 \le X \le 2|Y = 3) = \int_{x=3/2}^{2} \frac{1}{\ln 2}\frac{1}{x} \, dx = \frac{\ln 2 - \ln 3/2}{\ln 2}$$

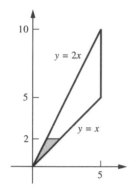

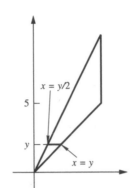

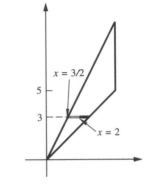

Figure P.4a   **Figure P.4b**

(c) If $0 \leq y \leq 5$, then

$$f(x|y) = \frac{1}{\ln 2} \frac{1}{x} \text{ for } \frac{1}{2}y \leq x \leq y \text{ (from part (b))}$$

If $5 \leq y \leq 10$, then

$$f(y) = \int_{x=y/2}^{5} \frac{1}{5} \frac{1}{x} \, dx = \frac{1}{5} \ln 5 - \frac{1}{5} \ln \frac{1}{2}y \text{ for } \frac{1}{2}y \leq x \leq y$$

$$f(x|y) = \frac{1/x}{\ln 5 - \ln \frac{1}{2}y} \text{ for } \frac{1}{2}y \leq x \leq 5$$

If $0 \leq y \leq 4$, $P(X \geq 4|Y = y) = 0$.
If $4 \leq y \leq 5$,
$$P(X \geq 4|Y = y) = \int_{4}^{y} \frac{1}{\ln 2} \frac{1}{x} \, dx = \frac{\ln y - \ln 4}{\ln 2}.$$
If $5 \leq y \leq 8$,
$$P(X \geq 4|Y = y) = \int_{4}^{5} \frac{1}{\ln 5 - \ln 1/2y} \, dx = \frac{\ln 5 - \ln 4}{\ln 2}.$$
If $8 \leq y \leq 10$, $P(X \geq 4|Y = y) = 1$.

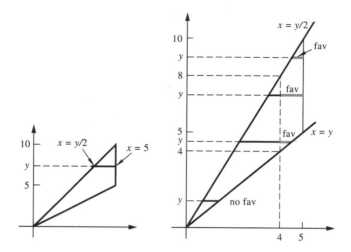

Figure P.4c

(d) *Method 1.* (theorem of total exp)

$Y|X$ is unif on $[x, 2x]$ so $E(Y|X = x) = 3x/2$ for $0 \le x \le 5$.

$$EY = \int_{x=0}^{5} E(Y|X = x) f(x) \, dx = \int_{x=0}^{5} \frac{3x}{2} \frac{1}{5} \, dx = \frac{15}{4}$$

Method 2.

$$EY = \int_{universe} y \, f(x, y) \, dA = \int_{x=0}^{5} \int_{y=x}^{2x} y \frac{1}{5} \frac{1}{x} \, dy \, dx$$

(e) $E(X|Y = y) = \int_{x=-\infty}^{\infty} x \, f(x|y) \, dx$

Case 1. $0 \le y \le 5$

$$E(X|Y = y) = \int_{x=y/2}^{y} x \frac{1}{\ln 2} \frac{1}{x} \, dx = \frac{y/2}{\ln 2}$$

Case 2. $5 \le y \le 10$

$$E(X|Y = y) = \int_{x=y/2}^{5} x \frac{1}{x} \frac{1}{\ln 5 - \ln \frac{1}{2}y} \, dx = \frac{5 - \frac{1}{2}y}{\ln 5 - \ln \frac{1}{2}y}$$

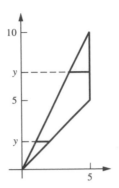

Figure P.4e

5. **(a)** $f(r) = 1/10, 0 \leq r \leq 10$

$$P(Z \leq 5 | R = r) = \frac{\text{fav area}}{\text{total}}$$

$$= \begin{cases} 1 & \text{if } 0 \leq r \leq 5 \\ \dfrac{25\pi}{\pi r^2} & \text{if } 5 \leq r \leq 10 \end{cases}$$

$$P(Z \leq 5) = \int_{r=0}^{10} P(Z \leq 5 | R = r) \, f(r) \, dr$$

$$= \int_{r=0}^{5} \frac{1}{10} \, dr + \int_{r=5}^{10} \frac{25}{r^2} \frac{1}{10} \, dr = \frac{3}{4}$$

$$EZ = \int_{r=0}^{10} E(Z | R = r) \, f(r) \, dr$$

total has radius r

fav has radius 5

Figure P.5a

Now we need $E(Z | R = r)$, which I'm going to call $E(Z | R = r_0)$ so that I can use the letter r later as a polar coordinate.

$$E(Z | R = r_0) = E\left(\sqrt{X^2 + Y^2}\right), \text{ where } (X, Y) \text{ is picked}$$

at random in a circle of radius r_0

$$= \int_{\text{universe}} \sqrt{x^2 + y^2} \ \underbrace{f(x,y)}_{1/\text{total area}} \ dA$$

$$= \int_{\theta=0}^{2\pi} \int_{r=0}^{r_0} r \frac{1}{\pi r_0^2} \ r \, dr \, d\theta = \frac{2}{3} r_0$$

$$EZ = \int_{r=0}^{10} \frac{2}{3} r \frac{1}{10} \, dr = \frac{10}{3}$$

(b) $F(z|r) = P(Z \le z | R = r) = \dfrac{\text{fav area}}{\text{total}}$

$$= \begin{cases} \dfrac{\pi z^2}{\pi r^2} & \text{for } 0 \le r \le 10, 0 \le z \le r \\[2mm] 1 & \text{for } 0 \le r \le 10, z \ge r \end{cases}$$

$$f(z|r) = D_z F(z|r) = \frac{2z}{r^2} \text{ for } 0 \le r \le 10, 0 \le z \le r$$

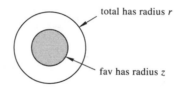

total has radius r

fav has radius z

Figure P.5b

(c) $f(r,z) = f(r)f(z|r) = \dfrac{1}{5}\dfrac{z}{r^2}$ for $0 \le r \le 10, 0 \le z \le r$

$$EZ = \int_{\text{universe}} z \, f(z,r) \, dz$$

$$= \int_{r=0}^{10} \int_{z=0}^{r} z \frac{1}{5} \frac{z}{r^2} \, dz \, dr = \frac{10}{3}$$

$$P(Z \le 5) = \int_{\text{fav}} \frac{1}{5} \frac{z}{r^2} \, dA$$

$$= \int_{z=0}^{5} \int_{r=z}^{10} \frac{1}{5} \frac{z}{r^2} \, dr \, dz = \frac{3}{4}$$

(d) $f(z) = \displaystyle\int_{r=z}^{10} \frac{1}{5} \frac{z}{r^2} \, dr = \frac{1}{5} - \frac{1}{50} z$ for $0 \le z \le 10$

$$f(r|z) = \frac{f(r,z)}{f(z)} = \frac{10z}{r^2(10-z)} \text{ for } 0 \le z \le 10, z \le r \le 10$$

$$P(R \leq 5 | Z = z) = \begin{cases} \displaystyle\int_{r=z}^{5} \frac{10z}{r^2(10-z)} \, dr & \text{if } 0 \leq z \leq 5 \\[3mm] 0 & \text{if } 5 \leq z \leq 10 \end{cases}$$

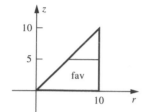

Figure P.5c

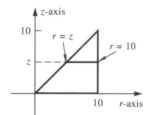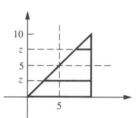

Figure P.5d

6. (a) V is uniform on $[0, v_0]$, so $f(v) = 1/v_0$ for $0 \leq v \leq v_0$.
If $V = v$, then X is exponential with $\lambda = 1/v^2$.
$E(X|V = v) = $ mean of the exp distribution $= 1/\lambda = v^2$.
By the theorem of total expectation,

$$EX = \int_{v=0}^{v_0} v^2 \frac{1}{v_0} \, dv = \frac{1}{3} v_0^2$$

(b) $f(x|v) = \dfrac{1}{v^2} e^{-x/v^2}$ for $x \geq 0$

$$f(v, x) = f(v) f(x|v) = \frac{1}{v_0} \frac{1}{v^2} e^{-x/v^2} \text{ for } 0 \leq v \leq v_0, x \geq 0$$

$$P(X \leq 3) = \int_{\text{fav}} f(v, x) \, dA = \frac{1}{v_0} \int_{v=0}^{v_0} \int_{x=0}^{3} \frac{e^{-x/v^2}}{v^2} \, dx \, dv$$

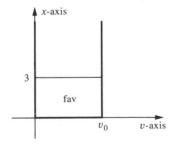

Figure P.6b

7. $f_X(x) = \dfrac{1}{10}$ for $0 \le x \le 10$

$$E(Y|X = x) = \frac{1}{2}x \text{ for } 0 \le x \le 10$$

$$EY = \int_{x=0}^{10} \frac{1}{2}x \, \frac{1}{10} \, dx = \frac{5}{2} \qquad \text{(theorem of total exp)}$$

$$f(x,y) = f(x)f(y|x) = \frac{1}{10}\frac{1}{x} \text{ for } 0 \le x \le 10, 0 \le y \le x$$

$$E(Y^2) = \int_{\text{universe}} y^2 \, f(x,y) \, dA$$

$$= \int_{x=0}^{10} \int_{y=0}^{x} y^2 \, \frac{1}{10}\frac{1}{x} \, dy \, dx = \frac{100}{9}$$

$$\text{Var } Y = E(Y^2) - (EY)^2 = \frac{100}{9} - \left(\frac{5}{2}\right)^2$$

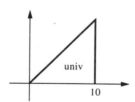

univ

10

Figure P.7

Solutions Section 8-3

1. (a) $N|X$ has a binomial distribution with parameters $n = 10, p = x$.

$$f(n|x) = P(N = n|X = x) = \binom{10}{n} x^n (1-x)^{10-n}$$

$$\text{for } 0 \le x \le 1; n = 0, 1, 2, \ldots, 10$$

$$f(x,n) = f(x)f(n|x) = 6\binom{10}{n} x^{n+1} (1-x)^{11-n}$$

$$\text{for } 0 \le x \le 1; n = 0, 1, 2, \ldots, 10$$

$$f(n) = \int_{x=0}^{1} f(x, n) \, dx$$

$$= 6\binom{10}{n} \int_{0}^{1} x^{n+1} (1-x)^{11-n} \, dx$$

$$= 6\binom{10}{n} \frac{(n+1)! \ (11-n)!}{13!} \quad \text{for } n = 0, 1, 2, \ldots, 10$$

$$f(x|n) = \frac{f(x, n)}{f(n)} = \frac{13! \ x^{n+1}(1-x)^{11-n}}{(n+1)! \ (11-n)!}$$

$$\text{for } n = 0, 1, \ldots, 10; 0 \le x \le 1$$

(b) *Method 1.*

$$P(10\,\text{H}) = P(N = 10) = f_N(10) = \frac{6 \cdot 11!}{13!} = \frac{1}{26}$$

Method 2. (theorem of total prob)

$$P(N = 10) = \int_{x=0}^{1} P(10\,\text{H}|X = x) \, f(x) \, dx$$

$$= \int_{0}^{1} x^{10} \, x(1-x) \, dx$$

$$= \int_{0}^{1} (x^{11} - x^{12}) \, dx = \frac{1}{26}$$

(c) (it's easiest to use theorem of total exp)

$N|X$ is binomial with parameters $n = 10, p = x$

so $E(N|X = x) = np = 10x$

$$E(N) = \int_{0}^{1} E(N|X = x) \, f(x) \, dx = 6 \cdot 10 \int_{0}^{1} x^2 (1-x) \, dx$$

$$= 60 \cdot \frac{2!}{4!} = 5$$

(d) $E(X|N = n) = \displaystyle\int_{0}^{1} x \, f(x|n) \, dx$

$$= \frac{13!}{(n+1)! \ (11-n)!} \int_{0}^{1} x^{n+2} (1-x)^{11-n} \, dx$$

$$= \frac{13!}{(n+1)! \ (11-n)!} \ \frac{(n+2)! \ (11-n)!}{14!}$$

$$= \frac{n+2}{14} \text{ for } n = 0, 1, 2, \ldots, 10$$

$$E(X|N = 10) = \frac{12}{14}$$

(e) $P\left(X \le \frac{1}{2}\middle| N = n\right) = \int_{x=0}^{1/2} f(x|n)\,dx$

$$= \int_{x=0}^{1/2} \frac{13! \ x^{n+1} \ (1-x)^{11-n}}{(n+1)! \ (11-n)!}\,dx$$

$$\text{for } n = 0, 1, \ldots, 10$$

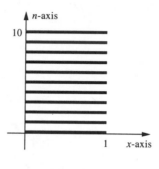

joint universe

Figure P.1   joint universe

2. $f(x) = 1$ for $0 \le x \le 1$

$N|X$ has a geometric dist with $p = x$

$P(N = n|X = x) = P$ (it takes n tosses to get a H)

$$= P(\text{T}^{n-1}\,\text{H}) = x(1-x)^{n-1}$$

$$\text{for } 0 \le x \le 1; n = 1, 2, 3, \ldots$$

$$E(N|X = x) = \frac{1}{p} = \frac{1}{x} \text{ for } 0 \le x \le 1$$

(a) *Method 1.* (theorem of total prob)

$$P(N = n) = \int_0^1 P(N = n|X = x) f(x) \, dx$$

$$= \int_0^1 x(1-x)^{n-1} \, dx = \frac{(n-1)!}{(n+1)!} = \frac{1}{n(n+1)}$$

$$P(N = 7) = \frac{1}{56}$$

Method 2. $P(N = n)$ is the marginal $f_N(n)$.

$$f(n|x) = P(N = n|X = x) = x(1-x)^{n-1}$$
$$\text{for } 0 \leq x \leq 1; n = 1, 2, 3, \ldots$$

$$f(x, n) = f(x)f(n|x) = x(1-x)^{n-1} \text{ for } 0 \leq x \leq 1; n = 1, 2, 3, \ldots$$

$$f(n) = \int_{x=-\infty}^{\infty} f(x, n) \, dx = \int_0^1 x(1-x)^{n-1} \, dx$$

$$= \frac{(n-1)!}{(n+1)!} = \frac{1}{n(n+1)} \text{ for } n = 1, 2, 3, \ldots$$

(b) $E(N) = \int_0^1 E(N|X = x) f(x) \, dx = \int_0^1 \frac{1}{x} \cdot 1 \, dx = \ln x \big|_0^1 = \infty$

(c) $f(x|n) = \dfrac{f(x, n)}{f(n)} = n(n+1) \, x(1-x)^{n-1}$

$$\text{for } 0 \leq x \leq 1; \; n = 1, 2, 3, \ldots$$

$$P(X \leq .5|N = n) \int_{x=0}^{.5} f(x|n) \, dx = n(n+1) \int_0^{.5} x(1-x)^{n-1} \, dx$$

(d) $E(X|N = n) = \int_0^1 x \, f(x|n) \, dx$

$$= n(n+1) \int_0^1 x \cdot x(1-x)^{n-1} \, dx$$

$$= n(n+1) \frac{2!(n-1)!}{(n+2)!} \text{ for } n = 1, 2, 3, \ldots$$

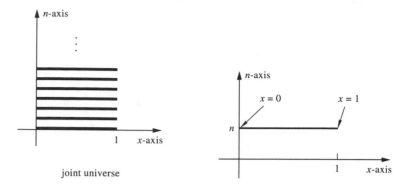

Figure P.2

3. (a) $f_N(n) = P(N = n) = \dfrac{1}{3}$ for $n = 1, 2, 3$

$X|N$ is uniform, so $f(x|n) = \dfrac{1}{n}$ for $0 \le x \le n; n = 1, 2, 3$

$f(n, x) = f(n) \, f(x|n) = \dfrac{1}{3n}$ for $0 \le x \le n; n = 1, 2, 3$

$f(x) = \displaystyle\sum_{n=1}^{3} f(n, x)$

But the number of non-zero terms in the sum depends on x.

Case 1. $\quad 0 \le x \le 1$

$$f(x) = f(1, x) + f(2, x) + f(3, x) = \dfrac{1}{3} + \dfrac{1}{3 \cdot 2} + \dfrac{1}{3 \cdot 3} = \dfrac{11}{18}$$

Case 2. $\quad 1 < x \le 2$

$$f(x) = f(2, x) + f(3, x) = \dfrac{1}{3 \cdot 2} + \dfrac{1}{3 \cdot 3} = \dfrac{5}{18}$$

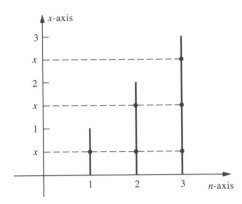

Figure P.3a   finding $f(x)$

Case 3. $2 < x \le 3$

$$f(x) = f(3, x) = \frac{1}{3 \cdot 3} = \frac{1}{9}$$

$$f(n|x) = \frac{f(n, x)}{f(x)} = \begin{cases} \dfrac{6}{11n} & \text{for } 0 \le x \le 1; n = 1, 2, 3 \\[2mm] \dfrac{6}{5n} & \text{for } 1 < x \le 2; n = 2, 3 \\[2mm] \dfrac{3}{n} = 1 & \text{for } 2 < x \le 3; n = 3 \end{cases}$$

(b) $N|X$ is discrete, so $f(n|x) = P(N = n|X = x)$. For each x in $[0, 1]$, we should have

$$P(N = 1|X = x) + P(N = 2|X = x) + P(N = 3|X = x) = 1$$

that is,

$$f_{N|X}(1|x) + f_{N|X}(2|x) + f_{N|X}(3|x) = 1 \text{ for each } x \text{ in } [0, 1]$$

Here's the check:

If $0 \le x \le 1$, then $f(1|x) + f(2|x) + f(3|x) = \dfrac{6}{11} + \dfrac{6}{22} + \dfrac{6}{33} = 1$.

If $1 < x \le 2$, then $f(1|x) + f(2|x) + f(3|x) = 0 + \dfrac{6}{10} + \dfrac{6}{15} = 1$.

If $2 < x \le 3$, then $f(1|x) + f(2|x) + f(3|x) = 0 + 0 + \dfrac{3}{3} = 1$.

(c) (easiest with theorem of total exp)

$$EX = \frac{1}{3} \cdot \frac{1}{2} + \frac{1}{3} \cdot 1 + \frac{1}{3} \cdot \frac{3}{2} = 1$$

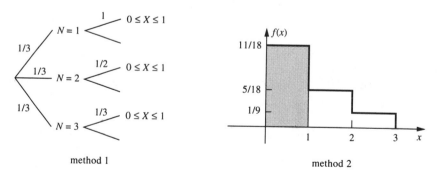

Figure P.3c

(d) *Method 1.* (theorem of total prob)

$$P(0 \leq X \leq 1) = \frac{1}{3} \cdot 1 + \frac{1}{3} \cdot \frac{1}{2} + \frac{1}{3} \cdot \frac{1}{3} = \frac{11}{18}$$

Method 2.

$$P(0 \leq X \leq 1) = \int_0^1 f(x)\, dx$$

$$= \int_0^1 \frac{11}{18}\, dx = \text{ area in the diagram } = \frac{11}{18}$$

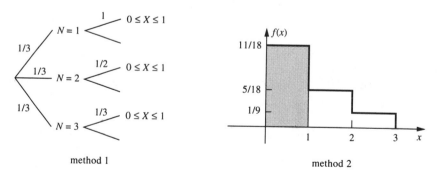

Figure P.3d

(e) $P(N = n | X = x)$ is just $f(n|x)$

$$E(N|X = x) = \sum_n n\, P(N = n | X = x)$$

$$= 1 \cdot f(1|x) + 2 \cdot f(2|x) + 3 \cdot f(3|x)$$

If $0 < x \leq 1$, then $E(N|X = x) = 1 \cdot \frac{6}{11} + 2 \cdot \frac{6}{22} + 3 \cdot \frac{6}{33} = \frac{18}{11}$.

If $1 < x \leq 2$, then $E(N|X = x) = 2 \cdot \dfrac{6}{10} + 3 \cdot \dfrac{6}{15} = \dfrac{12}{5}$.

If $2 < x \leq 3$, then $E(N|X = x) = 3 \cdot \dfrac{3}{3} = 3$ (obviously).

(f) $P(N = 2|X = x) = f(2|x)$

If $0 < x \leq 1$, then $f(2|x) = 6/22$.

If $1 < x \leq 2$, then $f(2|x) = 6/10$.

If $2 < x \leq 3$, then $f(2|x) = P(N = 2|X = x) = 0$.

(If you pick from $[0, n]$ and your pick is *above* 2, then n couldn't have been 2.)

4. (a) $f(y|n) = n/y^2$ for $n = 1, 2, 3, 4; y \geq n$

Legal because $f(y|n) \geq 0$ and

$$\int_{y=-\infty}^{\infty} f(y|n)\, dy = \int_{y=n}^{\infty} n/y^2 \, dy = 1$$

(b) *Method 1.* (theorem of total prob)

$$P(Y \geq \pi) = \sum_{n=1}^{4} P(N = n) P(Y \geq \pi | N = n)$$

$$= \frac{1}{4} \int_{\pi}^{\infty} \frac{1}{y^2} \, dy + \frac{1}{4} \int_{\pi}^{\infty} \frac{2}{y^2} \, dy + \frac{1}{4} \int_{\pi}^{\infty} \frac{3}{y^2} \, dy + \frac{1}{4} \cdot 1$$

$$= \frac{3}{2\pi} + \frac{1}{4}$$

Method 2. $f_N(n) = P(N = n) = 1/4$ for $n = 1, 2, 3, 4$.

$$f(n, y) = f(n)\, f(y|n) = \frac{1}{4} \frac{n}{y^2} \text{ for } n = 1, 2, 3, 4; y \geq n$$

To find $f_Y(y)$, sum out n; the number of terms in the sum depends on y.

If $1 \leq y < 2$, then $f_Y(y) = f(1, y) = 1/4y^2$.
If $2 \leq y < 3$, then $f(y) = f(1, y) + f(2, y) = 3/4y^2$.
If $3 \leq y < 4$, then $f(y) = f(1, y) + f(2, y) + f(3, y) = 6/4y^2$.
If $y \geq 4$, then $f(y) = f(1, y) + f(2, y) + f(3, y) + f(4, y) = 10/4y^2$.

$$P(Y \geq \pi) = \int_{y=\pi}^{\infty} f(y)\, dy = \int_{y=\pi}^{4} \frac{6}{4y^2}\, dy + \int_{y=4}^{\infty} \frac{10}{4y^2}\, dy$$

$$= \frac{3}{2\pi} + \frac{1}{4}$$

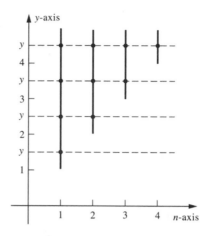

Figure P.4b   finding $f(y)$

(c) *Method 1. (theorem of total exp)*

$$E(Y|N = n) = \int_{y=-\infty}^{\infty} y\, f(y|n)\, dy = \int_{y=n}^{\infty} y \cdot \frac{n}{y^2}\, dy$$

$$= n \ln y|_{y=n}^{\infty} = \infty \text{ for } n = 1, 2, 3, 4$$

$$E(Y) = \sum_{n} E(Y|N = n)\, P(N = n) = \infty$$

Method 2.

$$E(Y) = \int_{y=-\infty}^{\infty} y\, f(y)\, dy$$

$$= \int_{y=1}^{2} y \cdot \frac{1}{4y^2}\, dy + \int_{2}^{3} y \cdot \frac{3}{4y^2}\, dy + \int_{3}^{4} y \cdot \frac{6}{4y^2}\, dy$$

$$+ \int_{4}^{\infty} y \cdot \frac{10}{4y^2}\, dy = \infty$$

(d) $f(n|y) = P(N = n|Y = y) = \dfrac{f(n, y)}{f(y)}$

(see part (b) method 2 for $f(n, y)$ and $f(y)$)

Case 1. $1 \leq y \leq 2$

$f(1|y) = 1$

Case 2. $2 \leq y \leq 3$

$$f(1|y) = \frac{f(1,y)}{f(y)} = \frac{(1/4)\,(1/y^2)}{3/4y^2} = \frac{1}{3}$$

$$f(2|y) = \frac{2}{3}$$

Case 3. $3 \leq y \leq 4$

$$f(1|y) = \frac{f(1,y)}{f(y)} = \frac{1}{6}$$

$$f(2|y) = \frac{f(2,y)}{f(y)} = \frac{1}{3}$$

$$f(3|y) = \frac{1}{2}$$

Case 4. $y \geq 4$

$$f(1|y) = \frac{1}{10}, \quad f(2|y) = \frac{2}{10}, \quad f(3|y) = \frac{3}{10}, \quad f(4|y) = \frac{4}{10}$$

$$E(N|Y = y) = 1 \cdot P(N = 1|Y = y) + 2 \cdot P(N = 2|Y = y)$$
$$+3 \cdot P(N = 3|Y = y) + 4 \cdot P(N = 4|Y = y)$$

If $1 \leq y \leq 2$, then $E(N|Y = y) = 1 \cdot 1 = 1$.

If $2 \leq y \leq 3$, then $E(N|Y = y) = 1 \cdot \dfrac{1}{3} + 2 \cdot \dfrac{2}{3} = \dfrac{5}{3}$.

If $3 \leq y \leq 4$, then $E(N|Y = y) = 1 \cdot \dfrac{1}{6} + 2 \cdot \dfrac{2}{6} + 3 \cdot \dfrac{3}{6} = \dfrac{14}{6}$.

If $y \geq 4$, then $E(N|Y = y) = 1 \cdot \dfrac{1}{10} + 2 \cdot \dfrac{2}{10} + 3 \cdot \dfrac{3}{10} + 4 \cdot \dfrac{4}{10} = 3$.

(e) $P(N = 3|Y = y) = f(3, y) = \begin{cases} \dfrac{3}{6} & \text{if } 3 \leq y < 4 \\[2mm] \dfrac{3}{10} & \text{if } y \geq 4 \end{cases}$

Solutions Review Problems for Chapters 6, 7, 8

1. (a) $EX = \displaystyle\int_{\text{plane}} x\, f(x,y)\, dA = \int_I \frac{3}{2}x\, dA + \int_{II} \frac{1}{2}x\, dA$

$\qquad = \dfrac{3}{2}\displaystyle\int_{x=0}^{1}\int_{y=0}^{1-x} x\, dy\, dx + \frac{1}{2}\int_{x=0}^{1}\int_{y=1-x}^{1} x\, dy\, dx$

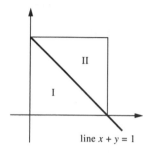

line $x + y = 1$

Figure P.1a

You can integrate directly or use this shortcut for $\displaystyle\int_R x\, dA$:

$\dfrac{\displaystyle\int_R x\, dA}{\text{area of } R} = \bar{x} = x\text{-coord of centroid}$

$\displaystyle\int_R x\, dA = \bar{x} \times \text{ area of } R$

This is useful if you can find $\bar{x}$ and the area of R easily. For a triangle, $\bar{x}$ is the average of the x-coords of the vertices, so

$$EX = \frac{3}{2}\cdot\frac{1}{2}\cdot\frac{1}{3} + \frac{1}{2}\cdot\frac{1}{2}\cdot\frac{2}{3} = \frac{5}{12}$$

$$E(X^2) = \frac{3}{2}\int_{x=0}^{1}\int_{y=0}^{1-x} x^2\, dy\, dx + \frac{1}{2}\int_{x=0}^{1}\int_{y=1-x}^{1} x^2\, dy\, dx = \frac{1}{4}$$

$$\text{Var } X = E(X^2) - (EX)^2 = \frac{1}{4} - \frac{25}{144}$$

(b) $f(y) = \displaystyle\int_{x=-\infty}^{\infty} f(x,y)\, dx = \int_{x=0}^{1-y}\frac{3}{2}\, dx$

$\qquad\qquad + \displaystyle\int_{x=1-y}^{1}\frac{1}{2}\, dx = \frac{3}{2} - y \text{ for } 0 \le y \le 1$

$$f(x|y) = \frac{f(x,y)}{f(y)} = \begin{cases} \dfrac{3}{3-2y} & \text{if } 0 \le y \le 1, 0 \le x \le 1-y \\[3mm] \dfrac{1}{3-2y} & \text{if } 0 \le y \le 1, 1-y \le x \le 1 \end{cases}$$

$$E(X|Y=y) = \int_{x=0}^{x=1-y} x\,\frac{3}{3-2y}\,dx + \int_{x=1-y}^{1} x\,\frac{1}{3-2y}\,dx$$

$$= \frac{1+2(1-y)^2}{2(3-2y)} \text{ for } 0 \le y \le 1$$

$$E\left(X|Y=\frac{1}{2}\right) = \frac{3}{8}$$

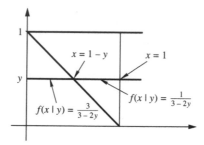

Figure P.1b

(c) $P\left(X \le \dfrac{3}{4} \middle| Y=y\right) = \displaystyle\int_{\text{fav}} f(x|y)\,dx$

Case 1. $0 \le y \le 1/4$

$$P\left(X \le \frac{3}{4} \middle| Y=y\right) = \int_{x=0}^{3/4} \frac{3}{3-2y}\,dx = \frac{3}{4}\,\frac{3}{3-2y}$$

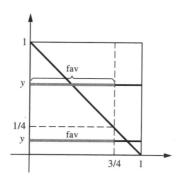

Figure P.1c

Case 2. $1/4 \le y \le 1$

$$P\left(X \le \frac{3}{4}\middle| Y = y\right) = \int_{x=0}^{1-y} \frac{3}{3-2y}\, dx + \int_{x=1-y}^{3/4} \frac{1}{3-2y}\, dx$$

$$= \frac{11/4 - 2y}{3 - 2y}$$

2. Let Y be the cost. Then

$$Y = \begin{cases} 2 & \text{if } 0 \le X \le 3 \\ 2 + \dfrac{1}{2}(X-3) & \text{if } X > 3 \end{cases}$$

$$EY = \int_{x=0}^{3} 2xe^{-x}\, dx + \int_{3}^{\infty}\left[2 + \frac{1}{2}(x-3)\right]xe^{-x}\, dx = 2 + \frac{5}{2}e^{-3}$$

3. $f(y) = \int_{x=0}^{y} ke^{-y}\, dx = kye^{-y}$ for $0 \le y \le 1$

$$f(x|y) = \frac{f(x,y)}{f(y)} = \frac{1}{y} \text{ for } 0 \le y \le 1, 0 \le x \le y$$

$f(x|y)$ has no x's in it, so $X|Y$ is uniform on $[0, y]$.

$E(X|Y) = \dfrac{1}{2}y$ for $0 \le y \le 1$

4. (a) $f(x) = \dfrac{1}{10}$ for $0 \le x \le 10$

$f(y|x) = \dfrac{2x^2}{y^3}$ for $0 \le x \le 10, y \ge x$

$f(x,y) = \dfrac{x^2}{5y^3}$ for $0 \le x \le 10, y \ge x$

$P(2 \le Y \le 3) = \displaystyle\int_{\text{fav}} f(x,y)\, dA$

$$= \int_{y=2}^{3}\int_{x=0}^{y} \frac{x^2}{5y^3}\, dx\, dy = \frac{1}{15}$$

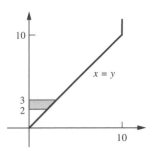

Figure P.4a

(b)
$$f(y) = \begin{cases} \displaystyle\int_{x=0}^{y} \frac{x^2}{5y^3}\,dx = \frac{1}{5} & \text{if } 0 \le y \le 10 \\[4mm] \displaystyle\int_{x=0}^{10} \frac{x^2}{5y^3}\,dx = \frac{1000}{15y^3} & \text{if } y \ge 10 \end{cases}$$

$$f(x|y) = \begin{cases} \displaystyle\frac{3x^2}{y^3} & \text{if } 0 \le y \le 10, 0 \le x \le y \\[4mm] \displaystyle\frac{3x^2}{1000} & \text{if } y \ge 10, 0 \le x \le 10 \end{cases}$$

If $0 \le y \le 2, P(X \ge 2|Y = y) = 0.$

If $2 \le y \le 10, P(X \ge 2|Y = y) = \displaystyle\int_{x=2}^{y} 3x^2/y^3\,dx = 1 - 8/y^3.$

If $y \ge 10, P(X \ge 2|Y = y) = \displaystyle\int_{x=2}^{10} 3x^2/1000\,dx = 1 - 8/1000.$

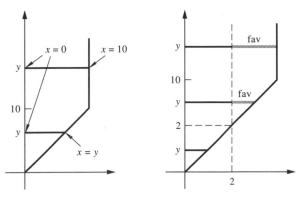

Figure P.4b

(c) $EY = \int_{\text{plane}} y f(x, y)\, dA = \int_{x=0}^{10} \int_{y=x}^{\infty} y\, \frac{x^2}{5y^3}\, dy\, dx = 10$

(d) *Case 1.* $0 \le y \le 10$

$$f(y) = \int_{x=0}^{y} \frac{x^2}{5y^3}\, dx = \frac{1}{15}$$

$$f(x|y) = \frac{3x^2}{y^3} \text{ for } 0 \le x \le y$$

$$E(X|Y = y) = \int_{x=0}^{y} x\, \frac{3x^2}{y^3}\, dx = \frac{3y}{4}$$

Case 2. $y \ge 10$

$$f(y) = \int_{x=0}^{10} \frac{x^2}{5y^3}\, dx = \frac{1000}{15y^3}$$

$$f(x|y) = \frac{3x^2}{1000} \text{ for } 0 \le x \le y$$

$$E(X|Y = y) = \int_{x=0}^{10} x\, \frac{3x^2}{1000}\, dx = \frac{15}{2}$$

5. $f(x) = e^{-x}, x \ge 0, E(Y|X = x) = 1/\lambda = x$ for $x \ge 0$

By theorem of total exp,

$$EY = \int_{x=0}^{\infty} E(Y|X = x)\, f(x)\, dx = \int_{x=0}^{\infty} x e^{-x}\, dx = 1.$$

To find Var Y, find $E(Y^2)$ first.

$$f(x, y) = f(x)\, f(y|x) = e^{-x}\, \frac{1}{x}\, e^{-y/x} \text{ for } x \ge 0, y \ge 0$$

$$E(Y^2) = \int_{x=0}^{\infty} \int_{y=0}^{\infty} y^2 e^{-x}\, \frac{1}{x}\, e^{-y/x}\, dy\, dx = 4$$

Var $Y = E(Y^2) - (EY)^2 = 3$

Figure P.5

6. Get the density $f_{\max}(x)$ of $\max(X_1, \cdots, X_n)$, and then use

$$E(\max) = \int_{-\infty}^{\infty} x f_{\max}(x)\, dx.$$

The random variables $X_1, \ldots, X_n$ have common density $f(x) = 1/3$ for $0 \le x \le 3$.

$$f_{\max}(x)\, dx = P(\max \approx x) = P(\text{1 of the } X_i\text{'s is} \approx x,\ \text{others are} \le x)$$

Use the multinomial where each trial has 3 outcomes:

$$P(\text{result is} \approx x) = f(x)\, dx = \frac{1}{3}\, dx$$

$$P(\text{result is} \le x) = \frac{\text{fav}}{\text{total}} = \frac{x}{3}$$

$$P(\text{result is} \ge x) = 1 - \frac{x}{3}$$

Then

$$f_{\max}(x)\, dx = \frac{n!}{1!(n-1)!}\, \frac{1}{3} dx \left(\frac{x}{3}\right)^{n-1}$$

$$f_{\max}(x) = \frac{n x^{n-1}}{3^n} \quad \text{for } 0 \le x \le 3$$

$$E(\max) = \int_{x=0}^{3} \frac{n x^n}{3^n}\, dx = \frac{3n}{n+1}$$

7. **(a)** Use theorem of total prob.

$$P(N = 0|X = x) = P(\text{no H} \mid \text{prob of H is } x)$$

$$= (1 - x)^{10} \text{ for } 0 \le x \le 1$$

$$P(N = 0) = \int_{x=0}^{1} P(N = 0|X = x) f(x) \, dx$$

$$= \int_{0}^{1} (1 - x)^{10} \cdot 1 \, dx = \frac{1}{11}$$

$$P(N = 1) = 1 - P(N = 0) = 10/11$$

(b) $f(x) = 1$ for $0 \le x \le 1$

$$f(0|x) = P(N = 0|X = x) = (1 - x)^{10} \text{ for } 0 \le x \le 1$$
$$f(1|x) = P(N = 1|X = x) = 1 - (1 - x)^{10} \text{ for } 0 \le x \le 1$$

$$f(x, n) = f(x)f(n|x) = \begin{cases} (1 - x)^{10} & \text{if } n = 0, 0 \le x \le 1 \\ 1 - (1 - x)^{10} & \text{if } n = 1, 0 \le x \le 1 \end{cases}$$

$$f_N(0) = P(N = 0) = 1/11$$

$$f(x|0) = \frac{f(x, 0)}{f_N(0)} = 11(1 - x)^{10} \text{ for } 0 \le x \le 1$$

$$P\left(0 \le X \le \frac{1}{2} \middle| N = 0\right) = \int_{x=0}^{1/2} f(x|0) \, dx$$

$$= \int_{x=0}^{1/2} 11(1 - x)^{10} \, dx = 1 - \left(\frac{1}{2}\right)^{11}$$

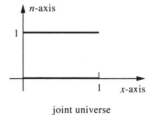

joint universe

Figure P.7

(c) $EN = 0 \cdot P(N = 0) + 1 \cdot P(N = 1) = 1 \cdot \frac{10}{11} = \frac{10}{11}$

(d) $E(X|N=0) = \displaystyle\int_{x=0}^{1} x f(x|0)\, dx$

$= \displaystyle\int_{x=0}^{1} 11x(1-x)\, dx$ (substitute $u = 1-x$)

$= 11 \displaystyle\int_{u=0}^{1} (u^{10} - u^{11})\, du = \dfrac{1}{12}$

8. $EY = \dfrac{1}{2}\cdot 3 + \dfrac{1}{2}\cdot\dfrac{5}{2} = \dfrac{11}{4}$ (theorem of total exp)

$E(Y^2) = \dfrac{1}{2}\, E(Y^2|H) + \dfrac{1}{2}\, E(Y^2|T)$

$E(Y^2|H) = 9,$

$E(Y^2|T) = \displaystyle\int_{y=-\infty}^{\infty} y^2\, f(y|T)\, dy = \displaystyle\int_{y=0}^{5} y^2\, \dfrac{1}{5}\, dy = \dfrac{25}{3}$

$E(Y^2) = \dfrac{1}{2}\cdot 9 + \dfrac{1}{2}\cdot\dfrac{25}{3} = \dfrac{26}{3}$

Var $Y = E(Y^2) - (EY)^2 = \dfrac{26}{3} - \dfrac{121}{16}$

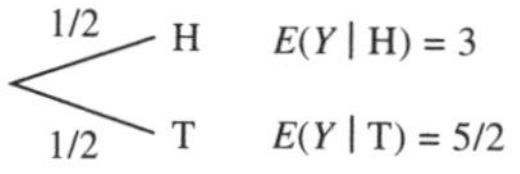

Figure P.8

9. $E(\max) = 1\cdot P(\max = 1) + 2P(\max = 2) + \cdots + 6P(\max = 6)$

$= 1P(1,1) + 2P(2, 1\text{ or } 1, 2\text{ or } 2, 2)$

$\quad + 3P(3, 3\text{ or } 3, 2\text{ or } 3, 1\text{ or } 2, 3\text{ or } 1, 3) + $ etc.

$= 1\cdot\dfrac{1}{36} + 2\cdot\dfrac{3}{36} + 3\cdot\dfrac{5}{36} + 4\cdot\dfrac{7}{36} + 5\cdot\dfrac{9}{36} + 6\cdot\dfrac{11}{36} = \dfrac{161}{36}$

$E(\max^2) = 1^2\cdot\dfrac{1}{36} + 2^2\cdot\dfrac{3}{36} + 3^2\cdot\dfrac{5}{36} + 4^2\cdot\dfrac{7}{36}$

$\quad\quad + 5^2\cdot\dfrac{9}{36} + 6^2\cdot\dfrac{11}{36} = \dfrac{791}{36}$

Var max $= E(\max^2) - (E\max)^2 = \dfrac{791}{36} - \left(\dfrac{161}{36}\right)^2$

10. $E(e^X) = \int_{-1}^{0} e^x \frac{1}{3}\, dx + \int_{0}^{1} e^x \frac{2}{3}\, dx = \frac{1}{3}(1 - e^{-1}) + \frac{2}{3}(e - 1)$

$E[(e^X)^2] = E(e^{2X}) = \int_{-1}^{0} e^{2x} \frac{1}{3}\, dx + \int_{0}^{1} e^{2x} \frac{2}{3}\, dx$

$$= \frac{1}{6}(1 - e^{-2}) + \frac{1}{3}(e^2 - 1)$$

$\text{Var } e^X = E[(e^X)^2] - (E\, e^X)^2$

11. (a) (theorem of total prob)

$$P(2 \le X \le 6) = \frac{2}{10} \cdot \frac{3}{5} + \frac{8}{10} \cdot \frac{4}{10} = \frac{11}{25}$$

(b) $EX = \frac{2}{10} \cdot \frac{5}{2} + \frac{8}{10} \cdot 5 = \frac{9}{2}$

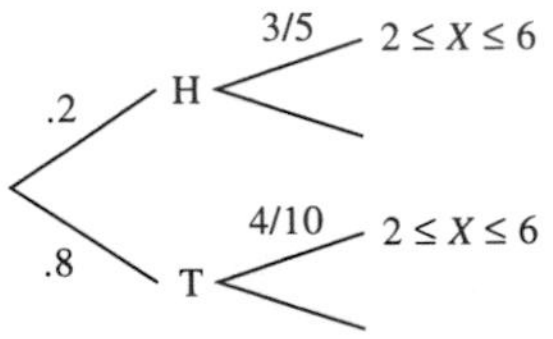

Figure P.11a 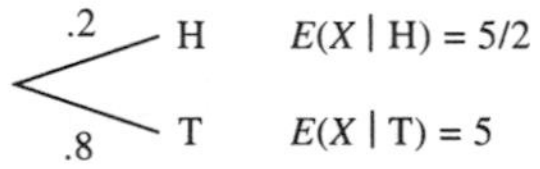   **Figure P.11b**

(c) Let's call the result of the coin toss N. Then N takes on values H and T (not entirely legal, since a random variable is supposed to take on numerical values only).

$$f_N(\text{H}) = P(N = \text{H}) = .2$$
$$f_N(\text{T}) = P(N = \text{T}) = .8$$
$$f(x|\text{H}) = \frac{1}{5} \text{ for } 0 \le x \le 5$$
$$f(x|\text{T}) = \frac{1}{10} \text{ for } 0 \le x \le 10$$
$$f(n, x) = f(n)f(x|n)$$
$$f(\text{H}, x) = (.2)\left(\frac{1}{5}\right) = \frac{1}{25} \text{ for } 0 \le x \le 5$$

$$f(\mathrm{T}, x) = (.8)\left(\frac{1}{10}\right) = \frac{2}{25} \text{ for } 0 \leq x \leq 10$$

$$f(x) = \sum_n f(n, x) = f(\mathrm{H}, x) + f(\mathrm{T}, x)$$

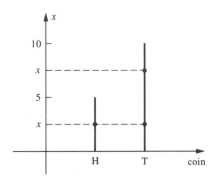

Figure P.11c

Case 1. $0 \leq x \leq 5$

$$f(x) = f(\mathrm{H}, x) + f(\mathrm{T}, x) = \frac{1}{25} + \frac{2}{25} = \frac{3}{25}$$

$$P(\mathrm{H}|X = x) = f(\mathrm{H}|x) = \frac{f(\mathrm{H}, x)}{f(x)} = \frac{1}{3}$$

$$P(\mathrm{T}|X = x) = \frac{2}{3}$$

Case 2. $5 \leq x \leq 10$

$$f(x) = f(\mathrm{T}, x) = \frac{2}{25}$$

$$P(\mathrm{H}|X = x) = f(\mathrm{H}|x) = 0 \text{ (by inspection)}$$

$$P(\mathrm{T}|X = x) = 1$$

All in all,

$$P(\mathrm{H}|X = x) = \begin{cases} \dfrac{1}{3} & \text{if } 0 \leq x \leq 5 \\[2mm] 0 & \text{if } 5 \leq x \leq 10 \end{cases}$$

$$P(\mathrm{T}|X = x) = \begin{cases} \dfrac{2}{3} & \text{if } 0 \leq x \leq 5 \\[2mm] 1 & \text{if } 5 \leq x \leq 10 \end{cases}$$

12. (a) $\text{Cov}(X, X) = E(X \cdot X) - (EX)(EX) = E(X^2) - (EX)^2$
$= \text{Var } X.$

(b) $E(X - \mu_X) = EX - E(\mu_X) = EX - \mu_X = 0.$

(c) $\text{Var}(X - \mu_X) = \text{Var } X$ since $\text{Var}(aX + b) = a^2 \text{Var } X.$

(d) $\text{Cov}(X, 3) = E(3X) - (EX)(E3) = 3EX - (EX) \cdot 3 = 0.$

Better still: X and 3 are independent random variables,
so $\text{Cov}(X, 3) = 0.$

13. (a) $\text{Cov}(X + Y, Z)$
$$= E\{[X + Y - E(X + Y)][Z - EZ]\}$$
$$= E[(X - EX + Y - EY)(Z - EZ)]$$
$$= E[(X - EX)(Z - EZ) + (Y - EY)(Z - EZ)]$$
$$= E[(X - EX)(Z - EZ)] + E[(Y - EY)(Z - EZ)]$$
$$= \text{Cov}(X, Z) + \text{Cov}(Y, Z)$$

(b) $\text{Cov}(X + Y, Z) = E[(X + Y)Z] - E(X + Y)E(Z)$
$$= E(XZ + YZ) - (EX + EY)EZ$$
$$= E(XZ) + E(YZ) - (EX)(EZ) - (EY)(EZ)$$
$$= E(XZ) - (EX)(EZ) + E(YZ) - (EY)(EZ)$$
$$= \text{Cov}(X, Z) + \text{Cov}(Y, Z)$$

14. (a) If $0 \le x \le 1$, then Y is uniform on $[0, 1 - x]$.

If $-1 \le x \le 0$, then Y is uniform on $[0, 1 + x]$.

$$E(Y|X = x) = \begin{cases} \dfrac{1 - x}{2} & \text{if } 0 \le x \le 1 \\[2mm] \dfrac{1 + x}{2} & \text{if } -1 \le x \le 0 \end{cases}$$

$$= \frac{1 - |x|}{2} \text{ for } -1 \le x \le 1$$

(b) If $Y = y$ where $0 \le y \le 1$, then X is uniform on $[y - 1, 1 - y]$, an interval centered at 0. So for $0 \le y \le 1$, $E(X|Y = y) = 0$.

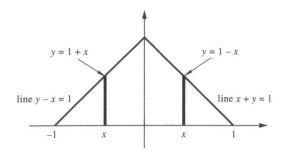

Figure P.14a

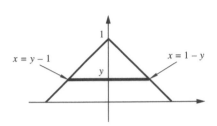

Figure P.14b

(c) *Method 1.* (theorem of total exp)

$$EY = \int_{x=-\infty}^{\infty} E(Y|X=x)\, f(x)\, dx = \int_{x=-1}^{0} \frac{1+x}{2} (1+x)\, dx$$

$$+ \int_{x=0}^{1} \frac{1-x}{2} (1-x)\, dx = \frac{1}{3}$$

Method 2.

$$EY = \int_{\text{plane}} y\, f(x,y)\, dA = \int_{y=0}^{1} \int_{x=y-1}^{1-y} y \cdot 1\, dx\, dy = \frac{1}{3}$$

(d) If $Y = \dfrac{1}{10}$, then X is unif on $\left[-\dfrac{9}{10}, \dfrac{9}{10} \right]$

$$P\left(X \le \frac{1}{2} \middle| Y = \frac{1}{10} \right) = \frac{\text{fav length}}{\text{total}} = \frac{1/2 - -9/10}{18/10} = 7/9$$

(e) $\min(x,y) = \begin{cases} x & \text{in region I} \\ y & \text{in region II} \end{cases}$

$$E(\min) = \int_{\text{I}} x\, f(x,y)\, dA + \int_{\text{II}} y\, f(x,y)\, dA$$

$$= \int_{x=-1}^{0} \int_{y=0}^{x+1} x\, dy\, dx + \int_{x=0}^{1/2} \int_{y=x}^{1-x} x\, dy\, dx$$

$$+ \int_{y=0}^{1/2} \int_{x=y}^{1-y} y\, dx\, dy = -\frac{1}{12}$$

(Shortcut: The last two integrals are equal, by symmetry.)

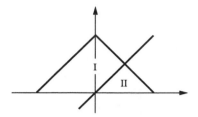

Figure P.14e

15. (a) $F(\rho|r) = P(\text{Dist} \le \rho|R = r) = \dfrac{\text{fav vol}}{\text{total}}$

$$= \frac{\frac{4}{3}\pi\rho^3}{\frac{4}{3}\pi r^3} = \frac{\rho^3}{r^3} \text{ for } 0 \le r \le 10, 0 \le \rho \le r$$

$$F(\rho|r) = 1 \quad \text{if } 0 \le r \le 10, \rho \ge r$$

$$f(\rho|r) = D_\rho F(\rho|r) = \frac{3\rho^2}{r^3} \text{ for } 0 \le r \le 10, 0 \le \rho \le r$$

(b) $f(r, \rho) = f(r) f(\rho|r) = \dfrac{1}{10} \dfrac{3\rho^2}{r^3} \text{ for } 0 \le r \le 10, 0 \le \rho \le r$

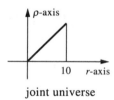

joint universe

Figure P.15b

(c) $P(\text{Dist} \le 5) = \displaystyle\int_{\text{fav}} f(r, \rho) \, dA$

$$= \int_{\rho=0}^{5} \int_{r=\rho}^{10} \frac{1}{10} \frac{3\rho^2}{r^3} \, dr \, d\rho = \frac{11}{16}$$

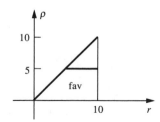

Figure P.15c

(d) $E(\text{Dist}) = \displaystyle\int_{\text{univ}} \rho f(r, \rho)\, dA$

$$= \frac{3}{10} \int_{\rho=0}^{10} \int_{r=\rho}^{10} \frac{\rho^3}{r^3}\, dr\, d\rho = \frac{15}{4}$$

(e) $f(\rho) = \displaystyle\int_{r=0}^{10} \frac{1}{10} \frac{3\rho^2}{r^3}\, dr$

$$= \frac{3}{20} - \frac{3\rho^2}{2000} \text{ for } 0 \le \rho \le 10$$

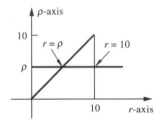

Figure P.15e

$$f(r|\rho) = \frac{f(r, \rho)}{f(\rho)} = \frac{200\rho^2}{r^3(100 - \rho^2)} \text{ for } 0 \le \rho \le 10, \rho \le r \le 10$$

$$E(R|\text{Dist} = \rho) = \int_{r=-\infty}^{\infty} r f(r|\rho)\, dr = \frac{200\rho^2}{1 - \rho^2} \int_{r=\rho}^{10} r \cdot \frac{1}{r^3}\, dr$$

$$= \frac{20\rho}{\rho + 10} \text{ for } 0 \le \rho \le 10$$

$$E(R|\text{Dist} = 7) = \frac{140}{17} \approx 8.24$$

Solutions Section 9-1

1. Let X_i be the service time for the ith customer. We'll assume that the X_i's are iid (one customer is just like another and they are independent). We want the probability that $X_1 + \cdots + X_{100} \le 120$ (minutes). The sum is approximately normal with mean 150 and variance 400. So

$$P(\text{sum } \le 120) = P\left(X^* \le \frac{120 - 150}{20}\right)$$

$$= P\left(X^* \le -\frac{3}{2}\right) = F^*\left(-\frac{3}{2}\right) = .067$$

2. Let X_i be the error in the ith gallon. We'll assume that the X_i's are iid. Then by (1), $X_i + \cdots + X_n$ is approximately normal with mean 0 and variance 4. So

$$P(|X_1 + \cdots + X_n| \leq 4) = P\left(\frac{-4}{2} \leq X^* \leq \frac{4}{2}\right)$$
$$= F^*(2) - F^*(-2) = .954 \text{ (approx)}$$

3. (a) The sum has a gamma distribution with parameters λ, n.

   (b) $X_1 + \cdots + X_n$ is approx normal with mean n/λ and variance n/λ^2. So

$$\frac{X_1 + \cdots + X_n - n/\lambda}{\sqrt{n/\lambda^2}}$$

has a unit normal distribution. So $a_n = n/\lambda, b_n = \sqrt{n/\lambda}$.

4. Let X_i be the number of tosses for the ith person to get 5 heads. Then the total number of tosses is $X_1 + \cdots + X_{100}$.

   X_i has a negative binomial distribution with parameters $r = 5$, $p = 1/3$, so

$$EX_i = \frac{r}{p} = 15, \text{ Var } X_i = \frac{rq}{p^2} = 30$$

The total is approximately normal with $\mu = 1500$ and $\sigma^2 = 3000$, so

$$P(\text{total} \leq 1550) = P\left(X^* \leq \frac{1550 - 1500}{\sqrt{3000}}\right)$$
$$= P(X^* \leq .91) = .816 \text{ (approx)}$$

5. Take a sample of size n. Let X_i be the height of the ith woman in the sample. The average sample height is $(X_1 + \cdots + X_n)/n$, where the X_i's are iid, so by (2) the sample average has (approximately) a normal distribution with mean 65 and variance $2/n$. We want n so that

$$P\left(64 \leq \frac{X_1 + \cdots + X_n}{n} \leq 66\right) = .9$$

$$P\left(-\frac{1}{\sqrt{2/n}} \le X^* \le \frac{1}{\sqrt{2/n}}\right) = .9$$

$$P\left(X^* \le \frac{1}{\sqrt{2/n}}\right) = .95 \text{ (by symmetry)}$$

$$\frac{1}{\sqrt{2/n}} = 1.65$$

$$n \approx 5.44$$

To be 90% sure, sample at least 6 women.

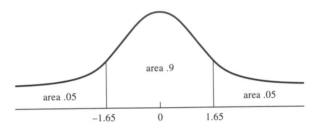

area .9

area .05   area .05

-1.65 0   1.65

Figure P.5

Solutions Review Problems for Chapters 4–9.

1. **(a)** $P(XY \le 2) = \int_{\text{fav}} f(x,y)\, dA$

$$= \int_{y=0}^{\sqrt{2}} \int_{x=y}^{2/y} f(x,y)\, dx\, dy.$$

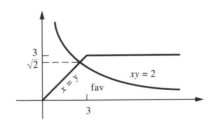

$\frac{3}{\sqrt{2}}$

$xy = 2$

fav

3

Figure P.1a

(b) Let $Z = X - Y$.

$F(z) = P(Z \le z)$

If $z \le 0$, then $F(z) = 0$. (X is always $\ge Y$, so $X - Y$ is never negative.)

If $z \geq 0$, then $F(z) = P(X - Y \leq z) = \displaystyle\int_{\text{fav}} f(x,y)\, dA$

$$= \int_{y=0}^{3} \int_{x=y}^{y+z} f(x,y)\, dx\, dy.$$

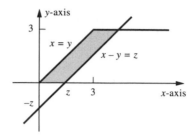

Figure P.1b

(c) $\displaystyle f(y) = \int_{x=y}^{\infty} f(x,y)\, dx \quad \text{for } 0 \leq y \leq 3$

$$f(x|y) = \frac{f(x,y)}{f(y)} \quad \text{for } 0 \leq y \leq 3, x \geq y$$

$$P(X \leq 2 | Y = y) = \begin{cases} 0 & \text{if } 2 \leq y \leq 3 \\[2mm] \displaystyle\int_{x=y}^{2} f(x|y)\, dx & \text{if } 0 \leq y \leq 2 \end{cases}$$

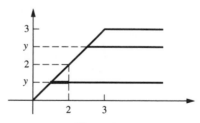

Figure P.1c

(d) $\displaystyle EX = \int_{y=0}^{3} \int_{x=y}^{\infty} x\, f(x,y)\, dx\, dy.$

(e) If $0 \leq x \leq 3$, then $f(x) = \displaystyle\int_{y=0}^{x} f(x,y)\, dy$. Call this $f_1(x)$.

If $x \geq 3$, then $f(x) = \displaystyle\int_{y=0}^{3} f(x,y)\, dy$. Call this $f_2(x)$.

$$f(y|x) = \frac{f(x,y)}{f_1(x)} \quad \text{if } 0 \le x \le 3, 0 \le y \le x$$

$$f(y|x) = \frac{f(x,y)}{f_2(x)} \quad \text{if } x \ge 3, 0 \le y \le 3$$

$$E(Y|X = x) = \int_{y=0}^{x} y\,\frac{f(x,y)}{f_1(x)}\,dy \quad \text{if } 0 \le x \le 3$$

$$E(Y|X = x) = \int_{y=0}^{3} y\,\frac{f(x,y)}{f_2(x)}\,dy \quad \text{if } x \ge 3$$

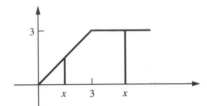

Figure P.1e

(f) $F_{\max}(z) = P(\max \le z) = \int_{\text{fav}} f(x,y)\,dA$

$$= \begin{cases} 0 & \text{if } z \le 0 \\[2mm] \int_{y=0}^{z}\int_{x=y}^{z} f(x,y)\,dx\,dy & \text{if } 0 \le z \le 3 \\[2mm] \int_{y=0}^{3}\int_{x=y}^{z} f(x,y)\,dx\,dy & \text{if } z \ge 3 \end{cases}$$

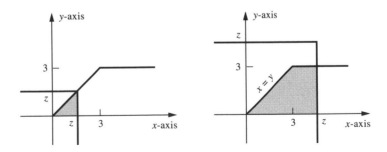

Figure P.1f

(g) $E(\max) = \displaystyle\int_{\text{universe}} \max(x, y)\, dA.$

But in this universe, the max is always x, so

$E(\max) = EX =$ answer to (d).

2. Let $Y = 1/(X + 1)$.

Distribution Function Method

We have $0 \le X \le 1$, so $\frac{1}{2} \le Y \le 1$.

If $y \le 1/2$, then $F(y) = 0$.

If $y \ge 1$, then $F(y) = 1$.

If $1/2 \le y \le 1$, then

$$F(y) = P(Y \le y) = P\left(X \ge \frac{1-y}{y}\right) = \text{fav area} = 2 - \frac{1}{y}$$

Density Method

We have $0 \le X \le 1$, so $\frac{1}{2} \le Y \le 1$, so $f(y) = 0$ for y outside $[\frac{1}{2}, 1]$.

Let $1/2 \le y \le 1$. Then $0 \le x \le 1$, so $f(x) = 1$. Also

$$y = \frac{1}{x+1}, \quad x = \frac{1-y}{y} = \frac{1}{y} - 1, \quad \frac{dx}{dy} = -\frac{1}{y^2}$$

$$f(y) = f(x)\left|\frac{dx}{dy}\right| = 1 \cdot \frac{1}{y^2}$$

So all in all, $f(y) = 1/y^2$ for $\frac{1}{2} \le y \le 1$.

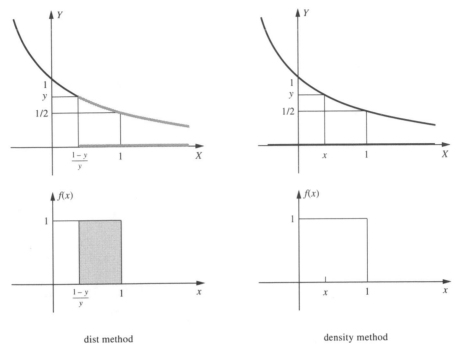

dist method   density method

Figure P.2

3. Let $X_1, \ldots, X_{100}$ be the lifetimes of the first 100 bulbs you use. There are 1095 days in 3 years, so we want $P(X_1 + \cdots + X_{100} \leq 1095)$. By the central limit theorem, $X_1 + \cdots + X_{100}$ is approximately normal with $\mu = 1020$ and $\sigma^2 = 900$. So

$$P(X_1 + \cdots + X_{100} \leq 1095) = P\left(X^* \leq \frac{1095 - 1020}{30}\right)$$

$$= P(X^* \leq 2.5) = F^*(2.5) = .994$$

Index